TOPICS IN COMPLEX FUNCTION THEORY

BY

C. L. SIEGEL

VOL. III

Abelian Functions and Modular Functions of Several Variables

TRANSLATED FROM THE ORIGINAL GERMAN BY

E. GOTTSCHLING

Johannes Gutenberg-Universität, Mainz

AND

M. TRETKOFF

Stevens Institute of Technology, New Jersey

WILEY-INTERSCIENCE

A DIVISION OF JOHN WILEY & SONS. INC.

NEW YORK . LONDON . SYDNEY . TORONTO

Library of Congress Cataloging in Publication Data:

Siegel, Carl Ludwig, 1896–
Topics in complex function theory.

(Interscience tracts in pure and applied mathematics, 25)
Vol. 2 translated by A. Shenitzer and M. Tretkoff; v. 3 translated by E. Gottschling and M. Tretkoff.
Bibliography: v. 3, p.
CONTENTS: 1. Elliptic functions and uniformization theory—v. 2. Automorphic functions and Abelian integrals.—v. 3. Abelian functions of several variables.
1. Functions of complex variables. I. Title.

QA331.S4713 517′.8 69–19931
ISBN 0–471–79070–2 (v. 1)

Printed in the United States of America

10 9 8 7 6 5 4 3 2 1

TOPICS IN COMPLEX FUNCTION THEORY

INTERSCIENCE TRACTS IN PURE AND APPLIED MATHEMATICS

1. *D. Montgomery and L. Zippin*—**Topological Transformation Groups**
2. *Fritz John*—**Plane Waves and Spherical Means Applied to Partial Differential Equations**
3. *E. Artin*—**Geometric Algebra**
4. *R. D. Richtmyer*—**Difference Methods for Initial-Value Problems**
5. *Serge Lang*—**Introduction to Algebraic Geometry**
6. *Herbert Busemann*—**Convex Surfaces**
7. *Serge Lang*—**Abelian Varieties**
8. *S. M. Ulam*—**A Collection of Mathematical Problems**
9. *I. M. Gel'fand*—**Lectures on Linear Algebra**
10. *Nathan Jacobson*—**Lie Algebras**
11. *Serge Lang*—**Diophantine Geometry**
12. *Walter Rudin*—**Fourier Analysis on Groups**
13. *Masayoshi Nagata*—**Local Rings**
14. *Ivan Niven*—**Diophantine Approximations**
15. *S. Kobayashi and K. Nomizu*—**Foundations of Differential Geometry. In two volumes**
16. *J. Plemelj*—**Problems in the Sense of Riemann and Klein**
17. *Richard Cohn*—**Difference Algebra**
18. *Henry B. Mann*—**Addition Theorems: The Addition Theorems of Group Theory and Number Theory**
19. *Robert Ash*—**Information Theory**
20. *W. Magnus and S. Winkler*—**Hill's Equation**
21. *Zellig Harris*—**Mathematical Operations in Linguistics**
22. *C. Corduneanu*—**Almost-Periodic Functions**
23. *Richard S. Bucy and Peter D. Joseph*—**Filtering and Its Applications**
24. *Paulo Ribenboim*—**Rings and Modules**
25. *C. L. Siegel*—**Topics in Complex Function Theory**
    - **Vol. I Elliptic Functions and Uniformization Theory**
    - **Vol. II Automorphic Functions and Abelian Integrals**
    - **Vol. III Abelian Functions and Modular Functions of Several Variables**

Preface

The present third and final part of my lectures on function theory is a revised edition of the work first published ten years ago. The original notes were recorded by E. Gottschling and H. Klingen. I have revised most of the work. Some of the less important material has been shortened; consequently I was able to expand the treatment of the foundations of the theory of modular functions of several variables.

C. L. Siegel

Preface to the English edition

This is the third and final part of Professor Carl L. Siegel's lectures on *Topics in Complex Function Theory*. The English translation is due to Erhard Gottschling and Marvin Tretkoff, both of whom were at the Courant Institute of Mathematical Sciences of New York University during the academic year 1969/70. The subsequent separation of the translators delayed publication of the English edition. Volume III is of particular importance for several reasons. Abelian functions and Siegel's modular functions (i.e., functions automorphic with respect to the symplectic modular group) constitute the best understood and most important classes of analytic functions of several complex variables. There exists no comparable rigorous introduction to these topics which is easily accessible to a second-year graduate student. The book should be especially welcome in view of the renewal of interest in Abelian functions and the growing literature on Siegel's modular functions.

A comprehensive bibliography of approximately 800 entries has been prepared for this volume by E. Gottschling. Special consideration has been given to the older literature. Moreover, the literature on Abelian functions has been supplemented by references to relevant papers on algebraic geometry. The bibliography also lists numerous papers on symmetric and homogeneous spaces and the theory of complex multiplication. In some cases papers have been included in the bibliography because they form the basis for work by their authors in fields related to this volume.

The bibliography contributes to the usefulness of Volume III of Professor Siegel's lectures and documents its timeliness.

Wilhelm Magnus
New York University

New York
May 1972

Contents

5

Abelian functions

1. Power series in several variables

The theory of functions of one complex variable can be established in two distinct fashions, one associated with Riemann and the other with Weierstrass. From Riemann's point of view considerations of conformal mapping and the determination of meromorphic functions by their singularities stand in the foreground. On the other hand, Weierstrass begins with function elements defined by power series. In the theory of functions of several complex variables the viewpoint of Weierstrass turns out to be advantageous. Therefore, we begin with a study of properties of power series of n complex variables $z_1, \ldots, z_n$.

Let

$$f = c + c_1 z_1 + \cdots + c_n z_n + c_{11} z_1^2 + c_{12} z_1 z_2 + \cdots \tag{1}$$

be such a power series, where the coefficients $c, c_1, \ldots, c_n, c_{11}, c_{12}, \ldots$ are complex constants. We assume that for a fixed arrangement of terms it converges for a certain system of values

$$z_k = \zeta_k \neq 0 \qquad (k = 1, \ldots, n). \tag{2}$$

Just as for power series of a single variable, one can show that the series (1) converges absolutely for

$$|z_k| < |\zeta_k| \qquad (k = 1, \ldots, n).$$

Given n positive real numbers $\rho_1, \ldots, \rho_n$ the region in n-dimensional complex space defined by the inequalities

$$|z_k| < \rho_k \qquad (k = 1, \ldots, n)$$

is called an *n-disk*. Furthermore, we denote the column formed from the n variables $z_1, \ldots, z_n$ by z, and we put

$$|z| = \max(|z_1|, \ldots, |z_n|).$$

It is clear that the power series (1) converges for a suitable system (2) if and only if it converges everywhere in the n-disk defined by

$$|z| < \rho \tag{3}$$

where ρ is a sufficiently small positive number. In this case we refer to the n-disk (3) as a *circle of convergence* of the power series f and we call ρ its *radius*. It is unnecessary for our future purposes to introduce the whole region of convergence of a power series in several variables. Furthermore, for a fixed f one can replace ρ by any smaller positive number. We should mention that it is possible to adopt a formal point of view at the outset and develop part of the theory without referring to the convergence or divergence of the series under consideration. However, we will not do this; instead we will consider throughout only those power series which possess a circle of convergence. Naturally, the radius, ρ, need not always have the same value.

If

$$g = d + d_1 z_1 + \cdots + d_n z_n + d_{11} z_1^2 + d_{12} z_1 z_2 + \cdots$$

is also a power series, we define $f = g$ if and only if the corresponding coefficients are equal. That is, all the conditions

$$c = d,\ c_1 = d_1, \ldots, c_n = d_n,\ c_{11} = d_{11},\ c_{12} = d_{12}, \ldots \tag{4}$$

are fulfilled. On the other hand, one could consider two power series to be equal if they assume the same numerical value at every point of a common circle of convergence. Beginning with the well-known case $n = 1$, it follows by induction on n that this definition of equality is equivalent to that given by (4).

If f and g are any two power series which converge, say, for $|z| < \rho$ and $|z| < \sigma$, then $f + g$, $f - g$, and fg are also power series which, in all three cases, have

$$|z| < \min(\rho, \sigma)$$

as a circle of convergence. Thus the set of all power series forms a ring which, by the way, contains all complex numbers.

Every power series, f, possesses exactly one decomposition of the form

$$f = f_0 + f_1 + f_2 + \cdots \tag{5}$$

where the homogeneous polynomial f_k $(k = 0, 1, \ldots)$ is formed out of all terms of total degree k. Clearly $f = 0$ implies that all $f_k = 0$. On the other hand, if $f \neq 0$ then there is a uniquely determined nonnegative integer m such that $f_k = 0$ $(k = 0, 1, \ldots, m - 1)$ and $f_m \neq 0$. We call m the *order of f* and write

$$m = \omega(f).$$

This suggests the definition

$$\omega(0) = \infty.$$

For two arbitrary power series f and g the following rules hold:

$$\omega(fg) = \omega(f) + \omega(g), \tag{6}$$
$$\omega(f+g) \geqslant \min(\omega(f), \omega(g)).$$

In particular if $fg = 0$, that is, $\omega(fg) = \infty$, it follows from (6) that at least one of the relations $\omega(f) = \infty$, $\omega(g) = \infty$ is valid; therefore, at least one of the power series f or g must be zero. In other words, the ring of power series has no zero divisors.

The concept of divisibility is now introduced as is customary in algebra. If f and g are two power series, then we call g a *divisor* of f and write $g \mid f$ if for a suitable third power series, h, the equation $f = gh$ holds. The relation $0 \mid f$ is fulfilled if and only if $f = 0$. But if $g \neq 0$ and $g \mid f$, then the power series h is uniquely determined by the equation $f = gh$; for, subtracting $f = gh^*$ from $f = gh$ yields $0 = g(h - h^*)$, $h - h^* = 0$, and $h = h^*$. We then define the *quotient* $f/g = h$.

A power series, g, is called a *unit* if $g \mid 1$. This means $gh = 1$, where h is a certain power series. According to (6) we have

$$\omega(g) + \omega(h) = \omega(gh) = \omega(1) = 0, \qquad \omega(g) = \omega(h) = 0.$$

Conversely, if g is any power series and $\omega(g) = 0$, then its constant term is $d \neq 0$ and

$$g = d + \cdots = d(1 + r)$$

where r is a power series whose constant term vanishes. If we write

$$h = \frac{d^{-1}}{1 + r} = d^{-1}(1 - r + r^2 - r^3 + \cdots)$$

then, upon rearrangement of the expression in the bracket on the right-hand side, we once again obtain a convergent power series. This yields

$$gh = 1, \qquad g \mid 1.$$

Now we define $g^{-1} = h$ and observe that the units form a group with respect to multiplication. It consists precisely of all those power series whose constant terms do not vanish.

If for two power series f and g we have $g \mid f$ and $f \mid g$ then either $f = 0$, $g = 0$ or $f \neq 0$, $g \neq 0$. In the second case both quotients $f/g = h$ and $g/f = j$ are power series, and from $f = gh$, $g = jf$ it follows that $f = jfh$. Thus $1 = jh$ and $h \mid 1$, so that $f = gh$ with h a unit. Conversely, if $f = gh$ with h a unit, then we also have $g = fh^{-1}$ with the inverse unit h^{-1}. Two such power series f and g will be called *associates*, and we indicate this by writing

$f \sim g$. It is clear that this concept of equivalence is reflexive, transitive, and symmetric.

We now consider all divisors, g, of a given power series, f. If $f = 0$, then g is obviously arbitrary. Now let $f \neq 0$ be a unit. From $\omega(f) = 0$ it follows that f has as divisors precisely all the units, u. If f is not a unit, then f obviously has as divisors all units, u, as well as all power series fu. If these are all the divisors of f, then we call f a *prime power series*. Otherwise there exists a decomposition $f = gh$ with $0 < \omega(g) < \omega(f)$, $0 < \omega(h) < \omega(f)$. By induction on the order it follows that f can be represented as a product of finitely many prime power series.

From considerations on the order we conclude that every power series of order 1 is a prime power series. In the case of power series of a single variable, z, there are no prime power series besides the series of order 1, and all prime power series are associates of z. If, on the other hand, $n > 1$ there are prime power series of order greater than 1. For example, $z_1^2 + z_2^3$ or, more generally, $z_1^2 + z_2^3 + r$ is a prime power series of order 2 when r is an arbitrary power series of order at least 4.

It is sometimes convenient to simplify a given power series, f, by introducing new variables, $w_1, \ldots, w_n$ by means of a suitably chosen invertible homogeneous linear transformation with complex coefficients of the variables $z_1, \ldots, z_n$. Let this transformation take f to ϕ. If we denote the images of the homogeneous polynomials f_k in decomposition (5) by the polynomials ϕ_k in the variables $w_1, \ldots, w_n$ then

$$\phi = \phi_0 + \phi_1 + \phi_2 + \cdots$$

is the decomposition of ϕ into its homogeneous components. The inverse transformation takes ϕ_k and ϕ to f_k and f. In particular, the order of a power series remains unchanged under every invertible homogeneous linear transformation of the variables $z_1, \ldots, z_n$.

2. *The preparation theorem*

Let f be a power series other than 0 and let s be its order, that is,

$$f = f_s + f_{s+1} + \cdots; \qquad f_s \neq 0$$

where f_k $(k = s, s+1, \ldots)$ stands for a homogeneous polynomial of total degree k in $z_1, \ldots, z_n$. We call the power series f *normalized* if the term z_1^s appears with a nonzero coefficient in f_s. When $n = 1$ it is evident that every nonzero power series is normalized. However, this is no longer true for $n > 1$; in this case, the variable z_1 is distinguished by the definition of a normalized power series.

Let

$$f = f_s(z_1, \ldots, z_n) + \cdots, \qquad g = g_t(z_1, \ldots, z_n) + \cdots,$$

be finitely many nonzero power series and $s, t, \ldots$ their orders. We will show that by a suitably chosen invertible homogeneous linear transformation

$$z_k = \sum_{l=1}^{n} a_{kl} w_l \qquad (k = 1, \ldots, n)$$

they can be simultaneously transformed into normalized power series in the variables $w_1, \ldots, w_n$ with w_1 distinguished. First, viewing the a_{kl} in the desired transformation as indeterminates, we find that upon application of this transformation to the power series $f, g, \ldots$ the monomials $w_1^s, w_1^t, \ldots$ appear with the respective coefficients $f_s(a_{11}, \ldots, a_{n1}), g_t(a_{11}, \ldots, a_{n1}), \ldots$. It is therefore sufficient for our purposes to select the entries $a_{11}, \ldots, a_{n1}$ of the first column of the coefficient matrix (a_{kl}) so that the nonidentically vanishing polynomial $f_s g_t \cdots$ is unequal to zero at the point

$$z_1 = a_{11}, \ldots, z_n = a_{n1}.$$

Next we complete this first column to a nonsingular matrix.

The investigation of divisibility of power series is based on the following important theorem which its discoverer Weierstrass called the *preparation theorem*.

Theorem 1: Let f be a normalized power series of order s. Then there is a unique unit, u, such that

$$fu = z_1^s + \alpha_1 z_1^{s-1} + \cdots + \alpha_s \tag{1}$$

where $\alpha_1, \ldots, \alpha_s$ are s power series in the $n - 1$ variables $z_2, \ldots, z_n$.

Proof: If the assertion of the theorem is valid for the power series f, then for an arbitrary constant, $c \neq 0$, one can replace the series f and u in (1) with cf and $c^{-1}u$ and deduce the validity of the statement with cf in place of f. Consequently, it is sufficient to prove the theorem under the assumption that the coefficient of the term z_1^s in f is 1.

We introduce $n - 1$ additional variables $\zeta_2, \ldots, \zeta_n$ and apply the transformation

$$z_k = z_1 \zeta_k \qquad (k = 2, \ldots, n). \tag{2}$$

Because of the assumption we can write

$$f = z_1^s(1 + r) \tag{3}$$

where r is a power series in the n variables $z_1, \zeta_2, \ldots, \zeta_n$ which has no constant term. Since f is a convergent power series in $z_1, \ldots, z_n$ there exists a positive real number $\rho < 1$ such that f converges absolutely everywhere in the n-disk $|z| < \rho$. Therefore, the series r is absolutely convergent at least when

$$(4) \qquad |z_1| < \rho, \qquad |z_1|\,|\zeta_k| < \rho \qquad (k = 2, \ldots, n).$$

If we replace every term of the power series r by its absolute value, then we obtain another series which we denote by $\lceil r \rceil$. Since r contains no constant term, we can choose a positive real number $\sigma < \rho$ such that the inequality

$$(5) \qquad \lceil r \rceil < 1$$

is satisfied whenever the conditions

$$(6) \qquad |z_1| < \sigma, \qquad |\zeta_k| < \sigma \qquad (k = 2, \ldots, n)$$

hold. These conditions imply (4).

The logarithmic series

$$(7) \qquad \log(1 + r) = r - \frac{r^2}{2} + \cdots = w$$

yields a power series in $z_1, \zeta_2, \ldots, \zeta_n$ which is absolutely convergent in the n-disk defined by (6) and possesses no constant term. Now for each integer $l = 0, \pm 1, \pm 2, \ldots$ form g_l, the subseries consisting of those monomials of w for which the exponents $k_1, k_2, \ldots, k_n$ of $z_1, \zeta_2, \ldots, \zeta_n$ satisfy the condition

$$(8) \qquad k_1 - (k_2 + \cdots + k_n) = l.$$

Putting

$$g_0 + g_1 + \cdots = \phi, \qquad g_{-1} + g_{-2} + \cdots = \psi$$

we obtain

$$w = \phi + \psi.$$

Equations (3) and (7) then imply

$$(9) \qquad fe^{-\phi} = z_1^s e^{\psi}.$$

In analogy with the definition of the power series g_l we define the subseries ϕ_l of

$$e^{-\phi} = 1 - \frac{\phi}{1!} + \cdots = \phi_0 + \phi_1 + \cdots,$$

$$e^{\psi} = 1 + \frac{\psi}{1!} + \cdots = 1 + \phi_{-1} + \phi_{-2} + \cdots$$

and the subseries f_l of

$$f = f(z_1, z_1\zeta_2, \ldots, z_1\zeta_n) = f_0 + f_1 + \cdots.$$

Expanding the left-hand side of (9), we only obtain terms for which the sum l, given by (8), satisfies $l \geqslant 0$. Thus the series $z_1^s\phi_l$ which appear on the right side of (9) must be zero whenever $l < -s$.

Finally, we put

$$(10) \qquad e^{-\phi} = u, \qquad z_1\phi_{-1} = \alpha_1, \ldots, z_1^s\phi_{-s} = \alpha_s$$

and again let $\zeta_k = z_k/z_1$ $(k = 2, \ldots, n)$. The $\alpha_1, \ldots, \alpha_s$ are then power series in the $n - 1$ variables $z_2, \ldots, z_n$ and u is a power series in $z_1, \ldots, z_n$ with constant term 1. Each power series in this collection is actually convergent, since by (5) and (6) they converge absolutely for

$$z_1 = \frac{\sigma}{2}, \qquad z_k = \zeta_k z_1 = \frac{\sigma^2}{4} \qquad (k = 2, \ldots, n)$$

and u is a unit. Assertion (1) follows from (9) and (10).

It remains for us to prove the uniqueness of u, which, by the way, we will not need in the future. Now if

$$fv = z_1^s + \beta_1 z_1^{s-1} + \cdots + \beta_s$$

holds for a unit, v, and s power series $\beta_1, \ldots, \beta_s$ in the variables $z_2, \ldots, z_n$, then β_k has order at least k. We expand

$$uv^{-1} = (1 + \alpha_1 z_1^{-1} + \cdots + \alpha_s z_1^{-s})(1 + \beta_1 z_1^{-1} + \cdots + \beta_s z_1^{-s})^{-1}$$

in powers of $z_1, \zeta_2, \ldots, \zeta_n$ assuming

$$|z_1| < \tau, \qquad |\zeta_k| < \tau \qquad (k = 2, \ldots, n)$$

which are conditions analogous to (6). Thus we obtain

$$(11) \qquad uv^{-1} = 1 + \chi_{-1} + \chi_{-2} + \cdots$$

where χ_l $(l = -1, -2, \ldots)$ consists of all terms whose exponents satisfy condition (8). On the other hand, since the left-hand side of (11) is a power series in $z_1, \ldots, z_n$, it cannot yield negative values of l upon application of substitution (2). Thus all $\chi_l = 0$ and, indeed, it follows that $u = v$.

As an example of the various applications of the preparation theorem we sketch a simple proof of the inverse function theorem for power series. Let

$$(12) \qquad w_k = \phi_k(z_1, \ldots, z_n) \qquad (k = 1, \ldots, n)$$

be a system of n power series where each

$$\phi_k = c_{k1}z_1 + \cdots + c_{kn}z_n + \cdots$$

has no constant term. We also assume that the determinant of the coefficients c_{kl} $(k, l = 1, \ldots, n)$ of the linear terms is nonzero. Then we must show that there is a unique system of n power series

$$(13)\quad z_k = \psi_k(w_1, \ldots, w_n) = d_{k1}w_1 + \cdots + d_{kn}w_n + \cdots \qquad (k = 1, \ldots, n)$$

which satisfies equations (12) identically in $w_1, \ldots, w_n$.

The matrix (c_{kl}) can be transformed to the unit matrix by means of a homogeneous linear transformation. Thus it suffices to consider this special case. First, let $n = 2$. Viewing z_1, z_2 and w_1 as three independent variables and applying the preparation theorem with $n = 3$ and $s = 1$ we obtain an identity

$$(\phi_1(z_1, z_2) - w_1)u_1(z_1, z_2, w_1) = z_1 - \chi(z_2, w_1) = z_1 - w_1 + \cdots.$$

Analogously, viewing z_2, w_1, w_2 as independent variables and applying the preparation theorem once again, we obtain

$$(\phi_2(\chi(z_2, w_1), z_2) - w_2)u_2(z_2, w_1, w_2) = z_2 - \psi_2(w_1, w_2) = z_2 - w_2 + \cdots.$$

Putting

$$\chi(\psi_2(w_1, w_2), w_1) = \psi_1(w_1, w_2)$$

we get (13) as a solution of (12); we also see that there can be no other solutions. The generalization to arbitrary n can easily be carried out by induction, but we will not do this. In this connection it should be pointed out that the associative law for substitution of power series is a consequence of the corresponding law for polynomials. The latter can be derived from the axioms of algebra.

We now come to the main topic, namely the investigation of the divisibility of power series by means of the preparation theorem. Let there be given finitely or countably many power series $h_1, h_2, \ldots$ in the variables $z_1, \ldots, z_n$. Another power series, q, in the same variables is called a *common divisor* of $h_1, h_2, \ldots$ provided $q \mid h_k$ $(k = 1, 2, \ldots)$. It is clear that this condition is always fulfilled for every unit q. In case there is no common divisor besides these trivial divisors, one says that $h_1, h_2, \ldots$ are *relatively prime* and indicates this by the notation

$$(14)\qquad (h_1, h_2, \ldots) \sim 1.$$

The other extreme case occurs when $h_1, h_2, \ldots$ are all 0, and then q is permitted to be any power series. If we now exclude this case and take, say, $h_1 \neq 0$, then $\omega(h_1)$ yields an upper bound for $\omega(q)$. We take any common divisor q selected so that its order is as large as possible and define

$$(h_1, h_2, \ldots) \sim q.$$

This is evidently a generalization of (14). Of course, we have to show that all such *greatest common divisors*, q, are actually associates of one another.

Henceforth, in this section, we will denote power series in the $n - 1$ variables $z_2, \ldots, z_n$ by lower case Greek letters. A power series

$$f = \alpha_0 + \alpha_1 z + \alpha_2 z^2 + \cdots,$$

which is ordered with respect to powers of z_1, is called *primitive* in case $n > 1$ provided $\alpha_0, \alpha_1, \ldots$ are relatively prime as power series in $z_2, \ldots, z_n$. That is, f is primitive if the relation

$$(\alpha_0, \alpha_1, \ldots) \sim 1 \tag{15}$$

holds in the ring of power series in the variables $z_2, \ldots, z_n$. Now we will carry over the well-known number-theoretic theorem of Gauss as well as his proof of it. Namely, we will show that the product of two primitive power series is again primitive.

If, in addition to f, the power series

$$g = \beta_0 + \beta_1 z_1 + \beta_2 z_1^2 + \cdots$$

is also primitive, that is if

$$(\beta_0, \beta_1, \ldots) \sim 1, \tag{16}$$

then we form the product series

$$fg = \gamma_0 + \gamma_1 z_1 + \gamma_2 z_1^2 + \cdots,$$

where the coefficients are given by

$$\gamma_k = \alpha_0 \beta_k + \alpha_1 \beta_{k-1} + \cdots + \alpha_k \beta_0 \qquad (k = 0, 1, \ldots). \tag{17}$$

Now, if the coefficients $\gamma_0, \gamma_1, \ldots$ are not relatively prime, then there exists as common divisor a power series, κ, which does not contain z_1 and is of smallest possible positive order. Then κ must be a prime series. As a result of (15) and (16) there are two nonnegative integers r and s such that

$$\kappa \mid \alpha_l \ (l = 0, \ldots, r - 1), \qquad \kappa \mid \beta_l \ (l = 0, \ldots, s - 1),$$

$$(\kappa, \alpha_r) \sim 1, \qquad (\kappa, \beta_s) \sim 1.$$

If we choose $k = r + s$ in (17) then each of the terms $\alpha_p \beta_q$ $(p + q = k)$ on the right side for which $p \neq r$ and $q \neq s$ contains a factor, α_p or β_q, divisible by κ. Since the left side is also divisible by κ it follows that

$$\kappa \mid \alpha_r \beta_s. \tag{18}$$

By the divisibility criterion for power series in $n - 1$ variables, (18) yields a contradiction of the assumptions $(\kappa, \alpha_r) \sim 1$ and $(\kappa, \beta_s) \sim 1$. This criterion will be proved forthwith, thereby completing the proof of the above assertion.

We will refer to the following statement as the *divisibility criterion.*

Theorem 2: Let f, g, h be three power series such that f and g are relatively prime and f divides the product gh. Then f is a divisor of h.

Proof: The proof will be carried out by means of induction on n. One is permitted to use the power series version of the theorem of Gauss for the transition from $n - 1$ to n when $n > 1$. This is because, in the proof of Gauss' theorem, we used the divisibility criterion for $n - 1$ rather than n.

First we dispose directly of the trivial cases in which one of the three series, f, g, h is a unit or 0. If $f = 0$, then it follows from $f \mid (gh)$ and $(f, g) \sim 1$ that $h = 0$. Thus $f \mid h$. If $g = 0$, then it follows from $(f, g) \sim 1$ that f is a unit and, therefore, a divisor of h. If $h = 0$ or f is a unit, then we again have $f \mid h$. If g is a unit, then from $f \mid (gh)$ it follows that f also divides $g^{-1}(gh) = h$. Finally, if h is a unit, then f also divides $h^{-1}(gh) = g$; on the other hand, f is relatively prime to g. Thus f is a unit which again divides h.

From now on we can assume that the orders

$$\omega(f) = k, \qquad \omega(g) = l, \qquad \omega(h) = m$$

are three natural numbers. Since the statement of the theorem remains unchanged under invertible homogeneous linear transformations of the variables $z_1, \ldots, z_n$, we can assume the three power series f, g, h are normalized. Furthermore, since divisibility properties are unchanged under multiplication by units, according to the preparation theorem we may take f, g, h in the form

$$\begin{aligned} f &= z_1^k + \alpha_1 z_1^{k-1} + \cdots + \alpha_k, \\ g &= z_1^l + \beta_1 z_1^{l-1} + \cdots + \beta_l, \\ h &= z_1^m + \gamma_1 z_1^{m-1} + \cdots + \gamma_m \end{aligned} \tag{19}$$

where $\alpha_1, \ldots, \gamma_m$ are power series in the $n - 1$ variables $z_2, \ldots, z_n$. In particular, when $n = 1$ we have $f = z_1^k$, $g = z_1^l$ which is impossible because $(f, g) \sim 1$. Thus the assertion is proved for $n = 1$. We now assume that $n > 1$ and that the assertion is proved for power series in $n - 1$ variables.

In addition to the concept of divisibility of power series we also utilize the corresponding notion for polynomials which is familiar from algebra. We form the resultant $\rho(f, g)$ of the two polynomials f and g in the variable z_1 defined in (19). It is a polynomial in $\alpha_1, \ldots, \alpha_k$ and $\beta_1, \ldots, \beta_l$ with integral coefficients; therefore it is a power series in $z_2, \ldots, z_n$. First, we consider the case $\rho(f, g) = 0$. We denote the ring of power series in $z_2, \ldots, z_n$ by Σ and the corresponding quotient field by Ω. It follows from the vanishing of the resultant that f and g have a common divisor in the ring of all polynomials in z_1 with coefficients from Ω. This divisor actually contains the variable z_1. Clearing denominators, we obtain two equations

$$\lambda f = t f_1, \qquad \mu g = t g_1 \tag{20}$$

where λ and μ are elements of Σ and $\lambda\mu \neq 0$. Furthermore, t, f_1, and g_1 are polynomials in z_1 with coefficients from Σ, and z_1 actually appears in t. Let r be the degree of t with respect to z_1. Since the highest power of z_1 occurring in f has coefficient 1, in view of (20) every common divisor of the coefficients of the polynomial f_1 also divides λ. Therefore, we can assume that f_1 and similarly g_1 and t are primitive. Since they are products of primitive power series, tf_1 and tg_1 are themselves primitive. Now λ and μ are units by (20); consequently $t \mid f$ and $t \mid g$. From the assumption $(f, g) \sim 1$ we find that t itself is a unit. Now, in descending powers of z_1, we have

$$t = \tau z_1^r + \cdots; \qquad \lambda^{-1} f_1 = \sigma z_1^{k-r} + \cdots; \qquad \tau\sigma = 1.$$

Since the power series σ is a unit, it contains a nonvanishing constant term. Therefore we obtain

$$\omega(\lambda^{-1} f_1) \leqslant k - r < k = \omega(f) = \omega(t) + \omega(\lambda^{-1} f_1).$$

This yields the contradiction $\omega(t) > 0$ and therefore, the absurdity of the assumption $\rho(f, g) = 0$.

Because $\rho(f, g) = \rho \neq 0$ we can determine two polynomials p and q in z_1 with coefficients from Σ such that

$$pf + qg = \rho, \qquad pfh + qgh = \rho h.$$

In the last equation the left-hand side is divisible by f because of the assumption $f \mid (gh)$. Therefore, we have

$$\rho h = fd \tag{21}$$

where d is a power series in the variables $z_1, \ldots, z_n$. Using the procedure of the uniqueness proof in Theorem 1, we observe that d is a polynomial in z_1. Arrange d with respect to powers of z_1 and let δ be a greatest common divisor of the coefficients. Since the highest power of z_1 appearing in h again has coefficient 1, it follows from (21) that δ divides ρ. The product $(\rho\delta^{-1})h$ is primitive because f and $\delta^{-1}d$ are primitive. Thus $\rho\delta^{-1}$ is a unit, and (21) yields the proof of the theorem.

The uniqueness theorem for prime factorization of a power series $f \neq 0$ now follows from the divisibility criterion as it does from the corresponding theorem in algebra and number theory. If f is not a unit, let p_1 be a divisor of f whose order, $\omega(p_1)$, is positive and as small as possible. If we apply the same procedure to the quotient f/p_1 and observe that its order is smaller than that of f, then after a finite number of steps we obtain a decomposition

$$f = p_1 p_2 \cdots p_r$$

where $p_1, p_2, \ldots, p_r$ are prime power series. Thus

$$f \sim p_1 p_2 \cdots p_r, \tag{22}$$

and this formula remains valid if $p_1, \ldots, p_r$ are replaced by associate power series. The uniqueness theorem to be proved states that the left-hand side of (22) determines the prime series $p_1, \ldots, p_r$ uniquely except for ordering and replacement by associate prime series.

Namely, if

$$p_1 p_2 \cdots p_r \sim q_1 q_2 \cdots q_s \tag{23}$$

where $q_1, q_2, \ldots, q_s$ are also prime series and p_1 is not an associate of any of these prime series, then p_1 is also relatively prime to each of them. On the other hand, the statement

$$p_1 \mid q_m q_{m+1} \cdots q_s$$

is correct for $m = 1$. Using Theorem 2 with $p_1 = f$, $q_m = g$, and $q_{m+1} \cdots q_s = h$, the validity of this statement for a natural number $m < s$ implies its validity for $m + 1$ in place of m. The statement is false, however, for $m = s$. Thus with a suitable ordering we can assume that $p_1 \sim q_1$ and correspondingly shorten relation (23). The uniqueness theorem then follows by induction.

It is now also easy to show that the greatest common divisor of finitely or infinitely many power series $h_1, h_2, \ldots$, which are not all zero, is uniquely determined by these series up to an arbitrary unit factor. Because of the uniqueness of prime factorization which has already been proved, we can find finitely or infinitely many nonassociate prime series p_l $(l = 1, 2, \ldots)$ such that

$$h_k \sim \prod_l p_l^{g_{kl}}$$

is the unique decomposition for every $h_k \neq 0$. For each value of k under consideration the exponents g_{kl} are nonnegative integers which are unequal to 0 for only finitely many l. A power series, q, is a common divisor of $h_1, h_2, \ldots$ if and only if

$$q \sim \prod_l p_l^{g_l}, \qquad g_l \leqslant \min_k g_{kl} \qquad (l = 1, 2, \ldots).$$

Moreover, the equality sign holds throughout only in the case of the greatest common divisor.

We pass to the field of quotients of the ring of convergent power series in the variables $z_1, \ldots, z_n$ using the method which is customary in algebra. We assign to any two power series p, q with $q \neq 0$ the symbol p/q, and we define equality of two such symbols

$$\frac{p}{q} = \frac{v}{w} \tag{24}$$

by the condition

$$pw = qv \tag{25}$$

which is understood to mean equality of power series. It should be observed that in defining the symbols p/q the variables $z_1, \ldots, z_n$ are considered as indeterminates. Therefore, the symbols will not presently be viewed as numerical functions of the variables. In this way the introduction of the numerical expression 0/0 is avoided. As a consequence of (24) and (25) every fraction p/q may be put in reduced form by dividing the numerator and the denominator by their greatest common divisor, (p, q). The numerator and denominator are then uniquely determined up to a common unit factor. The example $p = z_1$, $q = z_2$ shows that neither the numerator nor the denominator of a reduced fraction need be a unit when $n = 2$. This is in contradistinction to the case $n = 1$. It should also be noted that our previous considerations may be carried over to the ring of all power series, convergent and divergent. The proof of the preparation theorem even becomes somewhat shorter since the earlier convergence considerations can be omitted. We then show the corresponding uniqueness of prime factorization. In this connection there is the question of whether a convergent power series which is a prime series in the earlier sense may, perhaps, decompose into two divergent factors, neither of which is a unit. It can be shown that this cannot happen, but we will not do so.

3. *Regular functions*

We will deal with continuation of power series before taking up the definition of regular functions of n complex variables. In this connection we begin at the same point as in Section 3, of Chapter 1 (Volume I); however, for $n > 1$ new problems arise.

A point z in complex n-space $\mathfrak{Z}$ is given by a column of n complex quantities $z_1, \ldots, z_n$. The two rules

$$|a + b| \leqslant |a| + |b|, \qquad |\lambda a| = |\lambda|\,|a|,$$

with λ a complex scalar, hold exactly as in the case $n = 1$. Here

$$|z| = \max(|z_1|, \ldots, |z_n|)$$

is the absolute value already defined in Section 1. Accordingly, every point set given by the inequality $|z - a| < \rho$, with ρ a positive real number, is convex. We call it the *n-disk with radius* ρ *and center* a.

If a given power series, $p(z)$, in the variables $z_1, \ldots, z_n$ is convergent in the n-disk $|z| < \rho$, then every substitution

$$z = a + x \qquad (|a| < \rho)$$

gives rise to another power series $p_a(x)$ in the variables $x_1, \ldots, x_n$ by rearrangement of terms. The series $p_a(x)$ certainly converges in the n-disk $|x| < \rho - |a|$. In this n-disk we have

$$p(z) = p(a + x) = p_a(x)$$

and $p_a(x)$ is said to arise from $p(z)$ at a by *rearrangement of power series.* As in the case $n = 1$ it might be possible to come out of the original circle of convergence $|z| < \rho$ by means of a finite chain of rearrangements of power series

$$z = a^{(1)} + x^{(1)}, \qquad x^{(1)} = a^{(2)} + x^{(2)}, \ldots, x^{(m-1)} = a^{(m)} + x.$$

The considerations in Section 3 of Chapter 1 (Volume I) can be carried over without any new difficulties to the case of any number of variables. In particular, one can define continuation of a power series, $p(z - c)$, from c along a path in $\mathfrak{Z}$ in the same manner as indicated in that section. At this point it is not necessary to explain this once again. However, we must now investigate the behavior of certain statements about divisibility under continuation of power series.

Let $u(z)$ be a unit, thus $u(0) = \gamma \neq 0$ and $u - \gamma = v$ is a power series of positive order. Let a sufficiently small positive number ρ be selected so that $u(z)$ converges and $|v(z)| < |\gamma|$ for all z in the n-disk $|z| < \rho$. Putting $z = a + x$ with $|a| < \rho$ we have

$$u_a(x) = u(z) = \gamma + v(z), \qquad u_a(0) = \gamma + v(a) \neq 0;$$

therefore, $u_a(x)$ is also a unit. Thus we have shown that every unit remains a unit upon continuation along a sufficiently small path. We pass from this almost trivial remark to a deeper theorem.

Theorem 1: Let $p(z)$ and $q(z)$ be two relatively prime power series which converge for $|z| < \rho$. Then there is a positive number $\sigma \leqslant \rho$ such that for all a with $|a| < \sigma$ and $z = a + x$ the power series $p_a(x)$ and $q_a(x)$ obtained at a by rearrangement of power series are always relatively prime.

Proof: We proceed in a fashion similar to the derivation of the divisibility criterion in the above paragraph. In the case $n = 1$ the assumptions imply that at least one of the given power series is a unit and we come back to the nearly trivial case already treated. Thus we can assume $n > 1$ and neither $p(z)$ nor $q(z)$ is a unit.

By a properly selected homogeneous linear transformation of the variables $z_1, \ldots, z_n$ we first normalize p and q, and we note that the value of ρ may be changed by this procedure. The preparation theorem uniquely determines two units u and v such that

(1) $$pu = z_1^k + \alpha_1 z_1^{k-1} + \cdots + \alpha_k, \qquad qv = z_1^l + \beta_1 z_1^{l-1} + \cdots + \beta_l.$$

Here the natural numbers k and l are the orders of p and q. The radius ρ will eventually be further decreased in order that u and v converge. The coefficients $\alpha_1, \ldots, \alpha_k$ and $\beta_1, \ldots, \beta_l$ are power series in the $n-1$ variables $z_2, \ldots, z_n$.

The assumption $(p, q) \sim 1$ implies that $(pu, qv) \sim 1$. Now, if the theorem is correct with pu, qv in place of p, q, then the rearranged series

$$(pu)_a = p_a u_a, \qquad (qv)_a = q_a v_a \qquad (|a| < \sigma \leqslant \rho)$$

are also relatively prime; consequently, the same is true for the pair p_a, q_a. Thus it is sufficient to prove the theorem under the assumption that

(2) $$p = z_1^k + \alpha_1 z_1^{k-1} + \cdots + \alpha_k, \qquad q = z_1^l + \beta_1 z_1^{l-1} + \cdots + \beta_l$$

instead of (1). Let μ denote the resultant of these two polynomials with respect to z_1. Because $(p, q) \sim 1$ it is a nonzero power series in $z_2, \ldots, z_n$ as was already demonstrated in the proof of Theorem 2 of Section 2.

Now for two suitable power series r and s we have

$$rp + sq = \mu.$$

In particular, r and s are polynomials with respect to z_1 whose coefficients are polynomials in $\alpha_1, \ldots, \alpha_k$ and $\beta_1, \ldots, \beta_l$ with integral coefficients. Therefore, upon rearrangement of power series, we also have

(3) $$r_a p_a + s_a q_a = \mu_a \qquad (|a| < \sigma \leqslant \rho).$$

If we decompose the transformation $z = a + x$ into its n components $z_m = a_m + x_m$ $(m = 1, \ldots, n)$, then the relations

(4) $$p_a = x_1^k + p^*, \qquad q_a = x_1^l + q^*$$

follow from $z_1 = a_1 + x_1$ and (2). Here p^* and q^* are polynomials in x_1 of respective degrees at most $k - 1$ and $l - 1$, whose coefficients are power series in $x_2, \ldots, x_n$. Now, because x_1^k and x_1^l have 1 as coefficient in (4), we see that p_a and q_a are primitive power series in $x_1, x_2, \ldots, x_n$ with x_1 distinguished. We must show that if we put $(p_a, q_a) \sim h$, then the power series $h(x)$ is a unit. Since p_a is primitive h must also be primitive.

Now (3) yields a decomposition

(5) $$\mu_a = jh, \qquad j = \delta_0 + \delta_1 x_1 + \delta_2 x_1^2 + \cdots$$

with coefficients $\delta_0, \delta_1, \ldots$ which are power series in $x_2, \ldots, x_n$. Let

$$(\delta_0, \delta_1, \ldots) \sim \lambda.$$

Because h is primitive, (5) implies the relation

$$\lambda \sim \mu_a.$$

On the other hand, the orders satisfy

$$\omega(\mu_a) = \omega(j) + \omega(h), \qquad \omega(j) \geqslant \omega(\lambda) = \omega(\mu_a);$$

therefore,

$$\omega(h) = 0$$

which proves our assertion.

Let $\mathfrak{G}$ be a given domain in n-dimensional complex space $\mathfrak{Z}$, that is, a connected open set of points z. A single-valued complex function, $f(z)$, defined on $\mathfrak{G}$ is said to be *regular on* $\mathfrak{G}$ provided that for each point, a, of $\mathfrak{G}$ there is a convergent power series, $p_a(x)$, in the variables $x_1, \ldots, x_n$ and a circle of convergence $|x| < \rho_a$ which is carried into $\mathfrak{G}$ by the transformation $z = a + x$ so that the following numerical equality holds

$$f(z) = p_a(z - a) \qquad (|z - a| < \rho_a).$$

The series p_a is called the *function element of f at the point a* or, simply, the *local part of f at a*. It is clear that p_a is uniquely determined as a power series by $f(z)$ and a. Conversely, as in the case of a single variable, we also obtain the complete function, $f(z)$, at all points z of $\mathfrak{G}$ by means of analytic continuation of a given function element, $p_a(z - a)$, at an arbitrarily fixed point a.

If b is another point of $\mathfrak{G}$ and, correspondingly,

$$f(z) = p_b(z - b) \qquad (|z - b| < \rho_b),$$

then the relation

$$p_a(z - a) = p_b(z - b) \qquad (|z - a| < \rho_a;\ |z - b| < \rho_b)$$

holds at all points in the intersection of both circles of convergence. This formula will be referred to as the *coherence condition*. Obviously, we can also start out to define the function $f(z)$ by specifying the local part, $p_a(z - a)$, at each point, a, of $\mathfrak{G}$ subject to the coherence condition. In order to verify the coherence condition at given points a and b we must establish that rearrangement of the power series $p_a(z - a)$ and $p_b(z - b)$ at a common point, c, of their circles of convergence gives rise to two identical power series.

The definition of a regular function given above conforms to the viewpoint of Weierstrass. However, as will be shown presently, we can also begin by requiring complex differentiability; that is, we can follow the road taken by Cauchy in the case of a single variable. In this connection it is convenient to impose additional requirements upon the functions under consideration.

These functions are to be continuous with respect to all the variables $z_1, \ldots, z_n$ throughout the domain of definition, not merely continuous with respect to each individual variable.

Theorem 2: A complex function defined in $\mathfrak{G}$ is regular there if and only if it is continuous with respect to all the variables and differentiable in each separate variable at every point of $\mathfrak{G}$.

Proof: If the given function $f(z)$ is regular in $\mathfrak{G}$ then, just as in the well-known case of a single variable, the continuity and existence of partial derivatives of arbitrarily high order follow from the local expansion in a power series.

Conversely, let $f(z)$ be continuous in all variables and differentiable in each of them throughout $\mathfrak{G}$. First, in a neighborhood of a point a, we consider $f(z)$ as a function of z_1 alone by holding the variables $z_2, \ldots, z_n$ fixed. Now we obtain the representation

$$f(z_1, z_2, \ldots, z_n) = \frac{1}{2\pi i} \int_{C_1} \frac{f(\zeta_1, z_2, \ldots, z_n)}{(\zeta_1 - z_1)} \, d\zeta_1 \qquad (|z - a| < \rho)$$

from the Cauchy integral formula when the closed n-disk $|z - a| \leqslant \rho$ lies in $\mathfrak{G}$ and C_1 denotes the positively oriented circle $|\zeta_1 - a_1| = \rho$. If we view the integrand as a function of z_2 and, analogously, carry out the procedure of applying the Cauchy formula $n - 1$ additional times, then we obtain

$$(6) \quad f(z_1, \ldots, z_n) = (2\pi i)^{-n} \int_{C_1} \cdots \left(\int_{C_n} \frac{f(\zeta_1, \ldots, \zeta_n)}{(\zeta_1 - z_1) \cdots (\zeta_n - z_n)} \, d\zeta_n \right) \cdots d\zeta_1 \qquad (|z - a| < \rho).$$

Here the successive integrations range over the circles $C_n, C_{n-1}, \ldots, C_1$ defined by $|\zeta_k - a_k| = \rho$ $(k = n, \ldots, 1)$. Since the function $f(z)$ is, by assumption, continuous in all variables, we can replace the iterated integral on the right hand side of (6) by an n-fold integral. Finally, as in the case of a single variable, we expand each fraction, $1/(\zeta_k - z_k)$, as a geometric series in powers of $(z_k - a_k)/(\zeta_k - a_k)$. By interchanging integration and summation we obtain the desired function element; thereby, we have completed the proof.

We conclude immediately that the regular functions in a domain $\mathfrak{G}$ form a ring without zero-divisors. Accordingly, the definitions of divisibility and units carry over to this ring. In particular, f is called a *unit in* $\mathfrak{G}$ if f^{-1} is regular in $\mathfrak{G}$. It follows from Theorem 2 that the latter is the case if and only if f possesses no zeros in $\mathfrak{G}$.

We are now going to investigate which relations hold between local divisibility and divisibility in the large, that is, the relations between divisibility of power series and divisibility of regular functions. If f is a regular function on $\mathfrak{G}$, then we denote by f_a the local part of f at the point a. If the relation $g \mid f$ holds for a regular function, g, on $\mathfrak{G}$, then $f = gh$ for a function h likewise regular on $\mathfrak{G}$. Thus we also have $f_a = g_a h_a$, from which follows the formula $g_a \mid f_a$ in the sense of divisibility of power series.

Conversely, now let $g_a \mid f_a$ for each point a of $\mathfrak{G}$. Thus we also have

$$f_a = g_a h_a \qquad (|z - a| < \rho_a) \tag{7}$$

with h_a a power series and where ρ_a is properly selected. If the power series $g_a = 0$ for a given a, then by (7) f_a is also zero, and by continuation it follows that $g = 0, f = 0$. Thus we also have $g \mid f$. It will now be assumed that g does not vanish identically. If the circles of convergence $|z - a| < \rho_a$ and $|z - b| < \rho_b$ have a common point c, then by rearrangement of power series it follows that

$$(f_a)_c = (f_b)_c \qquad \text{and} \qquad (g_a)_c = (g_b)_c \text{ at } c.$$

Next from (7) it also follows that

$$0 = (g_a)_c((h_a)_c - (h_b)_c), \qquad (h_a)_c = (h_b)_c.$$

Therefore the power series h_a satisfy the coherence condition; consequently, they are the local parts of a function h which is regular on $\mathfrak{G}$ and satisfies the equation $f = gh$. Thus we have shown that the relation $g \mid f$ holds between two functions f and g which are regular on $\mathfrak{G}$ if and only if the corresponding relation $g_a \mid f_a$ holds for the local parts f_a and g_a at every point a of $\mathfrak{G}$.

Two functions f and g which are regular in $\mathfrak{G}$ are said to be *relatively prime on* $\mathfrak{G}$ if, except for units, there is no regular function on $\mathfrak{G}$ which is a common divisor of f and g. This will be expressed, once again, by the symbol $(f, g) \sim 1$. On the other hand, if f and g are not relatively prime, then the relations

$$f = hp, \qquad g = hq$$

hold on $\mathfrak{G}$, where h, p, q are regular and h is not a unit. For each point a of $\mathfrak{G}$ we also have

$$f_a = h_a p_a, \qquad g_a = h_a q_a, \qquad h_a \mid f_a, \qquad h_a \mid g_a.$$

Since the function h is not a unit on $\mathfrak{G}$, it has a zero at a certain point, a, of $\mathfrak{G}$. Likewise f_a is then a nonunit in the local ring of power series. Thus if the local parts f_a and g_a are relatively prime for all points, a, of $\mathfrak{G}$, then the functions f and g are relatively prime on $\mathfrak{G}$. We remark without proof that the converse of this theorem is not valid for arbitrary domains, as may be seen by a suitable example. However, in Section 5 we will prove the converse for the case where $\mathfrak{G}$ is the whole space $\mathfrak{Z}$, that is, when f and g are entire functions.

4. Meromorphic functions

At this point the theory of functions of one complex variable suggests that we extend the concept of a Riemann surface to the case of n variables and introduce the notions of holomorphic and meromorphic functions on the same level of generality. However, the introduction of ramification points requires a great deal of effort and, moreover, will not be needed in the sequel. Therefore, in the sequel we will restrict ourselves to the case of a domain $\mathfrak{G}$ in the space $\mathfrak{Z}$. Most of the time $\mathfrak{G}$ will coincide with $\mathfrak{Z}$.

We already introduced the field of quotients of the ring of convergent power series. Passing from the origin to an arbitrary point a and replacing z by $x = z - a$ we obtain the quotient

$$r(x) = \frac{p(x)}{q(x)} \tag{1}$$

where p and q are power series in $x_1, \ldots, x_n$, and $q(x)$ is not the zero series. We assume the fraction p/q to be reduced; consequently, the power series p and q are relatively prime. They are uniquely determined by r only up to a common unit factor. For fixed p and q let $|z - a| < \rho_a$ be a common circle of convergence denoted by $\mathfrak{K}_a$. Furthermore, let $\mathfrak{K}_a^*$ be the set of points, z, of $\mathfrak{K}_a$ where not both of the values $p(z - a)$ and $q(z - a)$ are equal to zero. Since $q(x)$ is not the zero series the set $\mathfrak{K}_a^*$ is everywhere dense in $\mathfrak{K}_a$. In addition, $\mathfrak{K}_a^*$ is an open subset of $\mathfrak{K}_a$. Then according to (1), at each point of $\mathfrak{K}_a^*$ the fraction $r(z - a)$ is either equal to a certain complex number or equal to ∞. The latter occurs in case the denominator is zero. Therefore, the points of $\mathfrak{K}_a^*$ are called *points of determinacy* of $r(x)$. It must be noted that for the purpose of this definition p, q, and ρ_a are understood to be given quantities.

We now come to the definition of meromorphic functions on $\mathfrak{G}$. To each point a of $\mathfrak{G}$ let there be assigned a quotient,

$$r_a(x) = \frac{p_a(x)}{q_a(x)} \qquad (x = z - a),$$

of two relatively prime power series p_a and $q_a \neq 0$. Let $\mathfrak{K}_a$ be a common circle of convergence of these series. If z is any point of $\mathfrak{G}$ which is a point of determinacy of $r_a(x)$ with respect to a suitable choice of a, then we assume the value

$$f(z) = r_a(z - a) \tag{2}$$

to be independent of the choice of a. Thus we can associate a unique value $f(z)$ with such a point z in $\mathfrak{G}$. No value, $f(z)$, is associated with any other point z in $\mathfrak{G}$. The function $f(z)$ is then said to be *meromorphic on* $\mathfrak{G}$, and the points z occurring in (2) are called its *points of determinacy*. The requirement that $f(z)$ be uniquely determined by (2) means that whenever the intersection $\mathfrak{K}_a^* \cap \mathfrak{K}_b^*$ is nonvoid we have

$$r_a(z-a) = r_b(z-a)$$

for every point z of $\mathfrak{K}_a^* \cap \mathfrak{K}_b^*$; or equivalently,

$$p_a(z-a)q_b(z-b) = q_a(z-a)p_b(z-b). \tag{3}$$

Furthermore, if c is any point of $\mathfrak{K}_a^* \cap \mathfrak{K}_b^*$, then the relation

$$(p_a)_c(q_b)_c = (q_a)_c(p_b)_c \tag{4}$$

is an identity of power series. Hence, relation (3) is valid throughout the intersection $\mathfrak{K}_a \cap \mathfrak{K}_b$. Again, this formula will be referred to as the *coherence condition*. Letting a and b be any two points of $\mathfrak{G}$, the coherence condition yields a necessary and sufficient condition for the existence of a meromorphic function f on $\mathfrak{G}$ whose local parts are given by

$$f_a = r_a(z-a) = \frac{p_a(z-a)}{q_a(z-a)}, \qquad |z-a| < \rho_a.$$

We must still verify that the concept of points of determinacy and the definition of the value of $f(z)$ at these points are independent of both the representation of the r_a as a reduced fraction, p_a/q_a, and the choice of the corresponding radius, ρ_a. Let b be a point of determinacy contained in $\mathfrak{K}_a^*$; then at least one of the values $p_a(b-a)$, $q_a(b-a)$ is unequal to zero. Now, if

$$r_b(z-b) = \frac{P_b(z-b)}{Q_b(z-b)} \qquad (|z-b| < \sigma_b) \tag{5}$$

is another reduced representation of

$$r_b(z-b) = \frac{p_b(z-b)}{q_b(z-b)} \qquad (|z-b| < \rho_b) \tag{6}$$

then

$$p_bQ_b = q_bP_b, \tag{7}$$

whereas, by (4),

$$(p_a)_bq_b = (q_a)_bp_b, \tag{8}$$

and at least one of the power series $(p_a)_b$, $(q_a)_b$ is a unit. It follows from (8) and (7) that one of the power series p_b, q_b as well as one of the power series P_b, Q_b is a unit. Thus b is also a point of determinacy with respect to the representations (5) and (6), and the value $f(b) = r_b(0)$ is, in fact, uniquely determined by (8). These considerations depend in an essential fashion on the fact that the fractions r_a are given in reduced form.

We have just seen that the points of determinacy, a, are characterized in a more concise fashion by saying that at least one of the power series p_a, q_a is a unit. In particular, if q_a is a unit then the value $f(a)$ is finite and the local part, f_a, is also a convergent power series. In this case the point a is said to be a *regular point* of the function. On the other hand, if q_a is not a unit, that is, $q_a(0) = 0$, then the value $f(a)$ is infinite and the point a is called a *pole* of the function. Finally, if p_a is not a unit, that is, $p_a(0) = 0$, then the value $f(a)$ vanishes and the point a is called a *zero* of the function.

Since every neighborhood of each point a of $\mathfrak{G}$ contains points z satisfying $q_a(z - a) \neq 0$, the regular points of f form an open subset of $\mathfrak{G}$ which is everywhere dense in $\mathfrak{G}$. On the other hand, the power series p_a is a unit at every pole a. Therefore, any pole, a, has a sufficiently small neighborhood containing neither zeros nor points of indeterminacy. Furthermore, assuming $z \to a$ the reciprocal value of $f(z)$ tends to 0. Thus $f(z)$ itself tends to ∞. Finally, if a neighborhood of any point, a, of $\mathfrak{G}$ consists only of zeros, then we have $p_a = 0$; applying the coherence condition, it follows that $p_b = 0$ for every point b of $\mathfrak{G}$ because $\mathfrak{G}$ is connected. Therefore, in this case, $f(z)$ vanishes identically.

Now we investigate the behavior of $f(z)$ when z, restricted to points of determinacy, tends to a pole or a point of indeterminacy. After a suitable homogeneous linear transformation of the variables, the preparation theorem yields a representation

$$(9) \qquad r_a(x) = \frac{p_a(x)}{q_a(x)}\, u_a(x), \qquad p_a(x) = x_1^k + \alpha_1 x_1^{k-1} + \cdots + \alpha_k,$$
$$q_a(x) = x_1^l + \beta_1 x_1^{l-1} + \cdots + \beta_l,$$

where $\alpha_1, \ldots, \beta_l$ are power series in $x_2, \ldots, x_n$ and $u_a(x)$ is a local unit. The integers k and l are the orders of p_a and q_a respectively, and, in particular, $l > 0$ because the point a is not regular. Since the coefficients $\beta_1, \ldots, \beta_l$ have positive order, their absolute values are smaller than any given positive number ε provided that $x_2, \ldots, x_n$ lie in a sufficiently small neighborhood of 0. Then, for given values of $x_2, \ldots, x_n$ the equation $q_a(x) = 0$ has l solutions x_1, each of which tends to 0 whenever ε tends to 0. In case $n = 1$ we obtain the unique solution $x_1 = 0$, thus proving the well known fact that a meromorphic function of one variable has only isolated poles. Moreover, we have $p_a(x) = 1$, which implies that there are no points of indeterminacy in this case.

In case $n > 1$ we still have $n - 1$ independent parameters $x_2, \ldots, x_n$. In particular, we can find a sequence of points, $b \neq a$, tending to the point a such that each b is a zero of $q_a(z - a)$. In this case if the point a is a pole of $f(z)$, we again have $p_a(x) = 1$. Therefore, the rearrangements of power series p_a and q_a at the points b are also relatively prime. Consequently, each point b is a pole of $f(z)$. Hence every pole is a limit point of poles.

If the point a is not a pole but is a point of indeterminacy, then we also have $k > 0$ in (9). Since the power series p_a and q_a are relatively prime the resultant μ of the polynomials p_a and q_a with respect to x_1 can be written in the form

$$\mu = sp_a + tq_a \tag{10}$$

as was done in Section 3. Here s and t are power series in $x_1, \ldots, x_n$ and μ does not vanish identically in $x_2, \ldots, x_n$. Subjecting the $n - 1$ variables $x_2, \ldots, x_n$ to the condition $\mu \neq 0$ and determining x_1 by the equation $q_a(x) = 0$, we obtain from (10) that $p_a(x) \neq 0$. Thus we can construct a sequence of poles b tending to the point a. Interchanging p_a and q_a we also obtain a sequence of zeros tending to the point a. Furthermore, for every complex constant λ the series $p_a - \lambda q_a$ and q_a are relatively prime because p_a and q_a are relatively prime. Therefore, the point a is also a limit point of zeros of $f - \lambda$. It follows that for every prescribed value λ there are arbitrarily small neighborhoods of the point a containing points of determinacy at which the function $f(z)$ assumes the value λ.

In view of (10), the resultant μ has positive order m since both integers k and l are positive. Applying the preparation theorem once again, we obtain

$$\mu\omega = x_2^m + \gamma_1 x_2^{m-1} + \cdots + \gamma_m \tag{11}$$

where $\gamma_1, \ldots, \gamma_m$ are power series without constant term in the $n - 2$ variables $x_3, \ldots, x_n$ and ω is a local unit depending upon $x_2, \ldots, x_n$. In case $n = 2$, the right-hand side of (11) vanishes only for $x_2 = 0$. Then, in view of (9), within a suitable neighborhood of a the power series p_a and q_a have only the trivial zero $z = a$ in common. Thus the point of indeterminacy, a, is isolated in this case. On the other hand, if $n > 2$ it is possible to assign complex numbers of arbitrarily small absolute value to the variables $x_2, \ldots, x_n$ such that $x_3 \cdots x_n \neq 0$ and $\mu = 0$. This yields a common zero of p_a and q_a as polynomials in x_1, which is arbitrarily near the origin. Using Theorem 1 in Section 3 it follows that the points of indeterminacy accumulate at a. More precisely, the totality of points of indeterminacy in suitable neighborhoods of a coincides with the set of common zeros of $p_a(z - a)$ and

$q_a(z - a)$. Here $x_3, \ldots, x_n$ can be considered locally as independent parameters; whereas, upon application of (9) and (10), x_1 and x_2 can assume finitely many values depending on $x_3, \ldots, x_n$. We will not treat the set of points of indeterminacy in greater detail. We remark without proof, however, that the set of points of determinacy, as well as the set of regular points, of any meromorphic function on $\mathfrak{G}$ is a connected subset of $\mathfrak{G}$.

The theory of functions of several variables turns out to be essentially more difficult than the theory of one variable because of the existence of points of indeterminacy. In the case $n > 1$, a mere glance at the poles already indicates a behavior which is completely different from that in the case $n = 1$. The reason is that, in case $n > 1$, the poles are not isolated and, in general, there does not exist a Laurent expansion. In a neighborhood of a nonregular point we are forced to view meromorphic functions as quotients of power series.

The meromorphic functions on $\mathfrak{G}$ form a field if we define the field operations locally by

$$(f + g)_a = f_a + g_a, \qquad (fg)_a = f_a g_a, \qquad (f^{-1})_a = (f_a)^{-1}.$$

This field contains the ring of regular functions on $\mathfrak{G}$ and, consequently, the field of quotients of this ring. The question arises whether, conversely, every meromorphic function on $\mathfrak{G}$ can be represented as the quotient of two regular functions. This question leads to interesting and complicated problems, problems whose solutions were initiated by Weierstrass and Poincaré, continued by Cousin, and completed mainly by Oka during the last thirty years. In the next section we will, for the most part, treat the special case $\mathfrak{G} = \mathfrak{Z}$ which is contained in the investigations of Cousin.

5. *The theorem of Weierstrass and Cousin*

At the outset we consider an arbitrary domain, $\mathfrak{G}$, in $\mathfrak{Z}$. Let $u(z)$ be a function which is a unit on $\mathfrak{G}$. Then u and u^{-1} are regular on $\mathfrak{G}$ and this holds if and only if the local part, u_a, is a unit for every point a of $\mathfrak{G}$. Under these assumptions, we can put

$$u_a = c(1 + w_a), \qquad c = u(a) \neq 0$$

where w_a is a power series of positive order in $x = z - a$. By rearrangement of terms, the expression

$$\log(1 + w_a) = w_a - \tfrac{1}{2}w_a^2 + \cdots$$

yields a convergent power series in the variables $x_1, \ldots, x_n$. Putting

$$v_a = \log c + \log(1 + w_a) = \log c + w_a - \tfrac{1}{2}w_a^2 + \cdots \tag{1}$$

with a specific choice of $\log c = \log(u(a))$, we obtain

$$\log u_a = v_a, \qquad u_a = e^{v_a} = 1 + v_a + \frac{v_a^2}{2!} + \cdots.$$

To begin with, the function $\log u(z)$ is uniquely determined by (1) within a certain circle of convergence of $v_a(z - a)$. Since the relation

$$\frac{\partial \log u(z)}{\partial z_k} = u^{-1} \frac{\partial u}{\partial z_k} \qquad (k = 1, \ldots, n)$$

holds in this circle, analytic continuation of the function element

$$(\log u)_a = \log u_a$$

from a to arbitrary points z of $\mathfrak{G}$ can be achieved by means of the integral

$$\log u - \log c = \int_a^z u^{-1}\left(\frac{\partial u}{\partial z_1}\, dz_1 + \cdots + \frac{\partial u}{\partial z_n}\, dz_n\right).$$

The analytic continuation is independent of the path connecting a with z when $\mathfrak{G}$ is simply connected, that is, when every closed path is homotopic to zero in $\mathfrak{G}$. In that case, once $\log c$ is fixed, for each b in $\mathfrak{G}$ the constant term, $\log b$, of the power series v_b formed analogously to (1) is uniquely determined. Furthermore, the power series v_b are the local parts of the single valued function $v = \log u$ which is regular on $\mathfrak{G}$ and satisfies the relation

$$u = e^v = 1 + v + \frac{v^2}{2!} + \cdots. \tag{2}$$

Therefore, when the domain $\mathfrak{G}$ is simply connected every unit, u, can be put in the form (2), where the regular function v is uniquely determined except for a fixed additive multiple of $2\pi i$. In general this is not true for an arbitrary, not necessarily simply connected, domain. This has already been seen in the case $n = 1$ by taking as a counter example the function $u = z$ in the annulus $1 < |z| < 2$. In fact the function z is a unit but $\log z$ is not single valued in this annulus. The converse, however, is true in any case. If v is any function, regular on an arbitrary domain $\mathfrak{G}$, then the function $u = e^v$ is regular and different from zero everywhere in $\mathfrak{G}$, thus a unit in $\mathfrak{G}$.

We now introduce the concept of *divisors* by a method somewhat analogous to that utilized in Section 6 of Chapter 4 (Volume II). Two functions f and g meromorphic on $\mathfrak{G}$ are called *associates* if the relation $g = fu$ holds, where u is a unit. The set of all functions which are associates of f is called a *principal divisor* or, more precisely, the *divisor of* f. It is designated by (f). If $f = 0$ we obtain the principal divisor (0) consisting only of the zero function. The principal divisors different from (0) obviously form a commutative group arising as a factor group from the multiplicative group of all functions $f \neq 0$ with respect to the group of units.

Local divisors can be defined correspondingly. Two elements

$$r_a = \frac{p_a(z - a)}{q_a(z - a)}, \qquad s_a = \frac{t_a(z - a)}{w_a(z - a)}$$

of the quotient field of all power series in $z - a$ are said to be in the same class if the relation $s_a = r_a u_a$ holds with a local unit u_a. The class containing r_a is then called a *local divisor at the point a* or, more precisely, the *divisor of* r_a, and, it is denoted by

$$(r_a) = \mathfrak{d}_a.$$

Obviously every principal divisor (f) determines the local divisors (f_a) arising from the local parts, f_a, of the function f. If we now have $(f_a) = (r_a)$ and $(f_b) = (r_b)$ we obtain

$$\text{(3)} \qquad f_a = r_a u_a \qquad (|z - a| < \rho_a), \qquad f_b = r_b v_b \qquad (|z - b| < \rho_b)$$

where u_a and v_b are local units at the respective points a and b. Here the radii ρ_a and ρ_b of the corresponding circles of convergence $\mathfrak{K}_a$ and $\mathfrak{K}_b$ shall be chosen so small that the values $u_a(z - a)$ and $v_b(z - b)$ are different from zero everywhere in $\mathfrak{K}_a$ and $\mathfrak{K}_b$ respectively. In case $\mathfrak{K}_a$ and $\mathfrak{K}_b$ have a nonvoid intersection, $\mathfrak{K}_{ab}$, the relation $f_a = f_b$ holds in $\mathfrak{K}_{ab}$; in view of (3), we have the coherence condition

$$\text{(4)} \qquad r_a = r_b u_{ab} \qquad (|z - a| < \rho_a, |z - b| < \rho_b)$$

with u_{ab} a unit on $\mathfrak{K}_{ab}$.

If $(f) = (0)$, that is, $f = 0$ it follows that $(f_a) = (0)$ and vice-versa. More generally, if f and g are any two meromorphic functions and $f \neq 0$, then the relation $(f_a) = (g_a)$ for every point a of $\mathfrak{G}$ implies that

$$(h_a) = (g_a)(f_a)^{-1} = (1).$$

Thus the function $h = gf^{-1}$ is a unit, and we have $(f) = (g)$. The principal divisor (f) is therefore uniquely determined by the local divisors (f_a).

Now instead of starting with a given principal divisor (f) we assign a local divisor $\mathfrak{d}_a = (r_a)$ to every point a of $\mathfrak{G}$. The representatives r_a are required to satisfy the coherence condition (4) throughout $\mathfrak{G}$. In this case we say a *divisor*, $\mathfrak{d}$, is defined on $\mathfrak{G}$ with *local parts* $\mathfrak{d}_a$. If the divisor (0) is excluded, both the local divisors at any point a as well as the divisors on $\mathfrak{G}$ form commutative multiplicative groups. The question arises whether or not an arbitrarily given divisor $\mathfrak{d}$ is always a principal divisor. That is, whether there is a meromorphic function f on $\mathfrak{G}$ whose local parts satisfy $(f_a) = \mathfrak{d}_a$. Cousin answered this question affirmatively if $\mathfrak{G}$ satisfies certain assumptions, and we therefore call this problem *Cousin's problem.* However, it was later shown by Oka that Cousin's problem is not necessarily solvable under certain topological conditions on $\mathfrak{G}$.

A local divisor $\mathfrak{d}_a$ is said to be *integral* if there exists a representative r_a which is a power series. In this case all representatives of $\mathfrak{d}_a$ are power series. Correspondingly, a divisor $\mathfrak{d}$ is said to be *integral* if all its local parts, $\mathfrak{d}_a$, are integral. Obviously any principal divisor (f) is integral if and only if the function f is regular everywhere in $\mathfrak{G}$. Two local divisors

$$\mathfrak{d}_a = (r_a), \qquad \mathfrak{c}_a = (s_a)$$

at any point a are called *relatively prime* if they are integral and the representatives r_a and s_a are relatively prime as power series. Furthermore, two divisors $\mathfrak{d}$ and $\mathfrak{c}$ defined on $\mathfrak{G}$ are called *relatively prime* if their local parts $\mathfrak{d}_a$ and $\mathfrak{c}_a$ are relatively prime at every point, a, of $\mathfrak{G}$. We now show that every divisor on $\mathfrak{G}$ can be written uniquely as a quotient of relatively prime divisors.

Let $\mathfrak{d}$ be a divisor with the local parts

$$\mathfrak{d}_a = (r_a), \qquad r_a = \frac{p_a}{q_a}, \qquad (p_a, q_a) \sim 1. \tag{5}$$

According to Theorem 1 of Section 3 the circles of convergence, $\mathfrak{K}_a$, given by $|z - a| < \rho_a$ can be chosen so small that the power series arising from p_a and q_a by rearrangement of power series at any point of $\mathfrak{K}_a$ are still relatively prime. Applying the coherence condition (4), we obtain

$$p_a q_b = p_b q_a u_{ab} \qquad (|z - a| < \rho_a, |z - b| < \rho_b)$$

where the function u_{ab} is a unit on the intersection, $\mathfrak{K}_{ab}$, of $\mathfrak{K}_a$ and $\mathfrak{K}_b$. It follows that the quotients q_a/q_b and q_b/q_a are both regular functions of z on $\mathfrak{K}_{ab}$. Consequently we obtain

$$q_a = q_b v_{ab} \qquad (|z - a| < \rho_a, |z - b| < \rho_b) \tag{6}$$

where v_{ab} is a unit on $\mathfrak{K}_{ab}$. Since q_b is not the zero series, we also obtain

$$p_a = p_b u_{ab} v_{ab} \qquad (|z - a| < \rho_a, |z - b| < \rho_b). \tag{7}$$

According to (6) and (7), the power series q_a and p_a satisfy the coherence condition (4) for divisors. Thus $\mathfrak{p}_a = (p_a)$ and $\mathfrak{q}_a = (q_a)$ are the local parts of two integral divisors $\mathfrak{p}$ and $\mathfrak{q}$ respectively. In view of (5), these divisors are relatively prime and satisfy the desired equation

$$\mathfrak{d} = \frac{\mathfrak{p}}{\mathfrak{q}}. \tag{8}$$

The result thus obtained also reduces the solution of Cousin's problem to the case of integral divisors. Assuming the solvability of Cousin's problem for integral divisors we consider in particular a principal divisor $\mathfrak{d} = (f)$. Then there exist two relatively prime principal divisors $\mathfrak{p} = (p)$ and $\mathfrak{q} = (q) \neq (0)$ such that the functions f and pq^{-1} differ from each other only by a unit factor. Absorbing this factor in p we obtain

$$f = \frac{p}{q} \tag{9}$$

where p and q are regular functions on $\mathfrak{G}$ and the local parts p_a, q_a are relatively prime at any point, a, of $\mathfrak{G}$. We apply this result to the case $f = gh^{-1}$ where g and $h \neq 0$ are regular and relatively prime functions on $\mathfrak{G}$. According to the divisibility criterion the equation

$$gq = hp$$

implies that $q_a \mid h_a$ at every point, a, of $\mathfrak{G}$, thus proving that the quotient

$$\frac{h}{q} = w$$

is regular on $\mathfrak{G}$. Then we also have $g = wp$ and the function w must be a unit since it is a common divisor of g and h. Therefore, at the outset we could have chosen $p = g$, $q = h$ in (9). Furthermore, the local parts g_a, h_a are relatively prime at every point, a, of $\mathfrak{G}$. Thus assuming the solvability of Cousin's problem for a given domain $\mathfrak{G}$, it is proved that any two regular functions on $\mathfrak{G}$ are relatively prime if and only if the same is true of their local parts at every point.

In the sequel it is useful to pass from the given domain $\mathfrak{G}$ to the closure, $\mathfrak{B}$, of $\mathfrak{G}$ by adding the boundary points. A complex valued function f on $\mathfrak{B}$ is called *meromorphic* or *regular* if it is defined on a domain containing $\mathfrak{B}$ and has the corresponding properties on this domain. The concept of *divisors on* $\mathfrak{B}$ is defined analogously. Therefore we can speak about Cousin's problem on $\mathfrak{B}$. For our purposes it is sufficient to assume $\mathfrak{B}$ to be a *closed box*, that is, the cartesian product $\mathfrak{R}_1 \times \mathfrak{R}_2 \times \cdots \times \mathfrak{R}_n$ of n closed rectangles each of which is contained in one of the complex planes represented by the variables $z_1, \ldots, z_n$. The following theorem will be called *Cousin's Lemma.*

Theorem 1: Let $\mathfrak{Q}_1$ and $\mathfrak{Q}_2$ be two boxes which have a common face of $2n - 1$ real dimensions and let $\mathfrak{Q}$ be the union of $\mathfrak{Q}_1$ and $\mathfrak{Q}_2$. Let $\mathfrak{d}$ be an integral divisor on $\mathfrak{Q}$. If Cousin's problem is solvable for each of the restrictions of $\mathfrak{d}$ to $\mathfrak{Q}_1$ and $\mathfrak{Q}_2$, then this problem is also solvable for $\mathfrak{Q}$ itself.

Proof: We can assume

$$\mathfrak{Q}_k = \mathfrak{R}_1^{(k)} \times \mathfrak{Q}^* \quad (k = 1, 2), \qquad \mathfrak{Q}^* = \mathfrak{R}_2 \times \cdots \times \mathfrak{R}_n$$

where $\mathfrak{R}_1^{(1)}$ and $\mathfrak{R}_1^{(2)}$ are two abutting rectangles in the z_1-plane whose common edge is L. Let $\mathfrak{d}$ be an integral divisor on $\mathfrak{Q}$ given by its local parts $\mathfrak{d}_a$ at all points, a, of $\mathfrak{Q}$. By restriction of $\mathfrak{d}$ to the subsets $\mathfrak{Q}_1$ and $\mathfrak{Q}_2$, we obtain two corresponding divisors $\mathfrak{d}^{(1)}$ and $\mathfrak{d}^{(2)}$. According to the assumptions, there exists a regular function f_k on $\mathfrak{Q}_k$ $(k = 1, 2)$ whose local parts satisfy the condition

$$(f_k)_a = \mathfrak{d}_a^{(k)} \tag{10}$$

everywhere on $\mathfrak{Q}_k$. In particular, for points a of $L \times \mathfrak{Q}^*$, the relation $\mathfrak{d}_a^{(1)} = \mathfrak{d}_a^{(2)}$ is valid. Then in the z_1-plane there exists a sufficiently small open neighborhood $\mathfrak{H}$ of L such that the functions f_1 and f_2 are regular on $\mathfrak{H} \times \mathfrak{Q}^*$ and satisfy the condition

$$f_1 = f_2 u \tag{11}$$

with u a unit on $\mathfrak{H} \times \mathfrak{Q}^*$. We can assume the domain $\mathfrak{H}$ to be bounded by a simple positively oriented closed rectifiable curve C.

Since $\mathfrak{H} \times \mathfrak{Q}^*$ is simply connected, the relations

$$u = e^v, \qquad v = \log u \tag{12}$$

hold with v a regular function on $\mathfrak{H} \times \mathfrak{Q}^*$. According to Cauchy's formula we obtain

$$v(z_1, \ldots, z_n) = \frac{1}{2\pi i} \int_C \frac{v(\zeta, z_2, \ldots, z_n)}{\zeta - z_1} \, d\zeta \tag{13}$$

where z_1 ranges over the interior of $\mathfrak{H}$ and $(z_2, \ldots, z_n)$ is any point of $\mathfrak{Q}^*$. The curve C can be decomposed into two arcs C_1 and C_2 such that C_k $(k = 1, 2)$ does not intersect $\mathfrak{R}_1^{(k)}$ (Figure 1). Putting

$$\frac{1}{2\pi i} \int_{C_k} \frac{v(\zeta, z_2, \ldots, z_n)}{\zeta - z_1} \, d\zeta = v_k(z) \qquad (k = 1, 2)$$

the function v_k obviously is regular throughout $\mathfrak{Q}_k$. Furthermore, according to (13) the relation

$$v = v_1 + v_2 \tag{14}$$

holds in the interior of $\mathfrak{H} \times \mathfrak{Q}^*$. We now define

$$f = f_1 e^{-v_1} \quad (z \text{ in } \mathfrak{Q}_1), \qquad f = f_2 e^{v_2} \quad (z \text{ in } \mathfrak{Q}_2). \tag{15}$$

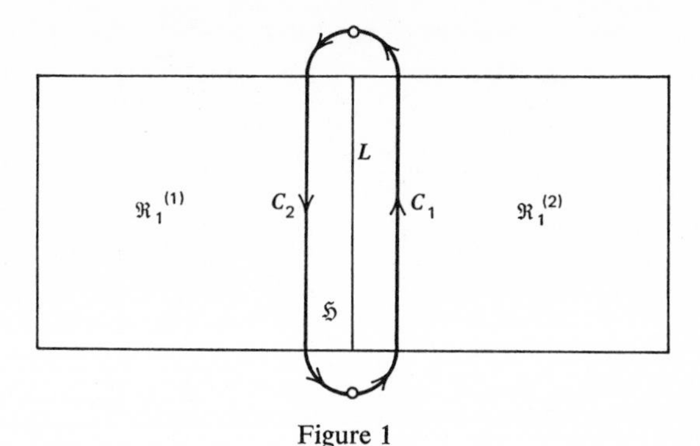

Figure 1

According to (11), (12), and (14) the functions $f_1e^{-v_1}$ and $f_2e^{v_2}$ agree on the interior of $\mathfrak{H} \times \mathfrak{Q}^*$. Therefore the function $f_2e^{v_2}$, defined and regular on $\mathfrak{Q}_2$, is the analytic continuation of the function $f_1e^{-v_1}$, defined and regular on $\mathfrak{Q}_1$. Hence the function f uniquely defined by (15) on the whole of $\mathfrak{Q}$ is a regular function on $\mathfrak{Q}$. It follows from (10) that

$$(f_a) = \mathfrak{d}_a$$

at each point a of $\mathfrak{Q}$ as desired.

The proof shows that the statement of the theorem can be carried over to more general domains than boxes, but we will not use this generalization in the sequel.

From Cousin's Lemma it easily follows that Cousin's problem is solvable for every box. Indeed, let $\mathfrak{d}$ be an integral divisor on a box $\mathfrak{Q}$. For every point, a, of $\mathfrak{Q}$ fix a representative p_a of the local part $\mathfrak{d}_a = (p_a)$ as well as a circle of convergence $|z - a| < \rho_a$ of p_a. Next, finitely many points, a, can be selected such that the whole of $\mathfrak{Q}$ is covered by the n-disks $|z - a| < \frac{1}{2}\rho_a$. Let t be any natural number. By dividing every edge of $\mathfrak{Q}$ into t equal parts, the box $\mathfrak{Q}$ is decomposed into t^{2n} congruent boxes $\mathfrak{Q}_k$ $(k = 1, \ldots, t^{2n})$. For sufficiently large t, each of these boxes $\mathfrak{Q}_k$ is contained in at least one n-disk $|z - a| < \rho_a$. Considering the local parts $\mathfrak{d}_b$ of $\mathfrak{d}$ and restricting b to the $\mathfrak{Q}_k$ we obtain a divisor $\mathfrak{d}^{(k)}$ on $\mathfrak{Q}_k$. The local parts $\mathfrak{d}_b^{(k)}$ arise from a single power series p_a by rearrangement of power series, as is seen from the coherence condition. Therefore, restricting the divisor $\mathfrak{d}$ to $\mathfrak{Q}_k$, Cousin's problem is solved by the power series $p_a(z - a)$.

We must now pass from the individual $\mathfrak{Q}_k$'s to $\mathfrak{Q}$ by repeated application of the Lemma. Again let $\mathfrak{Q} = \mathfrak{R}_1 \times \cdots \times \mathfrak{R}_n$. The above subdivision decomposes each rectangle $\mathfrak{R}_j$ $(j = 1, \ldots, n)$ into t^2 congruent rectangles $\mathfrak{R}_{kl}^{(j)}$ $(k, l = 1, \ldots, t)$ whose location in $\mathfrak{R}_j$ is given by the subscripts k, l in the same fashion as the entries of a matrix (Figure 2). Then each rectangle

$\mathfrak{R}_{11}$	$\mathfrak{R}_{12}$		$\mathfrak{R}_{1t}$
$\mathfrak{R}_{21}$	$\mathfrak{R}_{22}$		$\mathfrak{R}_{2t}$
$\mathfrak{R}_{t1}$	$\mathfrak{R}_{t2}$		$\mathfrak{R}_{tt}$

Figure 2

$\mathfrak{R}_j$ can be built up successively from its t^2 parts as follows. At the outset we fix the first index, k, and put together the t parts $\mathfrak{R}_{kl}^{(j)}$ $(l = 1, \ldots, t)$ starting by affixing $\mathfrak{R}_{k2}^{(j)}$ to $\mathfrak{R}_{k1}^{(j)}$, then affixing $\mathfrak{R}_{k3}^{(j)}$ to the union $\mathfrak{R}_{k2}^{(j)} \cup \mathfrak{R}_{k1}^{(j)}$ and so forth. For each $k = 1, \ldots, t$ we thus obtain a long rectangle, $\mathfrak{S}_k^{(j)}$, after $t - 1$ steps. In a similar fashion the rectangles $\mathfrak{S}_1^{(j)}, \ldots, \mathfrak{S}_t^{(j)}$ yield the rectangle $\mathfrak{R}_j$ by $t - 1$ successive pastings. First, we consider $j = 1$ and define $\mathfrak{R}_1^{(2)}$ successively as one of the $t^2 - 1$ rectangles $\mathfrak{R}_{12}^{(1)}, \ldots, \mathfrak{R}_{1t}^{(1)}$, $\mathfrak{R}_{22}^{(1)}, \ldots, \mathfrak{R}_{2t}^{(1)}, \ldots, \mathfrak{R}_{t2}^{(1)}, \ldots, \mathfrak{R}_{tt}^{(1)}$ and $\mathfrak{S}_2^{(1)}, \ldots, \mathfrak{S}_t^{(1)}$. On the other hand, in each case let $\mathfrak{R}_t^{(1)}$ be any subrectangle of $\mathfrak{R}_1$ already pasted together. Furthermore let $\mathfrak{Q}^*$ be the product of $n - 1$ subrectangles chosen from $\mathfrak{R}_2, \ldots, \mathfrak{R}_n$ and put

$$\mathfrak{Q}_k = \mathfrak{R}_1^{(k)} \times \mathfrak{Q}^* \qquad (k = 1, 2).$$

Applying Theorem 1 a total of $t^2 - 1$ times we derive the solvability of Cousin's problem for the restriction of $\mathfrak{d}$ to $\mathfrak{R}_1 \times \mathfrak{Q}^*$. Then we consider $j = 2$ and choose one of the t^2 subrectangles of $\mathfrak{R}_2$ as the first factor in $\mathfrak{Q}^*$, while the other $n - 2$ factors are kept fixed. By building up the rectangle $\mathfrak{R}_2$ in $t^2 - 1$ steps and repeatedly applying Theorem 1 it follows that Cousin's problem is solvable for the restriction of $\mathfrak{d}$ to the box $\mathfrak{R}_1 \times \mathfrak{R}_2 \times \mathfrak{Q}^{**}$ where $\mathfrak{Q}^{**}$ is the product of $n - 2$ subrectangles chosen from $\mathfrak{R}_3, \ldots, \mathfrak{R}_n$. Continuing this procedure of glueing together we obviously obtain, by induction, the solvability of Cousin's problem for $\mathfrak{Q}$.

Theorem 2: Cousin's problem is solvable for complex n-space $\mathfrak{Z}$.

Proof: Let t be a natural number and $\mathfrak{R}_t^{(j)}$ $(j = 1, \ldots, n)$ the square in the z_j-plane which is defined by

$$z_j = x_j + iy_j, \qquad -t \leqslant x_j \leqslant t, \qquad -t \leqslant y_j \leqslant t.$$

Form the box $\mathfrak{Q}_t = \mathfrak{R}_t^{(1)} \times \cdots \times \mathfrak{R}_t^{(n)}$. Then the sequence $\mathfrak{Q}_t$ $(t = 1, 2, \ldots)$ strictly increases, covering the whole space $\mathfrak{Z}$ as $t \to \infty$. Let $\mathfrak{d}$ be an integral divisor on $\mathfrak{Z}$ and $\mathfrak{d}_t$ the restriction of $\mathfrak{d}$ to $\mathfrak{Q}_t$. Then there exists a function f_t regular on $\mathfrak{Q}_t$ such that at any point, a, of $\mathfrak{Q}_t$ the local part of f_t is a representative of the local part of $\mathfrak{d}_t$ at a. Thus, in particular, the quotient f_{t+1}/f_t is a unit on the simply connected box $\mathfrak{Q}_t$. Consequently, we have

$$f_{t+1} = f_t e^{v_t} \qquad (t = 1, 2, \ldots)$$

where v_t is a regular function on $\mathfrak{Q}_t$. Since the box $\mathfrak{Q}_t$ contains the closed n-disk $|z| \leqslant t$, the power series expansion of v_t at the origin converges absolutely throughout this n-disk. If ε_t is any arbitrarily small positive number, then, by selecting a suitable partial sum of the power series expansion of v_t, there exists a polynomial w_t in the variables $z_1, \ldots, z_n$ for which the relation

$$|v_t - w_t| < \varepsilon_t \qquad (|z| \leqslant t) \tag{16}$$

holds.

We now put

$$v_t - w_t = d_t, \qquad g_t = f_t e^{-(w_1 + \cdots + w_{t-1})} \qquad (t = 1, 2, \ldots).$$

Then the quotient g_t/f_t obviously is a unit on $|z| \leqslant t$ and the relation

$$g_k = g_t e^{d_t + d_{t+1} + \cdots + d_{k-1}} \qquad (k = t, t+1, \ldots) \tag{17}$$

holds on this n-disk. Furthermore, let the numbers $\varepsilon_1, \varepsilon_2, \ldots$ be chosen so that their sum converges. According to (16) and (17) the sequence of functions $g_t, g_{t+1}, \ldots$ converges uniformly to a limit function, f, on $|z| < t$. Application of the Weierstrass convergence theorem shows that f is regular there. We see that f is defined everywhere independently of t and is regular; hence, it is an entire function. Furthermore, if we define

$$d_t + d_{t+1} + \cdots = c_t,$$

then the function c_t is regular on $|z| < t$ and the relation

$$f = g_t e^{c_t} = f_t e^{c_t - (w_1 + \cdots + w_{t-1})} \qquad (t = 1, 2, \ldots)$$

holds on this n-disk. This yields

$$(f) = \mathfrak{d}$$

and the proof is complete.

In the case of a single variable, Theorem 2 yields Weierstrass' theorem about the existence of entire functions with prescribed zeros. This is the reason for calling Theorem 2 the theorem of Weierstrass and Cousin. Actually Cousin proved a more general theorem which, in particular, implies the solvability of Cousin's problem for every domain

$$\mathfrak{G} = \mathfrak{G}_1 \times \mathfrak{G}_2 \times \cdots \times \mathfrak{G}_n$$

where $\mathfrak{G}_k$ $(k = 1, \ldots, n)$ is an arbitrary simply connected domain in the z_k-plane. However, we will not pursue this topic.

In Section 7 Jacobian functions will be investigated. This topic will require a refinement of Theorem 2 whose proof will be based on the ideas explained above.

6. *The period group*

Let a rectangular Cartesian coordinate system be established in m dimensional real Euclidean space. Each point will then be represented by means of its m coordinates $x_1, \ldots, x_m$ arranged in a column, $\mathfrak{x}$, which we also call a *vector*. It is then clear what we mean by $\mathfrak{x} + \mathfrak{y}$ and $\lambda\mathfrak{x}$ where λ is a real scalar. If we define the Euclidean distance $|\mathfrak{x} - \mathfrak{y}|$ from $\mathfrak{x}$ to $\mathfrak{y}$ in the usual fashion by

$$|\mathfrak{x}| = +\sqrt{x_1^2 + \cdots + x_m^2},$$

then the following rules hold: $|\mathfrak{x} + \mathfrak{y}| \leqslant |\mathfrak{x}| + |\mathfrak{y}|, |\lambda\mathfrak{x}| = |\lambda|\,|\mathfrak{x}|$. A vector $\mathfrak{x}_0$ is called a *limit point* of a set, $\mathfrak{G}$, of vectors, $\mathfrak{x}$, provided there is an infinite sequence of distinct vectors $\mathfrak{x}_n$ $(n = 1, 2, \ldots)$ of $\mathfrak{G}$ such that $|\mathfrak{x}_n - \mathfrak{x}_0| \to 0$. Furthermore, from now on we will assume that $\mathfrak{G}$ is a closed module, that is, $\mathfrak{G}$ contains all its limit points and $\mathfrak{x} - \mathfrak{y}$ belongs to $\mathfrak{G}$ whenever $\mathfrak{x}$ and $\mathfrak{y}$ belong to $\mathfrak{G}$. We call $\mathfrak{G}$ a *closed vector group*.

We now determine all closed vector groups. The simplest case occurs when $\mathfrak{G}$ is *discrete*, that is, has no limit points. In order to obtain an example of a discrete vector group we consider n linearly independent vectors $\mathfrak{x}_1, \ldots, \mathfrak{x}_n$. Thus the vector

$$\mathfrak{x} = \lambda_1\mathfrak{x}_1 + \cdots + \lambda_n\mathfrak{x}_n \tag{1}$$

is zero if and only if $\lambda_1 = 0, \ldots, \lambda_n = 0$, where the coefficients $\lambda_1, \ldots, \lambda_n$ are real scalars. The vector group generated by $\mathfrak{x}_1, \ldots, \mathfrak{x}_n$ is obtained if we only allow integers, chosen independently of one another, for the $\lambda_1, \ldots, \lambda_n$ in expression (1). This is also called the *lattice with basis* $\mathfrak{x}_1, \ldots, \mathfrak{x}_n$. The assumption that $\mathfrak{x}_1, \ldots, \mathfrak{x}_n$ are linearly independent implies that the matrix $(\mathfrak{x}_1 \cdots \mathfrak{x}_n)$ has rank n, whence $n \leqslant m$. If we require that $|\mathfrak{x}|$ be bounded in (1), then it follows that $\lambda_1, \ldots, \lambda_n$ are also bounded. Accordingly, every lattice is a discrete vector group. We will now demonstrate the converse of this statement.

Theorem 1: Every discrete vector group is a lattice.

Proof: We will use induction. We assume for a given natural number, k, there exist $k - 1$ elements $\mathfrak{x}_1, \dots, \mathfrak{x}_{k-1}$ of the discrete vector group $\mathfrak{G}$ with the property that a vector of the form

$$\mathfrak{x} = \lambda_1\mathfrak{x}_1 + \cdots + \lambda_{k-1}\mathfrak{x}_{k-1}; \qquad \lambda_1, \dots, \lambda_{k-1} \text{ real} \tag{2}$$

belongs to $\mathfrak{G}$ if and only if $\lambda_1, \dots, \lambda_{k-1}$ are all integers. This condition is vacuous for $k = 1$. It follows from the definition that $\mathfrak{x}_1, \dots, \mathfrak{x}_{k-1}$ are linearly independent because $\mathfrak{x} = 0$ implies that $\rho\,\mathfrak{x} = 0$ for arbitrary real scalars ρ. If $\mathfrak{x}_1, \dots, \mathfrak{x}_{k-1}$ generate all of $\mathfrak{G}$, then the assertion of the theorem is proved.

Now we assume that $\mathfrak{G}$ is not generated by $\mathfrak{x}_1, \dots, \mathfrak{x}_{k-1}$ and we select an element, $\mathfrak{y}_k$, of $\mathfrak{G}$ which does not admit an expression of the form (2). We then consider all the vectors of the type

$$\mathfrak{x} = \lambda_1\mathfrak{x}_1 + \cdots + \lambda_{k-1}\mathfrak{x}_{k-1} + \lambda_k\mathfrak{y}_k; \qquad \lambda_1, \dots, \lambda_{k-1}, \lambda_k \text{ real}$$

which lie in $\mathfrak{G}$. These clearly provide us with a subgroup, $\mathfrak{H}$, of $\mathfrak{G}$. In particular, we obtain one such vector by putting $\lambda_1 = 0, \dots, \lambda_{k-1} = 0,\ \lambda_k = 1$. Moreover, we now consider only $\mathfrak{x}$ in $\mathfrak{H}$ for which $0 \leqslant \lambda_1 < 1, \dots, 0 \leqslant \lambda_{k-1} < 1,\ 0 < \lambda_k \leqslant 1$. Since $\mathfrak{H}$ is discrete, among these $\mathfrak{x}$ there is one with smallest positive λ_k, and we denote this vector by $\mathfrak{x}_k$. The vectors $\mathfrak{x}_1, \dots, \mathfrak{x}_{k-1}, \mathfrak{y}_k$ are linearly independent because of the assumption about $\mathfrak{y}_k$, and it follows that the same is true for $\mathfrak{x}_1, \dots, \mathfrak{x}_{k-1}, \mathfrak{x}_k$.

Let

$$\mathfrak{x} = \mu_1\mathfrak{x}_1 + \cdots + \mu_{k-1}\mathfrak{x}_{k-1} + \mu_k\mathfrak{x}_k$$

with real $\mu_1, \dots, \mu_k$ be an element of $\mathfrak{G}$. In particular, for any integers $\gamma_1, \dots, \gamma_{k-1}, \gamma_k$

$$\mathfrak{x}_0 = \gamma_1\mathfrak{x}_1 + \cdots + \gamma_{k-1}\mathfrak{x}_{k-1} + \gamma_k\mathfrak{x}_k$$

is such an element; consequently, the same is true of the difference

$$\mathfrak{x} - \mathfrak{x}_0 = (\mu_1 - \gamma_1)\mathfrak{x}_1 + \cdots + (\mu_{k-1} - \gamma_{k-1})\mathfrak{x}_{k-1} + (\mu_k - \gamma_k)\mathfrak{x}_k.$$

On the other hand

$$\mathfrak{x}_k = \lambda_1\mathfrak{x}_1 + \cdots + \lambda_{k-1}\mathfrak{x}_{k-1} + \lambda_k\mathfrak{y}_k,$$

where λ_k is minimal. Thus we also have

$$\begin{aligned}\mathfrak{x} - \mathfrak{x}_0 = (\mu_1 - \gamma_1 + \lambda_1(\mu_k - \gamma_k))\mathfrak{x}_1 + \cdots \\ + (\mu_{k-1} - \gamma_{k-1} + \lambda_{k-1}(\mu_k - \gamma_k))\mathfrak{x}_{k-1} + \lambda_k(\mu_k - \gamma_k)\mathfrak{y}_k.\end{aligned}$$

Now we determine the integers $\gamma_1, \ldots, \gamma_k$ uniquely by the conditions

$$0 \leqslant (\mu_k - \gamma_k) < 1, \qquad 0 \leqslant \mu_l - \gamma_l + \lambda(\mu_k - \gamma_k) < 1 \qquad (l = 1, \ldots, k-1).$$

In particular, we obtain

$$0 \leqslant \lambda_k(\mu_k - \gamma_k) < \lambda_k.$$

From this it follows that $\mu_k = \gamma_k$ and, furthermore, $\mathfrak{x} = \mathfrak{x}_0$. Thus we have carried out the induction step from $k - 1$ to k. In any case $k \leqslant m$ because $\mathfrak{x}_1, \ldots, \mathfrak{x}_k$ are linearly independent. Therefore, after at most m steps, the procedure leads to the desired goal, and the theorem is proved.

Another simple example of a closed vector group arises from any linear subspace, $\mathfrak{L}$, of the whole space. If d is the dimension of $\mathfrak{L}$, then, upon selection of any d linearly independent vectors $\mathfrak{x}_1, \ldots, \mathfrak{x}_d$ in $\mathfrak{L}$, the points of $\mathfrak{L}$ have unique expressions of the form

$$\mathfrak{x} = \lambda_1\mathfrak{x}_1 + \cdots + \lambda_d\mathfrak{x}_d$$

where $\lambda_1, \ldots, \lambda_d$ are real variables. More generally, we can obtain closed vector groups in the following way. We take $d + n$ linearly independent vectors $\mathfrak{x}_1, \ldots, \mathfrak{x}_d$ and $\mathfrak{y}_1, \ldots, \mathfrak{y}_n$ where $d + n \leqslant m$ and form the linear space $\mathfrak{L}$ spanned by $\mathfrak{x}_1, \ldots, \mathfrak{x}_d$ and the lattice $\mathfrak{G}$ generated by $\mathfrak{y}_1, \ldots, \mathfrak{y}_n$. We consider all expressions $\mathfrak{x} + \mathfrak{y}$ with

$$\mathfrak{x} = \lambda_1\mathfrak{x}_1 + \cdots + \lambda_d\mathfrak{x}_d, \qquad \mathfrak{y} = \mu_1\mathfrak{y}_1 + \cdots + \mu_n\mathfrak{y}_n$$

where $\lambda_1, \ldots, \lambda_d$ are arbitrarily selected real numbers and $\mu_1, \ldots, \mu_n$ are arbitrarily chosen integers. We then obtain the elements of the direct sum $\mathfrak{L} + \mathfrak{G}$, and it is clear that this is again closed. We will now demonstrate the converse of this statement.

Theorem 2: Every closed vector group is a direct sum of a linear space and a lattice.

Proof: Let $\mathfrak{H}$ be a closed vector group. For every positive number ρ we consider the set, $\mathfrak{H}(\rho)$, of elements, $\mathfrak{x}$, of $\mathfrak{H}$ for which $|\mathfrak{x}| < \rho$. It is evident that for $0 < \sigma < \rho$ the set $\mathfrak{H}(\sigma)$ is contained in $\mathfrak{H}(\rho)$ and $\mathfrak{x} = 0$ is contained in $\mathfrak{H}(\rho)$ for all ρ. Let $d(\rho)$ be the maximal number of linearly independent vectors in $\mathfrak{H}(\rho)$. Thus $d(\rho) \geqslant 0$ and, in any case, for $0 < \sigma < \rho$ we have $d(\sigma) \leqslant d(\rho)$. Since $d(\rho)$ is integral, there is a positive number r such that

$$d(\rho) = d \qquad (0 < \rho \leqslant r) \tag{3}$$

where d is independent of ρ.

For each $\rho \leqslant r$ we select d linearly independent vectors from $\mathfrak{H}(\rho)$ and denote them by $\mathfrak{x}_k = \mathfrak{x}_k(\rho)$ $(k = 1, \ldots, d)$. It follows from (3) that these

d vectors span a linear space, $\mathfrak{L}$, of dimension d independent of ρ. If

$$\mathfrak{x} = \lambda_1\mathfrak{x}_1 + \cdots + \lambda_d\mathfrak{x}_d$$

is a point of $\mathfrak{L}$, then we put $\lambda_k = \gamma_k + \eta_k$ $(k = 1, \ldots, d)$ where γ_k is integral and $0 \leqslant \eta_k < 1$. For the vectors

$$\mathfrak{v} = \gamma_1\mathfrak{x}_1 + \cdots + \gamma_d\mathfrak{x}_d, \qquad \mathfrak{r} = \eta_1\mathfrak{x}_1 + \cdots + \eta_d\mathfrak{x}_d$$

we then have

$$\mathfrak{x} - \mathfrak{v} = \mathfrak{r}, \qquad |\mathfrak{x} - \mathfrak{v}| = |\mathfrak{r}| \leqslant |\eta_1|\,|\mathfrak{x}_1| + \cdots + |\eta_d|\,|\mathfrak{x}_d| \leqslant d\rho \leqslant m\rho.$$

Because the $\mathfrak{x}_k$ are dependent upon ρ so is $\mathfrak{v}$. Nevertheless, for a fixed $\mathfrak{x}$, $\mathfrak{r} \to 0$ and therefore $\mathfrak{v} \to \mathfrak{x}$ when ρ tends to zero. Now it follows that $\mathfrak{x}$ belongs to $\mathfrak{H}$ because $\mathfrak{v}$ belongs to the closed vector group $\mathfrak{H}$. Thus the entire linear space $\mathfrak{L}$ belongs to $\mathfrak{H}$.

If $\mathfrak{L}$ does not coincide with $\mathfrak{H}$, then we select a maximal number, n, of vectors $\mathfrak{y}_1, \ldots, \mathfrak{y}_n$ such that $\mathfrak{x}_1, \ldots, \mathfrak{x}_d$ and $\mathfrak{y}_1, \ldots, \mathfrak{y}_n$ are all linearly independent. All elements of $\mathfrak{H}$ then have the form $\mathfrak{x} + \mathfrak{y}$ where

$$\mathfrak{x} = \lambda_1\mathfrak{x}_1 + \cdots + \lambda_d\mathfrak{x}_d, \qquad \mathfrak{y} = \mu_1\mathfrak{y}_1 + \cdots + \mu_n\mathfrak{y}_n$$

for suitably selected real $\lambda_1, \ldots, \lambda_d$ and $\mu_1, \ldots, \mu_n$. Since $\mathfrak{L}$ is contained in $\mathfrak{H}$ it now follows that $\mathfrak{y}$ is in $\mathfrak{H}$ whenever $\mathfrak{x} + \mathfrak{y}$ belongs to $\mathfrak{H}$. These vectors $\mathfrak{y}$ therefore form a subgroup, $\mathfrak{G}$, of $\mathfrak{H}$ which is likewise closed. If $\mathfrak{G}$ is not discrete it contains a sequence $\mathfrak{y} \to 0$, $\mathfrak{y} \neq 0$ and, therefore, it contains an element of $\mathfrak{H}(r)$ which is not in $\mathfrak{L}$. This is a contradiction. Thus $\mathfrak{G}$ is discrete and, as a result of Theorem 1, a lattice. Since $\mathfrak{H} = \mathfrak{L} + \mathfrak{G}$ the assertion is proved.

We now apply the result just proved to the periods of meromorphic functions of n complex variables $z_1, \ldots, z_n$. In this application we will only consider the most important case for the sequel, that is, when $f(z)$ is meromorphic on the entire space $\mathfrak{Z}$. It then follows from Theorem 2 of the previous section that f is a quotient of two entire functions g and $h \not\equiv 0$, where g and h may be selected locally relatively prime at each point. In particular f is meromorphic in the sense of Section 12 of Chapter 4 (Volume II). A column, ω, formed from n complex numbers is called a *period* of $f(z)$ provided

$$f(z) = f(z + \omega) \tag{4}$$

holds everywhere. It is clear that the totality of periods of $f(z)$ forms an additive Abelian group, Ω. If we decompose ω into its real part, α, and imaginary part, β, and place the column β below the column α, then we obtain an interpretation of the period group as a vector group in real $2n$

dimensional space. If we write f as a quotient of entire functions g and $h \neq 0$, then (4) may be expressed in the form

$$g(z)h(z + \omega) = h(z)g(z + \omega).$$

It follows that Ω is closed.

The meromorphic function $f(z)$ is transformed to the meromorphic function

$$\phi(w) = f(Lw)$$

by introducing new variables $w_1, \ldots, w_n$ in place of $z_1, \ldots, z_n$ by means of an invertible homogeneous linear substitution $z = Lw$ with complex matrix L. The period group of $\phi(w)$ consists of all the columns $L^{-1}\omega$ where ω belongs to Ω. Now, as a result of Theorem 2, Ω is a direct sum of a linear subspace and a lattice in real $2n$ dimensional space. If Ω is not discrete, then it follows from Theorem 1 that there are nonzero periods, ω, such that $\lambda\omega$ are again periods of $f(z)$ for arbitrary real values of the scalar λ. If we select ω as the first column of an invertible matrix, L, then the column $L^{-1}\omega$ is formed from the numbers $1, 0, \ldots, 0$. It follows that the function $\phi(w)$, viewed as a function of the first variable w_1, has every real number λ as a period. But this shows that the partial derivative of ϕ with respect to w_1 is 0, and therefore $\phi(w)$ is independent of w_1. Thus if the original function $f(z)$ cannot be transformed into a meromorphic function of fewer than n variables by a suitable invertible linear transformation, then its period group is a lattice. Moreover, the converse of this statement is also valid. From now on, we will call a meromorphic function *degenerate* if its period group is not a lattice.

Now let $f(z)$ be nondegenerate and let Ω be its period group, which, consequently, is a lattice in real $2n$ dimensional space. Thus there is an integer m in the interval $0 \leqslant m \leqslant 2n$ as well as m periods $\omega^{(1)}, \ldots, \omega^{(m)}$ which are linearly independent with respect to the reals and form a basis of Ω. From now on we will only concern ourselves with the case $m = 2n$. If z is any point in $\mathfrak{Z}$, we consider the equation

$$z = \xi_1\omega^{(1)} + \cdots + \xi_{2n}\omega^{(2n)} \tag{5}$$

where the unknowns $\xi_1, \ldots, \xi_{2n}$ are real scalars. It follows by application of complex conjugation that

$$\bar{z} = \xi_1\bar{\omega}^{(1)} + \cdots + \xi_{2n}\bar{\omega}^{(2n)}. \tag{6}$$

Since, upon decomposition into real and imaginary parts, the $2n$ periods $\omega^{(1)}, \ldots, \omega^{(2n)}$ span the entire real space equations (5) and (6) possess a unique real solution $\xi_1, \ldots, \xi_{2n}$ for every choice of z. On the other hand, if the scalars $\xi_1, \ldots, \xi_{2n}$ form a complex solution of (5) and (6), then application of complex conjugation shows that they must be real. We introduce the matrices

$$(\omega^{(1)} \cdots \omega^{(2n)}) = C, \qquad \begin{pmatrix} C \\ \bar{C} \end{pmatrix} = P.$$

P is the matrix of the system of linear equations given by (5) and (6), and consequently the determinant $|P| \neq 0$ if and only if $\omega^{(1)}, \ldots, \omega^{(2n)}$ span all of $\mathfrak{Z}$ in the above sense.

Whereas we previously began with a meromorphic function and then considered its period group, we now modify the question by starting with a lattice Ω and a basis $\omega^{(1)}, \ldots, \omega^{(2n)}$. By an *Abelian function corresponding to* Ω or simply an *Abelian function* we mean a meromorphic function whose period group contains Ω. Clearly the Abelian functions corresponding to Ω form a field K. Furthermore, it is obvious that K also contains degenerate functions, in particular, all the constants. We will show later that K may even consist entirely of constants for certain period groups Ω. We call K itself *degenerate* if all the functions belonging to K are degenerate. The purpose of the following sections is to derive necessary and sufficient conditions for the existence of a nondegenerate Abelian function corresponding to Ω. If such a function exists, we shall even be able to construct one whose period group coincides with Ω.

The Jacobi–Abel functions introduced in Section 12 of the previous chapter (Volume II) are subsumed in the definition given here for $n = p$. Namely, there Ω was the period lattice of the normalized integrals of the first kind of an algebraic function field of genus p. At that time the period parallelotope $\mathfrak{P}$ had already been defined in Section 8, and we again define it to be the totality of points

$$z = \xi_1\omega^{(1)} + \cdots + \xi_{2n}\omega^{(2n)}$$

with $0 \leqslant \xi_1 < 1, \ldots, 0 \leqslant \xi_{2n} < 1$. Thus $\mathfrak{P}$ is a fundamental region for the representation of Ω as a group of translations given by $z \to z + \omega$.

We say that two points a, b of $\mathfrak{Z}$ are *congruent* and write $a \equiv b$ if their difference, $a - b$, lies in Ω. Thus $\mathfrak{P}$ contains one representative of each congruence class. In the case $n = 1$ the torus arises from the identification of opposite sides of the period parallelogram. Correspondingly, for arbitrary n, we obtain a compact space by identification of congruent boundary points of $\mathfrak{P}$. This space is called the *period torus* and is denoted by $\mathfrak{T}$. The points of $\mathfrak{T}$ are the classes of congruent points $a + \omega$ of $\mathfrak{Z}$, and we can take as their neighborhoods the congruence classes of neighborhoods $|z - a - \omega| < \rho$ in $\mathfrak{Z}$. Here the radius ρ must be smaller than the minimum of the values $\frac{1}{2}|\omega|$, $\omega \neq 0$. This definition has the advantage of being independent of the choice of basis for Ω. We may indicate this by the occasional use of the notation $\mathfrak{T} = \mathfrak{Z}/\Omega$.

7. *Jacobian functions*

With a view to future needs, we begin with more general considerations before continuing our investigation of Abelian functions. In place of $\mathfrak{Z}$ we once again consider an arbitrary domain $\mathfrak{H}$ in $\mathfrak{Z}$ for which Cousin's problem is always solvable. In order that we be able to write the units on $\mathfrak{H}$ in the form e^w, where w is a regular function, we will also assume that $\mathfrak{H}$ is simply connected.

Let there be given a system of n functions

$$z_k^* = \gamma_k(z_1, \ldots, z_n) \qquad (k = 1, \ldots, n) \tag{1}$$

which are regular on $\mathfrak{H}$ and whose functional determinant is nonzero throughout $\mathfrak{H}$. Furthermore, we assume that (1) defines an invertible mapping of $\mathfrak{H}$ onto itself. Clearly the totality of such invertible regular mappings of $\mathfrak{H}$ onto itself forms a group G. Then we consider any discontinuous subgroup, Γ, of G, and for the sake of brevity, instead of (1) we write

$$z^* = \gamma z \tag{2}$$

where γ is an element of Γ. In doing this we interpret the symbol γ in (2) as an operator. If a meromorphic function $f(z)$ on $\mathfrak{H}$ satisfies the functional equation

$$f(\gamma z) = f(z) \tag{3}$$

for all γ in Γ, then we say that it is an *automorphic function with respect to* Γ, or simply, an *automorphic function.* We see immediately that the automorphic functions of one variable studied in Chapter 3 (Volume II) form a special case of this definition.

Since, by assumption, the Cousin problem is solvable on $\mathfrak{H}$ we have the representation

$$f(z) = \frac{g(z)}{h(z)} \tag{4}$$

where g and $h \neq 0$ are regular functions on $\mathfrak{H}$ which are everywhere locally relatively prime. It follows from (3) and (4) that

$$g(z)h(\gamma z) = h(z)g(\gamma z), \qquad g(\gamma z) = j_\gamma(z)g(z), \qquad h(\gamma z) = j_\gamma(z)h(z) \tag{5}$$

where $j_\gamma(z)$ is a regular function on $\mathfrak{H}$ which depends on the choice of the element γ of Γ. Consequently, we also have

$$h(\gamma^{-1}z) = j_{\gamma^{-1}}(z)h(z), \qquad h(z) = j_{\gamma^{-1}}(\gamma z)h(\gamma z)$$

whence

$$j_{\gamma^{-1}}(\gamma z)j_\gamma(z) = 1. \tag{6}$$

It follows that the function $j_\gamma(z)$ is a unit on $\mathfrak{H}$ which may be expressed in the form

$$(7) \qquad j_\gamma(z) = e^{s_\gamma(z)}$$

where $s_\gamma(z)$ is a regular function on $\mathfrak{H}$.

Relation (6) has a generalization; namely, for any other element, δ, of Γ we use the equations $h(\delta z) = j_\delta(z)h(z)$, $h(\gamma\delta z) = j_{\gamma\delta}(z)h(z) = j_\gamma(\delta z)h(\delta z)$ to obtain

$$(8) \qquad j_{\gamma\delta}(z) = j_\gamma(\delta z)j_\delta(z).$$

Disregarding the previous definition of the $j_\gamma(z)$, for each fixed element γ of Γ let there be defined a regular function $j_\gamma(z)$ on $\mathfrak{H}$ such that not all $j_\gamma(z)$ are identically zero and condition (8) is satisfied for all γ, δ in Γ. We say that the functions $j_\gamma(z)$ form a *factor system*. If ε is the identity element of the group Γ then, as a special case of (8), we find that the function $j_\varepsilon(z) = 1$ and condition (6) holds. Thus every function $j_\gamma(z)$ is a unit. The collection of $s_\gamma(z)$ in (7) is then called an *exponent system*. Those solutions, $g(z)$, of the functional equations

$$(9) \qquad g(\gamma z) = j_\gamma(z)g(z)$$

which are regular in $\mathfrak{H}$ are called *automorphic forms with factor system* $j_\gamma(z)$. Since $\lambda_1 g_1 + \lambda_2 g_2$ is a solution of (9) whenever g_1 and g_2 are solutions and λ_1, λ_2 are arbitrary complex constants, it follows that for a fixed factor system the automorphic forms constitute a complex vector space. The quotient of any two automorphic forms with the same factor system whose denominator is not 0 is an automorphic function. Conversely, it also follows from (3), (4), and (5) that every automorphic function is such a quotient for a suitable factor system.

In this connection there arises the problem of determining all factor systems for a given group, Γ, as well as the additional question of the existence of a nonzero automorphic form corresponding to a given factor system. Both questions have been solved only in special cases. The most important of these, the case of Jacobian functions, will be considered in the sequel. These investigations will be simplified by the introduction of the concept of equivalence of factor systems. The factor system $j_\gamma(z)$ was originally derived from the representation (4) of a given automorphic function, $f(z)$, as a quotient of two locally relatively prime regular functions $g(z)$ and $h(z)$. Now if

$$f(z) = \frac{g^*(z)}{h^*(z)}$$

is another such representation, then we have

$$g^*(z) = g(z)q(z), \qquad h^*(z) = h(z)q(z)$$

where $q(z)$ is a unit on $\mathfrak{H}$. The corresponding factor system $j_\gamma^*(z)$ then satisfies

$$g^*(\gamma z) = j_\gamma^*(z)g^*(z), \qquad h^*(\gamma z) = j_\gamma^*(z)h^*(z),$$

$$q(\gamma z)j_\gamma(z) = j_\gamma^*(z)q(z).$$

Thus we have

$$j_\gamma^*(z) = \frac{q(\gamma z)}{q(z)} j_\gamma(z). \tag{10}$$

On the other hand, if we are given an arbitrary factor system $j_\gamma(z)$ which must satisfy (8), then we can define new functions $j_\gamma^*(z)$ by (10) where $q(z)$ is now an arbitrarily given unit on $\mathfrak{H}$. Using (8) and the identity

$$\frac{q(\gamma\delta z)}{q(z)} = \frac{q(\gamma\delta z)}{q(\delta z)} \frac{q(\delta z)}{q(z)} \tag{11}$$

we verify that the j_γ^* likewise form a factor system. The two factor systems j_γ and j_γ^* are then said to be *equivalent*. This clearly defines an equivalence relation. We must now add another problem to the two given above. Namely, we must select a simplest possible representative for each equivalence class of factor systems by means of suitable reduction conditions. Passing to the exponent system $s_\gamma(z)$ by (7), we obtain the additive relation

$$s_{\gamma\delta}(z) - s_\gamma(\delta z) - s_\delta(z) = 2\pi i c_{\gamma,\delta} \tag{12}$$

in place of (8). Here the integer $c_{\gamma,\delta}$ depends on γ and δ but, because of the continuity with respect to z of the left side of (12), is a not necessarily zero constant. Furthermore, putting

$$q(z) = e^{t(z)}, \qquad j_\gamma^*(z) = e^{s_\gamma^*(z)}$$

on $\mathfrak{H}$ we obtain the relation

$$s_\gamma^*(z) = t(\gamma z) - t(z) + s_\gamma(z)$$

from (10). This formula is used to define an equivalence relation between exponent systems. From the additive identity for $t(z)$ corresponding to (11), we conclude that the integers $c_{\gamma,\delta}$ appearing in (12) depend only on the equivalence class of the exponent system $s_\gamma(z)$.

We will no longer pursue the case where the domain $\mathfrak{H}$ and the group Γ are arbitrary. Instead we now consider the special case which arises in the study of Abelian functions, namely, where $\mathfrak{H}$ is the entire space $\mathfrak{Z}$ and Γ is a lattice Ω with basis $\omega^{(1)}, \ldots, \omega^{(2n)}$. The mappings (1) are then merely the translations $z^* = z + \omega$ where ω is an arbitrary element of Ω. Correspondingly, we write j_ω, s_ω instead of j_γ, s_γ. Now, for a given factor system, we consider the corresponding automorphic forms, that is, the entire functions $g(z)$ which are solutions of the functional equations

$$g(z + \omega) = j_\omega(z)g(z) \tag{13}$$

for all ω. In accordance with the terminology introduced by Frobenius we will call these forms *Jacobian functions with factor system* j_ω because they are generalizations of the elliptic theta function introduced by Jacobi. We are now interested in the case where the vector space of Jacobian functions does not consist only of the zero function.

A *trivial factor system* is given by the conditions $j_\omega = 1$ for all ω. In view of (13), $g(z)$ is then an Abelian function whence, by periodicity, bounded on $\mathfrak{Z}$. Since Liouville's theorem is obviously valid in the case of several variables, the function $g(z)$ is constant. More generally, if j_ω and 1 are equivalent factor systems, then by (10) and (13) we see that every solution of (13) which is not identically zero is a unit on $\mathfrak{Z}$, that is, a nowhere zero entire function. Conversely, if (13) has a solution which is a unit, then j_ω and 1 are equivalent factor systems. Thus if we assume that $g(z)$ has zeros but does not vanish identically, then j_ω and 1 are not equivalent factor systems.

Let $\mathfrak{d}$ be the divisor of the Jacobian function $g(z)$, that is,

$$\mathfrak{d} = (g).$$

Then, as a consequence of (13), we have the relation

$$\mathfrak{d}_{a+\omega} = \mathfrak{d}_a \tag{14}$$

for the local parts, $\mathfrak{d}_a$, of the divisor $\mathfrak{d}$ at any point a of $\mathfrak{Z}$ and any ω in Ω. Conversely, let $\mathfrak{d}$ be a divisor defined on $\mathfrak{Z}$ whose local parts satisfy conditions (14) and which for present purposes need not be integral. Then, in view of Theorem 2 of Section 5, $\mathfrak{d}$ is a principal divisor (g); furthermore the function $g(z + \omega)$ is associated with $g(z)$ on $\mathfrak{Z}$. Hence (13) is satisfied for an appropriate factor system j_ω. Since the divisor $\mathfrak{d}$ determines g only up to a unit factor, it uniquely determines only the class of the factor system j_ω. A divisor $\mathfrak{d}$ on $\mathfrak{Z}$ is said to be *periodic* if it satisfies (14) for all points, a, of $\mathfrak{Z}$ and elements ω of Ω. Clearly, every periodic divisor is uniquely defined on the period torus $\mathfrak{T}$. Our aim is to obtain sharper versions of Theorems 1 and 2 of Section 5 for the case of periodic integral divisors $\mathfrak{d} \neq (0)$. For this purpose it is necessary to examine their proofs in greater detail and make certain estimates.

In view of (14), we can assume that the representatives $p_a(x)$ and $p_{a+\omega}(x)$ of $\mathfrak{d}_a$ and $\mathfrak{d}_{a+\omega}$ satisfy

$$p_{a+\omega}(x) = p_a(x). \tag{15}$$

Let $\mathfrak{K}_a$ be a circle of convergence of $p_a(z - a)$ defined by $|z - a| < \rho_a$ where $\rho_{a+\omega} = \rho_a$ and let $\mathfrak{K}_a^*$ be a concentric n-disk defined by $|z - a| < \frac{1}{2}\rho_a$. Because of the compactness of the period torus $\mathfrak{T}$ we can select finitely many points η of $\mathfrak{Z}$ so that the closure of the period parallelotope $\mathfrak{P}$ is completely covered by the $\mathfrak{K}_\eta^*$. It is clear that the entire space $\mathfrak{Z}$ is covered by the totality of n-disks $\mathfrak{K}_\zeta^*$ obtained from the corresponding translations $\zeta = \eta + \omega$. Furthermore, every compact subset of $\mathfrak{Z}$ intersects only finitely many $\mathfrak{K}_\zeta^*$. On the intersection $\mathfrak{K}_{\zeta_1\zeta_2}$ of two n-disks $\mathfrak{K}_{\zeta_1}$ and $\mathfrak{K}_{\zeta_2}$ the coherence condition takes the form

$$p_{\zeta_1}(z - \zeta_1) = p_{\zeta_2}(z - \zeta_2)e^{w(z)}, \qquad w(z) = w_{\zeta_1\zeta_2}(z) \tag{16}$$

where $w_{\zeta_1\zeta_2}(z)$ is a regular function on $\mathfrak{K}_{\zeta_1\zeta_2}$. By selecting ρ_a sufficiently small we can also arrange that $w_{\zeta_1\zeta_2}(z)$ is regular everywhere on the boundary of $\mathfrak{K}_{\zeta_1\zeta_2}$. Since $\mathfrak{P}$ meets only finitely many $\mathfrak{K}_\zeta$, as a consequence of (15) we can assume the inequality

$$|w_{\zeta_1\zeta_2}(z)| < \gamma \tag{17}$$

where γ is a positive number independent of ζ_1, ζ_2, and z.

If the variable z is replaced by λz, where λ is a positive constant, then the radii ρ_a are multiplied by λ. Since the investigations of the present section are independent of such coordinate transformations, we may assume throughout that $\rho_\eta > 3$ and therefore that $\rho_\zeta > 3$. By a "lattice point" we now mean a point, a, in $\mathfrak{Z}$ whose n coordinates a_k $(k = 1, \ldots, n)$ are Gaussian integers $l + mi$ $(l, m = 0, \pm 1, \ldots)$. In general these lattice points have nothing to do with the period lattice. Now for each lattice point, a, we select a point $\zeta = \zeta_a$ such that a lies in $\mathfrak{K}_\zeta^*$. If $|z - a| < 3/2$, then

$$|z - \zeta| \leqslant |z - a| + |a - \zeta| < 3/2 + \tfrac{1}{2}\rho_\zeta < \rho_\zeta;$$

therefore, z lies in $\mathfrak{K}_\zeta$. We define $\mathfrak{S}_a$ as the closed cube with center a and edge length 2. The coordinates of its points, z, must satisfy the condition that the real and imaginary parts of $z_k - a_k$ $(k = 1, \ldots, n)$ lie between -1 and 1. Since $|z - a| \leqslant \sqrt{2} < 3/2$, the closed set $\mathfrak{S}_a$ is entirely contained in the open set $\mathfrak{K}_\zeta$. Now we define

$$p_{\zeta_a}(z - \zeta_a) = \phi_a(z) \tag{18}$$

on $\mathfrak{S}_a$. If two cubes $\mathfrak{S}_a$ and $\mathfrak{S}_b$ intersect, then by (16) and (17) we have the relation

$$\phi_a(z) = \phi_b(z)e^{v_{ab}(z)}, \qquad v_{ab} = w_{\zeta_a\zeta_b}, \qquad |v_{ab}| < \gamma \tag{19}$$

on the intersection.

Shrinking the sides by the ratio of 1 to 2 and keeping the centers fixed, we obtain from $\mathfrak{S}_a$ closed cubes $\mathfrak{Q}_a$ for which the real and imaginary parts of $z_k - a_k$ $(k = 1, \ldots, n)$ lie between $-\frac{1}{2}$ and $\frac{1}{2}$. Obviously the $\mathfrak{Q}_a$ cover $\mathfrak{Z}$ completely with neither gaps nor overlaps. Letting $\mathfrak{Z}_k$ denote the plane of the complex variable z_k and $\mathfrak{S}_{a_k}$ and $\mathfrak{Q}_{a_k}$ the intersections of $\mathfrak{Z}_k$ with $\mathfrak{S}_a$ and $\mathfrak{Q}_a$, we have

$$\mathfrak{S}_a = \mathfrak{S}_{a_1} \times \cdots \times \mathfrak{S}_{a_n}, \qquad \mathfrak{Q}_a = \mathfrak{Q}_{a_1} \times \cdots \times \mathfrak{Q}_{a_n}.$$

Furthermore, let t be a natural number. If we select $a_k = l + mi$ with $l, m = 0, \pm 1, \ldots, \pm t$, then the $(2t + 1)^2$ corresponding squares $\mathfrak{S}_{a_k}$ form a square $\mathfrak{W}_k$ with side of length $2t + 2$ and center point $z_k = 0$. Therefore, a is a lattice point lying in the cube $\mathfrak{W} = \mathfrak{W}_1 \times \cdots \times \mathfrak{W}_n$. Furthermore, let α_k denote the sequence $a_{k+1}, \ldots, a_n$ of $n - k$ numbers and

$$\mathfrak{S}_{\alpha_k} = \mathfrak{S}_{a_{k+1}} \times \cdots \times \mathfrak{S}_{a_k}, \qquad \mathfrak{Q}_{a,k} = \mathfrak{Q}_{a_1} \times \cdots \times \mathfrak{Q}_{a_k} \times \mathfrak{S}_{\alpha_k},$$

$$\mathfrak{W}_{\alpha_k} = \mathfrak{W}_1 \times \cdots \times \mathfrak{W}_k \times \mathfrak{S}_{\alpha_k}.$$

Let $\gamma_1, \ldots, \gamma_{15}$ be positive numbers which depend only on γ, n; in particular, they are independent of t.

Theorem 1: For each α_k there exists a regular function $f_{\alpha_k}(z)$ defined on $\mathfrak{W}_{\alpha_k}$ such that $\mathfrak{d}$ restricted to $\mathfrak{W}_{\alpha_k}$ is the divisor of f_{α_k} and the following relations hold on $\mathfrak{Q}_{a,k}$:

$$f_{\alpha_k}(z) = \phi_a(z)e^{v_{a,k}(z)} \tag{20}$$

$$|v_{a,k}(z)| < \gamma_1 t^{5k} \tag{21}$$

where $v_{a,k}(z)$ is a regular function on $\mathfrak{Q}_{a,k}$.

Proof: We will use induction with respect to k. We remark that expression (20) results from the solvability of Cousin's problem in an arbitrary box, which was proved in Section 5. However, inequality (21) is essential for our later purposes. In order to derive (21) we must now use sharper versions of our previous arguments. In case $k = 0$, we define

$$\alpha_k = a, \qquad \mathfrak{S}_{\alpha_k} = \mathfrak{S}_a = \mathfrak{Q}_{a,k} = \mathfrak{W}_{\alpha_k}.$$

We then define

$$f_{\alpha_k}(z) = f_a(z) = \phi_a(z), \qquad v_{a,0}(z) = 0.$$

Clearly, as a consequence of (18), Theorem 1 is valid for $k = 0$. Assume that the result is valid for k satisfying $0 \leqslant k < n$. We must now derive the result for $k + 1$ in place of k. For a given α_{k+1} the region $\mathfrak{W}_{\alpha_{k+1}}$ is the union of the $(2t + 1)^2$ regions

$$(22)\quad \mathfrak{W}_{\alpha_k} = \mathfrak{W}_1 \times \cdots \times \mathfrak{W}_k \times \mathfrak{S}_{a_{k+1}} \times \mathfrak{S}_{\alpha_{k+1}} \qquad (a_{k+1} = l + mi;\ l, m = 0, \pm 1, \ldots, \pm t).$$

By the induction assumption, there exists a regular function $f_{\alpha_k}(z)$ on each of these regions whose divisor coincides with $\mathfrak{d}$ suitably restricted. Furthermore, on each of the corresponding subregions $\mathfrak{Q}_{a,k}$, the induction hypothesis yields a regular function $v_{a,k}(z)$ which satisfies (20) and (21). Let

$$\mathfrak{W}_{\beta_k} = \mathfrak{W}_1 \times \cdots \times \mathfrak{W}_k \times \mathfrak{S}_{b_{k+1}} \times \mathfrak{S}_{\alpha_{k+1}}$$

be another of the regions in (22). If the intersection of $\mathfrak{W}_{\alpha_k}$ and $\mathfrak{W}_{\beta_k}$ is nonempty, then $f_{\alpha_k}(z)$ and $f_{\beta_k}(z)$ have the same divisor on it. Thus on the intersection the relation

$$(23)\qquad f_{\alpha_k}(z) = f_{\beta_k}(z)e^{v(z)}, \qquad v(z) = v_{\alpha_k\beta_k}(z)$$

holds for a regular function, $v_{\alpha_k\beta_k}(z)$, whose absolute value will now be estimated from above.

Let b be the lattice point with coordinates $a_1, \ldots, a_k, b_{k+1}, a_{k+2}, \ldots, a_n$. If the intersection of $\mathfrak{W}_{\alpha_k}$ and $\mathfrak{W}_{\beta_k}$ is nonempty, then for every subregion $\mathfrak{Q}_{a,k}$ of $\mathfrak{W}_{\alpha_k}$ the intersection with $\mathfrak{Q}_{b,k}$ is also nonempty and expressions (20), (21), and

$$(24)\qquad f_{\beta_k}(z) = \phi_b(z)e^{v_{b,k}(z)}, \qquad |v_{b,k}(z)| < \gamma_1 t^{5k}$$

are all valid. Since $\mathfrak{Q}_{a,k}$ and $\mathfrak{Q}_{b,k}$ are contained respectively in $\mathfrak{S}_a$ and $\mathfrak{S}_b$, expression (19) must hold on the intersection. Since $\mathfrak{d} \neq (0)$, $\phi_a(z)$ is not identically zero, and from relations (19), (20), (21), (23), and (24) we therefore have

$$(25)\qquad \begin{aligned} &v_{\alpha_k\beta_k}(z) = v_{a,k}(z) - v_{b,k}(z) + v_{ab}(z) + 2\pi i\nu, \qquad \nu = \nu_{ab}, \\ &|v_{\alpha_k\beta_k}(z) - 2\pi i\nu_{ab}| < 2\gamma_1 t^{5k} + \gamma < \gamma_2 t^{5k}, \end{aligned}$$

with ν an integer depending on a and b. Because it is a continuous function of z, the integer ν must be constant. Every cube $\mathfrak{Q}_{a_1} \times \cdots \times \mathfrak{Q}_{a_k}$ lying in $\mathfrak{W}_1 \times \cdots \times \mathfrak{W}_k$ can be joined to the distinguished cube $\mathfrak{Q}_0 \times \cdots \times \mathfrak{Q}_0$ by means of at most kt pairwise abutting cubes of this type. Since the function $v_{\alpha_k\beta_k}(z)$ is single-valued on the whole intersection of $\mathfrak{W}_{a_k}$ and $\mathfrak{W}_{\beta_k}$, relation (25) implies that, at every common point of $\mathfrak{Q}_{a,k} \cap \mathfrak{Q}_{b,k}$ and any other analogously defined intersection $\mathfrak{Q}_{c,k} \cap \mathfrak{Q}_{d,k}$, the estimate

$$2\pi\,|\nu_{ak} - \nu_{cd}| < 2\gamma_2 t^{5k}$$

holds. Explicitly, we have

$$c = (c_1, \ldots, c_k, a_{k+1}, a_{k+2}, \ldots, a_n),$$
$$d = (c_1, \ldots, c_k, b_{k+1}, a_{k+2}, \ldots, a_n).$$

In the definition of $v_{\alpha_k\beta_k}(z)$ we may still select an arbitrary multiple of $2\pi i$. Therefore, we may normalize so that $\nu_{ab} = 0$ for $a_1 = b_1 = 0, \ldots, a_k = b_k = 0$. Thus it follows generally that

$$(26)\qquad \begin{aligned} 2\pi\,|\nu_{ab}| &< 2kt\gamma_2 t^{5k} \\ |v_{\alpha_k\beta_k}(z)| &< \gamma_3 t^{5k+1}, \end{aligned}$$

which is the desired estimate.

The construction of the function $f_{\alpha_{k+1}}(z)$ is analogous to that in the proof of Theorem 1 of Section 5 and the subsequent considerations. The square $\mathfrak{W}_{k+1}$ is built up of $(2t+1)^2$ unit squares $\mathfrak{Q}_{a_{k+1}}$ where $a_{k+1} = l + mi$ $(l, m = 0, \pm 1, \ldots, \pm t)$. If we first fix m, then we obtain $2t + 1$ squares corresponding to the values of l, and we denote them by $\mathfrak{Q}_l$. We denote the entire strip by $\mathfrak{G}_m$. Let $\mathfrak{S}_l$ be the square $\mathfrak{S}_{a_{k+1}}$ corresponding to $\mathfrak{Q}_l$. If we also fix α_{k+1}, then for each value $l = 0, \pm 1, \ldots, \pm t$ we denote the corresponding function $f_{\alpha_k}(z)$ by $f_l(z)$. If the points $(z_1, \ldots, z_k)$ and $(z_{k+2}, \ldots, z_n)$ of the respective cubes $\mathfrak{W}_1 \times \cdots \times \mathfrak{W}_k$ and $\mathfrak{S}_{\alpha_{k+1}}$ are held fixed, then $f_l(z)$ is a regular function of the single variable z_{k+1} on $\mathfrak{S}_l$. Letting $l = -t, -t+1, \ldots, t-1$ we obtain from (23) the following formulae on the hatched intersection of $\mathfrak{S}_l$ and $\mathfrak{S}_{l+1}$ in Figure 3:

$$(27)\qquad f_l(z) = f_{l+1}(z)e^{v_l(z_{k+1})}.$$

In this intersection $v_l(z_{k+1})$ is a regular function of z_{k+1} alone which, in addition, is also a regular function of the remaining $n - 1$ variables on

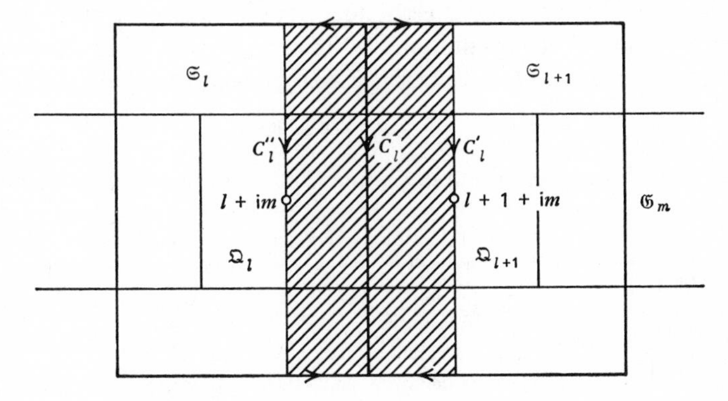

Figure 3

$\mathfrak{W}_1 \times \cdots \times \mathfrak{W}_k \times \mathfrak{S}_{\alpha_{k+1}}$. By (26) we then have

$$|v_l(z_{k+1})| < \gamma_3 t^{5k+1}. \tag{28}$$

Now we let C_l be the directed line segment which is equidistant from $l + im$ and $(l + 1) + im$, belongs to the intersection of $\mathfrak{S}_l$ and $\mathfrak{S}_{l+1}$, and runs from top to bottom. We form the expression

$$\sigma(z_{k+1}) = \sum_{l=-t}^{t-1} \frac{1}{2\pi i} \int_{C_l} \frac{v_l(\zeta)}{\zeta - z_{k+1}} \, d\zeta \tag{29}$$

where, at first, z_{k+1} does not lie on any segment C_l. Clearly $\sigma(z_{k+1})$ is a regular function of z_{k+1} which, in addition, is a regular function of the remaining $n - 1$ variables $z_1, \ldots, z_k, z_{k+2}, \ldots, z_n$ on the given regions. In particular, $\sigma(z_{k+1})$ is regular in the interior of the square $\mathfrak{Q}_\lambda$, where λ is one of the indices $0, \pm 1, \ldots, \pm t$. If z_{k+1} is an interior point of $\mathfrak{Q}_\lambda$, we set

$$\sigma(z_{k+1}) = \sigma_\lambda(z_{k+1}). \tag{30}$$

Now, as indicated in the diagram, let C_l' and C_l'' be the right and left parts respectively of the boundary of $\mathfrak{S}_l \cap \mathfrak{S}_{l+1}$. If z_{k+1} lies in the interior of $\mathfrak{Q}_\lambda$, then by Cauchy's theorem we can replace the integration along C_λ in (29) by integration along C_λ' when $\lambda < t$. Similarly, integration along $C_{\lambda-1}''$ can be substituted for integration along $C_{\lambda-1}$ when $\lambda > -t$. Accordingly, $\sigma_\lambda(z_{k+1})$ can be continued analytically to the whole interior of $\mathfrak{S}_\lambda$. If z_{k+1} is in the interior of the intersection of $\mathfrak{S}_l$ and $\mathfrak{S}_{l+1}$, then both functions $\sigma_l(z_{k+1})$ and $\sigma_{l+1}(z_{k+1})$ are regular there. We obtain the following formulae from (29) by subtraction

$$\sigma_{l+1}(z_{k+1}) - \sigma_l(z_{k+1}) = \frac{1}{2\pi i} \int_{C_l''} \frac{v_l(\zeta)}{\zeta - z_{k+1}} \, d\zeta - \frac{1}{2\pi i} \int_{C_l'} \frac{v_l(\zeta)}{\zeta - z_{k+1}} \, d\zeta.$$

We obtain a positive circuit of the boundary of $\mathfrak{S}_l \cap \mathfrak{S}_{l+1}$ by first traversing C_l'' in the direction of its orientation and then traversing C_l' in the direction opposite to its orientation. Hence the Cauchy integral formula yields

$$\sigma_{l+1}(z_{k+1}) - \sigma_l(z_{k+1}) = v_l(z_{k+1})$$

in the interior of the intersection of $\mathfrak{S}_l$ and $\mathfrak{S}_{l+1}$. Expression (27) then yields

$$f_l e^{\sigma_l(z_{k+1})} = f_{l+1} e^{\sigma_{l+1}(z_{k+1})} \qquad (l = -t, \ldots, t - 1) \tag{31}$$

in this domain.

In order to estimate the function σ_λ, we cut off a strip of width $\frac{1}{4}$ which runs around $\mathfrak{S}_\lambda$ and we call the remainder $\mathfrak{F}_\lambda$. The $2t + 1$ closed squares $\mathfrak{F}_\lambda$ ($\lambda = 0, \pm 1, \ldots, \pm t$) then cover a rectangle $\mathfrak{H}_m$ which arises from $\mathfrak{G}_m$ by addition of a boundary strip of width $\frac{1}{4}$. Now we view the indeterminates

$z_1, \ldots, z_k$ and $z_{k+2}, \ldots, z_n$ as variables and define a function $F_m(z)$ on the region

$$\mathfrak{B}_m = \mathfrak{W}_1 \times \cdots \times \mathfrak{W}_k \times \mathfrak{H}_m \times \mathfrak{S}_{\alpha_{k+1}}$$

by setting

$$F_m(z) = f_\lambda e^{\sigma_\lambda(z_{k+1})} \tag{32}$$

for all z_{k+1} in $\mathfrak{F}_\lambda$. Expression (31) now shows that $F_m(z)$ is uniquely defined on all of $\mathfrak{B}_m$ and that it is regular there. Since, by the induction hypothesis, the function f_λ has divisor $\mathfrak{d}$ on the region $\mathfrak{W}_1 \times \cdots \times \mathfrak{W}_k \times \mathfrak{S}_\gamma \times \mathfrak{S}_{\alpha_{k+1}}$ it follows that $\mathfrak{d}$ must also be the divisor of F_m on $\mathfrak{B}_m$. We once again change the paths of integration in (29) from C_λ to C'_λ for $\lambda < t$ and from $C_{\lambda-1}$ to $C''_{\lambda-1}$ for $\lambda > -t$ and observe that then $|\zeta - z_{k+1}| \geqslant 1/4$ on all paths of integration. Since the length of each of these paths remains $\leqslant 3$, we obtain from (28) the following estimate on the square $\mathfrak{F}_\lambda$

$$|\sigma_\lambda(z_{k+1})| < 2t \cdot \frac{1}{2\pi} \cdot 3 \cdot 4\gamma_3 t^{5k+1} = \gamma_4 t^{5k+2}. \tag{33}$$

We now pass from the $2t + 1$ rectangles $\mathfrak{H}_m$ ($m = 0, \pm 1, \ldots, \pm t$) to the large square $\mathfrak{W}_{k+1}$ in a manner analogous to the passage from $\mathfrak{S}_l$ ($l = 0, \pm 1, \ldots, \pm t$) to $\mathfrak{H}_m$. Since, by construction, the functions F_m have the divisor $\mathfrak{d}$ on $\mathfrak{B}_m$, we have

$$F_m(z) = F_{m+1}(z) e^{w_m(z)} \tag{34}$$

on the intersection of $\mathfrak{B}_m$ and $\mathfrak{B}_{m+1}$ for $m < t$. The function $w_m(z)$, which is regular there, will be estimated in the same way as $v_{\alpha_k\beta_k}(z)$ in (26). Namely, we again freeze all variables except z_{k+1} and consider $w_m(z)$ as a function of z_{k+1} on the intersection of $\mathfrak{H}_m$ and $\mathfrak{H}_{m+1}$. This intersection is itself the union of the $2t + 1$ intersections $\mathfrak{F}_{l,m} \cap \mathfrak{F}_{l,m+1}$ ($l = 0, \pm 1, \ldots, \pm t$). Here $\mathfrak{F}_{l,m}$ and $\mathfrak{F}_{l,m+1}$ are the previous squares $\mathfrak{F}_l$ corresponding to the indices m and $m + 1$. If, with greater precision, we write f_{lm}, σ_{lm} instead of f_l, σ_l, then from (32) and (33) we have the following relations in each intersection

$$F_m = f_{lm} e^{\sigma_{lm}}, \; F_{m+1} = f_{l,m+1} e^{\sigma_{l,m+1}}, \; |\sigma_{lm}| < \gamma_4 t^{5k+2}, \; |\sigma_{l,m+1}| < \gamma_4 t^{5k+2}. \tag{35}$$

On the other hand, we can apply (23) and (26) to f_{lm} and $f_{l,m+1}$ and obtain

$$f_{lm} = f_{l,m+1} e^{v_{lm}}, \qquad |v_{lm}| < \gamma_3 t^{5k+1}, \tag{36}$$

where each v_{lm} is regular on the region under consideration. It follows from (34), (35), and (36) that

$$w_m = \sigma_{lm} - \sigma_{l,m+1} + v_{lm} + 2\pi i \nu, \qquad \nu = \nu_{l,m}$$

where $\nu_{l,m}$ is an integer which is dependent on l and m. Again, referring to z, we see that $\nu_{l,m}$ is constant, and we can still normalize by putting $\nu_{0,m} = 0$.

From (35) and (36) it follows that

$$|w_m - 2\pi i \nu_{l,m}| < 2\gamma_4 t^{5k+2} + \gamma_3 t^{5k+1} < \gamma_5 t^{5k+2} \tag{37}$$

on $\mathfrak{F}_{l,m} \cap \mathfrak{F}_{l,m+1}$. Thus we also have

$$|\nu_{l,m} - \nu_{l+1,m}| < 2\gamma_5 t^{5k+2}$$

for $l = -t, \ldots, t-1$; therefore,

$$|\nu_{l,m}| < 2\gamma_5 t^{5k+3} \qquad (l = 0, \pm 1, \ldots, \pm t). \tag{38}$$

From (37) and (38) we obtain the estimate

$$|w_m| < \gamma_6 t^{5k+3} \qquad (m = -t, \ldots, t-1) \tag{39}$$

on the intersection of $\mathfrak{V}_m$ and $\mathfrak{V}_{m+1}$.

We can now carry over the considerations which led to (31). In place of $\mathfrak{S}_l, f_l, v_l$ there appear $\mathfrak{H}_m, F_m, w_m$, and in analogy with (29) and (30) we define a function $\tau(z_{k+1})$ and regular functions $\tau_\mu(z_{k+1}), \mu = 0, \pm 1, \ldots, \pm t$. Each $\tau_\mu(z_{k+1})$ is defined in the interior of $\mathfrak{H}_\mu$ and is regular in the remaining $n-1$ variables on the region $\mathfrak{W}_1 \times \cdots \times \mathfrak{W}_k \times \mathfrak{S}_{\alpha_{k+1}}$. Here the former line segment C_l has to be replaced by the horizontal line segment bisecting the intersection of $\mathfrak{H}_m$ and $\mathfrak{H}_{m+1}$ ($m = -t, \ldots, t-1$) and running from left to right. The length of this line segment is $2t + 3/2$. By (39), we obtain the analogue of (33), that is, the estimate

$$|\tau_\mu(z_{k+1})| < 2t \cdot \frac{1}{2\pi}(2t+2) \cdot 4\gamma_6 t^{5k+3} < \gamma_7 t^{5k+5} \tag{40}$$

on the rectangle $\mathfrak{G}_\mu$ which arises from $\mathfrak{H}_\mu$ by cutting off a boundary strip of width $1/4$. In the interior of $\mathfrak{H}_m \cap \mathfrak{H}_{m+1}$ we have

$$\tau_{m+1} - \tau_m = w_m$$

as well as

$$F_m e^{\tau_m} = F_{m+1} e^{\tau_{m+1}} \qquad (m = -t, \ldots, t-1), \tag{41}$$

which is the desired analogue of (31).

We define the desired function $f_{\alpha_{k+1}}(z)$ on the region $\mathfrak{W}_{\alpha_{k+1}}$ by setting

$$f_{\alpha_{k+1}} = F_\mu e^{\tau_\mu} \tag{42}$$

on $\mathfrak{V}_\mu$ ($\mu = 0, \pm 1, \ldots, \pm t$). This definition is the analogue of (32). It follows from (41) that $f_{\alpha_{k+1}}(z)$ is then uniquely determined and regular on $\mathfrak{W}_{\alpha_{k+1}}$. Furthermore, since the restriction of $\mathfrak{d}$ to $\mathfrak{V}_\mu$ is the divisor of F_μ there, we see that, in the same sense, $\mathfrak{d}$ is the divisor of $f_{\alpha_{k+1}}$ on $\mathfrak{W}_{\alpha_{k+1}}$. For $a_{k+1} = l + mi$ the region $\mathfrak{Q}_{\alpha_{k+1}}$ is the subsquare of $\mathfrak{G}_m$ previously denoted by $\mathfrak{Q}_l$. Since the region $\mathfrak{Q}_{a,k+1}$ is contained in $\mathfrak{Q}_{a,k}$, the formula

$$v_{a,k+1} = v_{a,k} + \sigma_{lm} + \tau_m$$

defines a regular function of $z_1, \ldots, z_n$ on $\mathfrak{Q}_{a,k+1}$ which, as a result of (20), (32), (42), and (21), (33), (40) satisfies the relation

$$f_{\alpha_{k+1}}(z) = \phi_a(z)e^{v_{a,k+1}(z)}, \quad |v_{a,k+1}(z)| < \gamma_1 t^{5k} + \gamma_4 t^{5k+2} + \gamma_7 t^{5k+5} < \gamma_8 t^{5(k+12)}.$$

Thus the induction step is carried out and the proof of Theorem 1 is complete.

We only require Theorem 1 in the case $k = n$, where $\mathfrak{W}_{\alpha_n}$ is then the entire cube with edge length $2t + 2$ and $\mathfrak{Q}_{a,n}$ is the single subcube $\mathfrak{Q}_a$. Consequently, there is a regular function, f, on $\mathfrak{W}$ with the prescribed divisor, $\mathfrak{d}$, there and which satisfies the condition

$$f = \phi_a e^{v_a}, \qquad |v_a| < \gamma_9 t^{5n} \tag{43}$$

on $\mathfrak{Q}_a$ for some function, v_a, which is regular there.

Theorem 2: Let $\mathfrak{d}$ be a periodic integral divisor. Then there exists a Jacobian function with divisor $\mathfrak{d}$ whose exponent system consists entirely of polynomials.

Proof: We begin with a consideration related to the method of proof of Schwarz' lemma. Let $f(z)$ be a regular function whose absolute value is less than or equal to M in the closed n-disk $|z| \leqslant \rho$. We expand $f(z)$ in a Taylor series in powers of $z_1, \ldots, z_n$ and let f_k $(k = 0, 1, \ldots)$ be the partial sum formed by all terms of degree k, and we set

$$f(z) = g(z) + r(z), \quad g(z) = f_0 + \cdots + f_k, \quad r(z) = f_{k+1} + f_{k+2} + \cdots. \tag{44}$$

Now fix z. If λ is a new complex variable, then the function

$$f(\lambda z) = h(\lambda)$$

is a regular function of the single variable λ in the disk $|\lambda| \leqslant \rho\,|z|^{-1}$; there we have

$$h(\lambda) = \sum_{k=0}^{\infty} f_k(z)\lambda^k.$$

If C is the positively oriented boundary circle, then by Cauchy's formula

$$f_k(z) = \frac{1}{2\pi i}\int_C \frac{h(\lambda)}{\lambda^{k+1}}\,d\lambda \qquad (k = 0, 1, 2, \ldots)$$

and therefore

$$|f_k(z)| \leqslant M\left(\frac{|z|}{\rho}\right)^k.$$

Thus we have the estimates

$$(45)\qquad |g(z)| \leqslant M \sum_{l=0}^{k} \left(\frac{|z|}{\rho}\right)^l, \qquad |r(z)| \leqslant M \sum_{l=k+1}^{\infty} \left(\frac{|z|}{\rho}\right)^l \leqslant 2M\left(\frac{|z|}{\rho}\right)^{k+1},$$

where z is arbitrary in the first estimate and is assumed to satisfy $|z| \leqslant \frac{1}{2}\rho$ in the second.

We assume that $\mathfrak{d} \neq (0)$ and apply the result (43) in the following fashion. Choose $t = t_l = 2^l$ $(l = 0, 1, \ldots)$. If we now write $\mathfrak{W}^{(l)}, f^{(l)}, v_a^{(l)}$ in place of the earlier $\mathfrak{W}, f, v_a$, then $f^{(l)}$ becomes a regular function on the cube $\mathfrak{W}^{(l)}$ with the divisor $\mathfrak{d}$ restricted to $\mathfrak{W}^{(l)}$, and $v_a^{(l)}$ is a regular function on the subcube $\mathfrak{Q}_a$ of $\mathfrak{W}^{(l)}$ for which

$$(46)\qquad f^{(l)} = \phi_a e^{v_a^{(l)}}, \qquad |v_a^{(l)}| < \gamma_9 t^{5n} \qquad (t = 2^l).$$

If we replace l by $l + 1$, then we also obtain

$$(47)\qquad f^{(l+1)} = \phi_a e^{v(l+1)}, \qquad |v_a^{(l+1)}| < \gamma_9 (2t_l)^{5n};$$

therefore,

$$(48)\qquad f^{(l+1)} = f^{(l)} e^{v_a^{(l+1)} - v_a^{(l)}}$$

holds on $\mathfrak{Q}_a$. Furthermore, since $f^{(l+1)}$ and $f^{(l)}$ have the same divisor $\mathfrak{d}$ on $\mathfrak{W}^{(l)}$ it follows that

$$(49)\qquad f^{(l+1)} = f^{(l)} e^{s^{(l)}}$$

for some regular function, $s^{(l)}$, on $\mathfrak{W}^{(l)}$. Expressions (48) and (49) yield the formula

$$(50)\qquad s^{(l)} = v_a^{(l+1)} - v_a^{(l)} + 2\pi i \nu, \qquad \nu = \nu_a^{(l)}$$

on $\mathfrak{Q}_a$ for some integer ν. Finally, if we normalize $\nu_0^{(l)} = 0$ and apply (46), (47), and (50) to abutting cubes $\mathfrak{Q}_a$, the previous method of proof yields the following inequality on $\mathfrak{W}^{(l)}$:

$$(51)\qquad |s^{(l)}| < \gamma_{10} t^k,$$

where $k = 5n + 1$.

In order to make the sequence of functions $f^{(l)}$ $(l = 0, 1, \ldots)$ converge, we will, as in the proof of Theorem 2 in Section 5, adjoin certain unit factors which are still to be defined. But now it is a question of making a more precise estimate. Corresponding to (44) we introduce the decomposition

$$(52)\qquad s^{(l)} = g^{(l)} + r^{(l)}$$

where $g^{(l)}$ contains all terms of degree less than or equal to k in the series expansion of $s^{(l)}$ in powers of $z_1, \ldots, z_n$, and $r^{(l)}$ consists of the remaining terms. Let $\mathfrak{K}_l$ be the n-disk $|z| \leqslant t_l$ of radius $t_l = 2^l$ which is contained within $\mathfrak{W}^{(l)}$. Then, since the radius of $\mathfrak{K}_{l-1}$ is $\frac{1}{2}t_l = 2^{l-1}$, that is, half the radius of $\mathfrak{K}_l$, it follows from (45) and (51) that for $l = 1, 2, \ldots$ both estimates

$$(53)\qquad \begin{aligned} |g^{(m)}| &< \gamma_{10} t_l^k \qquad (m = 1, \ldots, l-1), \\ |r^{(l)}| &< 2\gamma_{10} t_l^k \left(\frac{|z|}{t}\right)^{k+1} = 2\gamma_{10} |z|^{k+1} 2^{-l} \end{aligned}$$

are valid on $\mathfrak{K}_{l-1}$. By (49) and (52) we obtain $f^{(l+1)} e^{-g^{(l)}} = f^{(l)} e^{r^{(l)}}$ on $\mathfrak{K}_l$. Thus on $\mathfrak{K}_l$ we also have

$$(54)\qquad f^{(m+1)} e^{-(g^{(l)} + \cdots + g^{(m)})} = f^{(l)} e^{r^{(l)} + \cdots + r^{(m)}}$$

for each integer $m \geqslant l$, and in particular

$$f^{(m+1)} e^{-(g^{(1)} + \cdots + g^{(m)})} = f^{(1)} e^{r^{(1)} + \cdots + r^{(m)}}, \qquad (m = 1, 2, \ldots)$$

on $\mathfrak{K}_1$. Since, by (53), the series $r^{(1)} + r^{(2)} + \cdots$ converges uniformly in the disk $\mathfrak{K}_0$, the Weierstrass convergence theorem implies the existence of the limit

$$\lim_{m \to \infty} \{f^{(m+1)} e^{-(g^{(1)} + \cdots + g^{(m)})}\} = f(z)$$

which is a regular function of z on the interior of $\mathfrak{K}_0$. Furthermore, if we put

$$(55)\qquad \sigma_l = -(g^{(1)} + \cdots + g^{(l-1)}) + r^{(l)} + r^{(l+1)} + \cdots \qquad (l = 1, 2, \ldots),$$

then the expression (54) yields the following representation on the interior of $\mathfrak{K}_0$

$$(56)\qquad f(z) = f^{(l)} e^{\sigma_l}.$$

However for each fixed value of l, expressions (53) and (55) imply that σ_l is a uniformly convergent series of regular functions on $\mathfrak{K}_{l-1}$ and therefore regular on the interior of $\mathfrak{K}_{l-1}$. Clearly $f(z)$ is then analytically continued to the interior of $\mathfrak{K}_{l-1}$ by means of (56), and it is an associate of $f^{(l)}$ there. Since the restriction of $\mathfrak{d}$ to $\mathfrak{K}_l$ is the divisor of $f^{(l)}$ for any l we see that f is an entire function that has $\mathfrak{d}$ as divisor on $\mathfrak{Z}$.

Since the divisor $\mathfrak{d}$ is periodic, f is a Jacobian function. Let

$$j_\omega(z) = e^{s_\omega(z)}$$

be the factor system for $f(z)$, and let $s_\omega(z)$ be the corresponding exponent system. We have to show that for every period, ω, the function $s_\omega(z)$ is a polynomial in the variables $z_1, \ldots, z_n$. Since $l < 2^l = t$, expressions (53) and (55) yield the following estimate

$$|\sigma_l| < t\gamma_{10} t^k + 2\gamma_{10} t^k < \gamma_{11} t^{5n+2}$$

on $\mathfrak{R}_{l-1}$ ($l = 1, 2, \ldots$). Thus it follows from (46) and (56) that with the definition

$$\sigma_{al} = v_a^{(l)} + \sigma_l$$

the following formulas

$$(57) \qquad f(z) = \phi_a(z)e^{\sigma_{al}(z)}, \qquad |\sigma_{al}(z)| < \gamma_{12}t^{5n+2}$$

are valid on every cube $\mathfrak{Q}_a$ in the interior of $\mathfrak{R}_{l-1}$.

For a given period, ω, assume that l is so large that all cubes intersecting both n-disks $|z| \leqslant t/4$ and $|z + \omega| \leqslant t/4$ lie entirely in the interior of $\mathfrak{R}_{l-1}$. If z and $z + \omega$ are points of $\mathfrak{Q}_a$ and $\mathfrak{Q}_b$ respectively, then in addition to (57) we have the corresponding formula

$$(58) \qquad f(z + \omega) = \phi_b(z + \omega)e^{\sigma_{bl}(z+\omega)}, \qquad |\sigma_{bl}(z + \omega)| < \gamma_{12}t^{5n+2}.$$

On the other hand, by (15) and (18) we obtain

$$(59) \qquad \begin{aligned} \phi_a(z) &= p_{\zeta_a}(z - \zeta_a), \\ \phi_b(z + \omega) &= p_{\zeta_b}(z + \omega - \zeta_b) = p_{\zeta_b - \omega}(z - (\zeta_b - \omega)) \end{aligned}$$

where $\mathfrak{Q}_a$ and $\mathfrak{Q}_b$ are taken to lie in the respective circles of convergence $\mathfrak{R}_{\zeta_a}$ of p_{ζ_a} and $\mathfrak{R}_{\zeta_b}$ of p_{ζ_b}. The cube, $\mathfrak{Q}_{b-\omega}$, which arises from $\mathfrak{Q}_b$ by translation through $-\omega$ then lies in the circle of convergence $\mathfrak{R}_{\zeta_b-\omega}$ of $p_{\zeta_b-\omega}$. Since z is also a point in $\mathfrak{Q}_a$, we see that z belongs to the intersection of $\mathfrak{R}_{\zeta_a}$ and $\mathfrak{R}_{\zeta_b-\omega}$. Therefore, (16) and (17) yield

$$(60) \qquad p_{\zeta_a}(z - \zeta_a) = p_{\zeta_b-\omega}(z - (\zeta_b - \omega))e^{w(z)}, \qquad |w(z)| < \gamma$$

where $w(z)$ is a regular function on $\mathfrak{R}_{\zeta_a} \cap \mathfrak{R}_{\zeta_b-\omega}$. From the equation

$$f(z + \omega) = f(z)e^{s_\omega(z)}$$

we derive, by means of (57), (58), (59), and (60), the relation

$$(61) \qquad s_\omega(z) = \sigma_{bl}(z + \omega) - \sigma_{al}(z) - w(z) + 2\pi i \nu; \qquad \nu = \nu_{abl\omega}$$

with ν an integer independent of the point z in $\mathfrak{Q}_a \cap \mathfrak{Q}_{b-\omega}$. Next we determine c so that ω is a point of $\mathfrak{Q}_c$. Let l_0 be the smallest integer such that the interior of $\mathfrak{R}_{l_0-1}$ contains all cubes which meet either the n-disk $|z| \leqslant t/4$ or the n-disk $|z + \omega| \leqslant t/4$. Then we may fix $\nu = 0$ for $a = 0$, $b = c$, and $l = l_0$. In particular, the value $s_\omega(0)$ is hereby determined. Now use equation (61) to define ν for $a = 0$, $b = c$ and arbitrary l; whence (57), (58), and (60) imply that, as a function of t, $\nu_{0cl\omega}$ does not grow more rapidly than t^{5n+2}. Since the number of cubes $\mathfrak{Q}_{b-\omega}$ which intersect any $\mathfrak{Q}_a$ has an upper bound independent of a, l, and ω we can join any two points of the n-disk $|z| \leqslant t/4$ by a chain of intersections, $\mathfrak{Q}_a \cap \mathfrak{Q}_{b-\omega}$, whose length is less than $\gamma_{13}t$. By the same argument which we have employed repeatedly, it follows from (57),

(58), and (60) that the estimate

$$(62) \qquad |s_\omega(z)| < \gamma_{14} t^{5n+3}, \qquad \left(|z| \leqslant \frac{t}{4}, |z+\omega| \leqslant \frac{t}{4}\right)$$

is valid.

Finally, we expand the entire function $s_\omega(z)$ in a power series

$$s_\omega(z) = g(z) + r(z)$$

where $g(z)$ consists of all terms of total degree less than or equal to $5n + 3$, and $r(z)$ is composed of the remaining terms. In view of (45) and (62) we obtain the inequality

$$|r(z)| < 2\gamma_{14} t^{5n+3} \left(\frac{4|z|}{t}\right)^{5n+4} = \gamma_{15} |z|^{5n+4} t^{-1}$$

for $|z| \leqslant t/8$ and $|z + \omega| \leqslant t/4$. If, for fixed z, we let $t = 2^l$ tend to ∞, then it follows that $r(z) = 0$. Hence, $s_\omega(z)$ is a polynomial whose degree is, by the way, at most $5n + 3$. This completes the proof of Theorem 2.

8. Linearization of the exponent system

Our immediate goal is to prove the following sharpened version of Theorem 2 of the preceding section.

Theorem 1: Let $\mathfrak{d}$ be a periodic integral divisor. Then, there exists a Jacobian function with divisor $\mathfrak{d}$ whose exponent system consists entirely of linear functions.

Proof: We may again assume $\mathfrak{d} \neq (0)$. Let f and g be two Jacobian functions with the same divisor, $\mathfrak{d}$, and exponent systems s_ω and t_ω respectively. Then, there exists an entire function w, satisfying

$$(1) \qquad g = fe^w,$$

and we conclude that

$$(2) \qquad t_\omega(z) = s_\omega(z) + w(z + \omega) - w(z).$$

Here we should actually add an integral multiple of $2\pi i$, but this can be absorbed by $t_\omega(z)$. Conversely, given the Jacobian function f with divisor $\mathfrak{d}$ and exponent system s_ω and any arbitrary entire function $w(z)$, the relation (1) defines a Jacobian function $g(z)$ with divisor $\mathfrak{d}$ whose exponent system t_ω is determined by (2). Now we must construct the entire function $w(z)$ so that each $t_\omega(z)$ becomes a linear function. This problem was solved by Appell for $n = 1, 2$ and by Conforto for an arbitrary n. In view of Theorem 2 of Section 7, it suffices for our purposes to start with polynomials $s_\omega(z)$. Then, by simple algebraic considerations we will construct a polynomial solution, $w(z)$.

In the sequel we denote a fixed basis of the lattice Ω by $\omega_1, \ldots, \omega_{2n}$ and, for simplicity, put $s_{\omega_r}(z) = s_r(z)$, $t_{\omega_r}(z) = t_r(z)$, $r = 1, \ldots, 2n$. Assuming all $s_r(z)$ are given polynomials, then in view of (2) we have to determine a polynomial $w(z)$ such that each of the $2n$ polynomials

$$s_r(z) + w(z + \omega_r) - w(z) = t_r(z), \qquad (r = 1, \ldots, 2n) \tag{3}$$

turns out to be linear. Consequently, relation (8) in Section 7 implies that all the $t_\omega(z)$ are linear. According to relation (12) in Section 7 we put

$$s_{\omega_k+\omega_l}(z) - s_k(z + \omega_l) - s_l(z) = 2\pi i c_{kl}, \qquad (k, l = 1, \ldots, 2n). \tag{4}$$

Then each c_{kl} is an integer. We write

$$c_{lk} - c_{kl} = a_{kl}, \qquad 2\pi i a_{kl} = \alpha_{kl}.$$

Interchanging k and l in (4) and subtracting yields the formula

$$s_k(z + \omega_l) - s_l(z + \omega_k) - s_k(z) + s_l(z) = \alpha_{kl}, \qquad (k, l = 1, \ldots, 2n). \tag{5}$$

The matrix $(a_{kl}) = A$ turns out to be skew symmetric. Furthermore, the integers c_{kl} and a_{kl} remain unchanged when passing from s_r to t_r. Equations (5) will be essential in determining the linear functions $t_r(z)$ in (3). The matrix A will also be of great importance in the sequel.

In place of the system of $2n$ linear difference equations for the unknown function $w(z)$ given by (3), we first treat the single equation

$$\phi(u + 1) - \phi(u) = s(u). \tag{6}$$

Here

$$s(u) = a_1 u^{m-1} + a_2 u^{m-2} + \cdots + a_m$$

is a given polynomial of degree $m - 1$ in the single variable u with complex coefficients $a_1, \ldots, a_m$, and the polynomial $\phi(u)$ is still to be determined. Equation (6) is obtained from (3) when z, ω_r, s_r, w, t_r are replaced by u, 1, $-s$, ϕ, 0 respectively. Evidently the difference $\phi(u + 1) - \phi(u)$ has degree $m - 1$ whenever the polynomial $\phi(u)$ has degree m. Therefore, we try to solve (6) by putting

$$\phi(u) = b_0 u^m + b_1 u^{m-1} + \cdots + b_m$$

with unknown coefficients $b_0, b_1, \ldots, b_m$. Inserting this expression in (6) and comparing coefficients, we derive the following system of linear equations

$$\sum_{l=0}^{k-1} \binom{m-l}{m-k} b_l = a_k, \qquad (k = 1, \ldots, m).$$

These equations determine $b_0, b_1, \ldots, b_{m-1}$ recursively, whereas b_m remains arbitrary. Furthermore, requiring $\phi(0) = 0$ implies that $b_m = 0$ and consequently that (6) has a uniquely determined polynomial $\phi(u)$ as solution. We let ξ be a natural number and sum the equations obtained from (6) by inserting $u = 0, 1, \ldots, \xi - 1$. Thus we derive the relation

$$\phi(\xi) = \sum_{\lambda=0}^{\xi-1} s(\lambda), \qquad (\xi = 0, 1, 2, \ldots) \tag{7}$$

which remains true for $\xi = 0$ if, in that case, the sum is defined to be zero. In particular, if the polynomial $s(u)$ ranges over the successive powers of u, we obtain the Bernoulli polynomials for $\phi(u)$. However, for our purposes we need not introduce them explicitly.

Let t be the column formed by the n variables $t_1, \ldots, t_n$ and let k be an integer between 1 and $2n$. If λ stands for a scalar, the expression $s_k(t + \lambda\omega_k)$ is a polynomial in $t_1, \ldots, t_n$, and a single term of this polynomial is of the form $t_1^{g_1} \cdots t_n^{g_n} s(\lambda)$, where the coefficient $s(\lambda)$ turns out to be a polynomial in λ. In view of (7), summing over the values $\lambda = 0, \ldots, \xi - 1$ yields the expression $t_1^{g_1} \cdots t_n^{g_n}\phi(\xi)$, where $\phi(\xi)$ is the polynomial determined above. Hence the sum

$$\sum_{\lambda=0}^{\xi-1} s_k(t + \lambda\omega_k) = \phi_k(t, \xi), \qquad (\xi = 0, 1, \ldots) \tag{8}$$

is also a polynomial in $t_1, \ldots, t_n$, and ξ, which satisfies $\phi_k(t, 0) = 0$ and

$$\phi_k(z, 1) = s_k(z). \tag{9}$$

We now apply (5) to $z = t + \lambda\omega_k + \mu\omega_l$. For given natural numbers ξ, η we sum over $\lambda = 0, \ldots, \xi - 1$ and $\mu = 0, \ldots, \eta - 1$, thus obtaining

$$\sum_{\lambda=0}^{\xi-1} \{s_k(t + \eta\omega_l + \lambda\omega_k) - s_k(t + \lambda\omega_k)\}$$
$$= \sum_{\mu=0}^{\eta-1} \{s_l(t + \xi\omega_k + \mu\omega_l) - s_l(t + \mu\omega_l)\} + \alpha_{kl}\xi\eta.$$

In view of (8), we conclude that

$$\phi_k(t + \eta\omega_l, \xi) - \phi_k(t, \xi) = \phi_l(t + \xi\omega_k, \eta) - \phi_l(t, \eta) + \alpha_{kl}\xi\eta, \qquad (k, l = 1, \ldots, 2n). \tag{10}$$

This relation is valid for $\xi = 0, 1, \ldots,$ and $\eta = 0, 1, \ldots.$ But, since both sides of (10) are polynomials in ξ and η, this formula holds identically in ξ and η. Furthermore, replacing ξ by $\xi + \eta$ in (8), where $\eta = 0, 1, \ldots,$ we obtain

$$\phi_k(t, \xi + \eta) = \phi_k(t, \xi) + \sum_{\lambda=\xi}^{\xi+\eta-1} s_k(t + \lambda\omega_k)$$

$$= \phi_k(t, \xi) + \sum_{\lambda=0}^{\eta-1} s_k(t + \xi\omega_k + \lambda\omega_k).$$

Hence, we have

(11) $\phi_k(t, \xi + \eta) = \phi_k(t, \xi) + \phi_k(t + \xi\omega_k, \eta), \qquad (k = 1, \ldots, 2n).$

Relation (11) holds identically in the variables ξ and η for the same reason as above.

Since the $2n$ vectors $\omega_1, \ldots, \omega_{2n}$ are linearly independent over the real numbers, there are among them n which are linearly independent over the complex numbers. With a suitable reordering we may assume them to be $\omega_1, \ldots, \omega_n$. Then every point x of $\mathfrak{Z}$ has a unique representation as

(12) $$x = \xi_1\omega_1 + \cdots + \xi_n\omega_n.$$

The complex coefficients $\xi_1, \ldots, \xi_n$ will be called the "coordinates of x." We have to distinguish them from the elements of the column x, that is, the components of the vector x which we now denote by $x^{(1)}, \ldots, x^{(n)}$. Furthermore, we introduce the vectors

(13) $$x_k = \xi_1\omega_1 + \cdots + \xi_{k-1}\omega_{k-1}, \qquad (k = 1, \ldots, n + 1).$$

In particular, we obtain $x_1 = 0$, $x_{n+1} = x$, and

(14) $$x_{k+1} = x_k + \xi_k\omega_k, \qquad (k = 1, \ldots, n).$$

In view of (12) and (13), the coordinates, ξ_k, as well as the components of x_k are homogeneous linear functions of the components $x^{(1)}, \ldots, x^{(n)}$ of x with complex coefficients. We define

(15) $$\psi(t, x) = \sum_{k=1}^{n} \phi_k(t + x_k, \xi_k)$$

and will show that the function $-\psi(0, z)$ has exactly the properties required of $w(z)$ in (3). According to (15), it is obvious that $\psi(t, x)$ is a polynomial with respect to the components of t and x. Hence, $\psi(0, z)$ is also such a polynomial.

In (10) we replace t, ξ by $t + x_k$, ξ_k and sum over $k = 1, \ldots, n$. This yields

$$\sum_{k=1}^{n} \{\phi_k(t + x_k + \eta\omega_l, \xi_k) - \phi_k(t + x_k, \xi_k)\}$$

$$= \sum_{k=1}^{n} \{\phi_l(t + x_k + \xi_k\omega_k, \eta) - \phi_l(t + x_k, \eta) + \alpha_{kl}\xi_k\eta\}.$$

The left-hand side is treated by use of (15) and the right-hand side by use of (14). Furthermore, replacing l by r we obtain

$$\psi(t + \eta\omega_r, x) - \psi(t, x)$$
$$= \phi_r(t + x, \eta) - \phi_r(t, \eta) + \sum_{k=1}^{n} \alpha_{kr}\xi_k\eta, \qquad (r = 1, \ldots, 2n),$$

identically with respect to the vectors t, x and the scalar variable η. We choose $t = 0$, $x = z$, $\eta = 1$, and denote the coordinates of z by $\zeta_1, \ldots, \zeta_n$. In view of (9), it follows that

$$(16) \quad \psi(\omega_r, z) - \psi(0, z) = s_r(z) - s_r(0) + \sum_{k=1}^{n} \alpha_{kr}\zeta_k, \qquad (r = 1, \ldots, 2n).$$

In order to obtain the desired result, we need another formula for the expression $\psi(\omega_r, z)$. Let y be a vector with coordinates $\eta_1, \ldots, \eta_n$ and define the vectors $y_1, \ldots, y_{n+1}$ analogously to (13). In (10) we replace t, η, ξ by $t + x_k + y_l$, η_l, ξ_k and sum over all pairs k, l satisfying $1 \leqslant l < k \leqslant n$. The left-hand side yields the expression

$$\sum_{k=2}^{n} \sum_{l=1}^{k-1} \{\phi_k(t + x_k + y_l + \eta_l\omega_l, \xi_k) - \phi_k(t + x_k + y_l, \xi_k)\}$$
$$= \sum_{k=2}^{n} \{\phi_k(t + x_k + y_k, \xi_k) - \phi_k(t + x_k, \xi_k)\}$$

where summation over k could also start with $k = 1$. Correspondingly, the right-hand side yields the expression

$$\sum_{l=1}^{n-1} \sum_{k=l+1}^{n} \{\phi_l(t + x_k + y_l + \xi_k\omega_k, \eta_l) - \phi_l(t + x_k + y_l, \eta_l) + \alpha_{kl}\xi_k\eta_l\}$$
$$= \sum_{l=1}^{n-1} \{\phi_l(t + x + y_l, \eta_l) - \phi_l(t + x_{l+1} + y_l, \eta_l)\} + \sum_{1 \leqslant l < k \leqslant n} \alpha_{kl}\xi_k\eta_l,$$

where on the right the index l in the first sum could also assume the value $l = n$. Hence we obtain

$$\sum_{k=1}^{n} \{\phi_k(t + x_k + y_k, \xi_k) - \phi_k(t + x_k, \xi_k)\}$$
$$= \sum_{k=1}^{n} \{\phi_k(t + x + y_k, \eta_k) - \phi_k(t + x_{k+1} + y_k, \eta_k)\} + \sum_{1 \leqslant l < k \leqslant n} \alpha_{kl}\xi_k\eta_l.$$

In (11) we replace t, ξ, η by $t + x_k + y_k$, ξ_k, η_k and derive

$$\phi_k(t + x_k + y_k, \xi_k + \eta_k) = \phi_k(t + x_k + y_k, \xi_k) + \phi_k(t + x_{k+1} + y_k, \eta_k),$$
$$(k = 1, \ldots, 2n).$$

The preceding relation then yields the equation

$$\sum_{k=1}^{n}\{\phi_k(t + x_k + y_k, \xi_k + \eta_k) - \phi_k(t + x_k, \xi_k)\}$$
$$= \sum_{k=1}^{n}\phi_k(t + x + y_k, \eta_k) + \sum_{1\leqslant l<k\leqslant n} \alpha_{kl}\xi_k\eta_l,$$

which, in view of (15), is equivalent to

$$\psi(t, x + y) - \psi(t, x) = \psi(t + x, y) + \sum_{1\leqslant l<k\leqslant n} \alpha_{kl}\xi_k\eta_l.$$

In particular, we put $t = 0$, $x = \omega_r$, $y = z$ and denote the coordinates of ω_r by γ_{kr}, $k = 1, \ldots, n$. Then we obtain

$$\psi(0, z + \omega_r) - \psi(0, \omega_r) = \psi(\omega_r, z) + \sum_{1\leqslant l<k\leqslant n} \alpha_{kl}\gamma_{kr}\zeta_l, \quad (r = 1, \ldots, 2n) \tag{17}$$

which is the desired formula. Obviously, the expression

$$s_r(0) - \psi(0, \omega_r) - \sum_{k=1}^{n}\alpha_{kr}\zeta_k - \sum_{1\leqslant l<k\leqslant n} \alpha_{kl}\gamma_{kr}\zeta_l = t_r(z)$$

is a linear function of the components of z. Since, in view of (16) and (17), the relation

$$\psi(0, z + \omega_r) - \psi(0, z) = s_r(z) - t_r(z), \qquad (r = 1, \ldots, 2n)$$

holds, equation (3) is in fact satisfied by

$$w(z) = -\psi(0, z).$$

Although the preceding computations may not, at first glance, appear so obvious, they are made more transparent by the following observation. Our solution of the difference equation (3) is suggested by the corresponding problem concerning differential equations. More precisely, let $s_1, \ldots, s_m$ be m functions of m real variables $x_1, \ldots, x_m$ such that the relations

$$\frac{\partial s_k}{\partial x_l} - \frac{\partial s_l}{\partial x_k} = \alpha_{kl}, \qquad (k, l = 1, \ldots, m) \tag{18}$$

hold with some constants $\alpha_{kl} = -\alpha_{lk}$. We seek a function w of $x_1, \ldots, x_m$ such that each expression

$$s_r + \frac{\partial w}{\partial x_r} = t_r, \qquad (r = 1, \ldots, m) \tag{19}$$

turns out to be linear. Then by simple integration it is easy to write an explicit solution. The connection with the problem treated above is seen by putting $m = 2n$, identifying $x_1, \ldots, x_m$ with the real and imaginary parts of the components of z and considering (3) and (4) instead of (19) and (18).

If $v(z)$ is any polynomial of degree 2 in the variables $z_1, \ldots, z_n$ the difference

$$\delta_\omega(z) = v(z + \omega) - v(z)$$

turns out to be linear for each period ω. Hence, the function $w(z) + v(z)$ is a solution of (3) whenever $w(z)$ has this property, where the linear function $t_r(z)$ is to be selected suitably in each case. Conversely, the function $\delta_\omega(z)$ is linear for every solution $w + v$. Since, in that case, all partial derivatives of $v(z)$ of second order are entire Abelian functions they must be constant. Hence $v(z)$ is a polynomial of degree 2. Therefore, the general solution of our problem is obtained by adding an arbitrary quadratic polynomial to the specific solution constructed above.

Poincaré gave another proof of Theorem 1 based on the theory of harmonic functions of $2n$ real variables. In so doing one can avoid Theorem 2 of the preceding section and obtain directly a Jacobian function with linear exponent system and given periodic integral divisor. Our procedure is more involved technically, but conceptually simpler.

9. The period relations

Let $f(z) \not\equiv 0$ be a Jacobian function. In view of the result of the preceding section, we restrict ourselves to the case of a linear exponent system. The exponents $s_l(z)$, $l = 1, \ldots, 2n$ corresponding to the fundamental periods ω_l may, therefore, be written as

$$s_l(z) = 2\pi i \sum_{k=1}^{n} b_{kl} z_k + \beta_l, \qquad (l = 1, \ldots, 2n) \tag{1}$$

with constant coefficients b_{kl}, $k = 1, \ldots, n$, and β_l. These constants will now be treated in more detail and put into certain normal forms by suitable transformations.

We form the matrix $(b_{kl}) = B$ with n rows and $2n$ columns. Denoting its columns by b_l, $l = 1, \ldots, 2n$, we have

$$s_l(z) = 2\pi i b_l' z + \beta_l, \qquad (l = 1, \ldots, 2n). \tag{2}$$

Equation (5) in Section 8 implies

$$b_k' \omega_l - b_l' \omega_k = a_{kl}, \qquad (k, l = 1, \ldots, 2n). \tag{3}$$

Furthermore, introducing the period matrix $(\omega_1 \cdots \omega_{2n}) = C$ and the skew symmetric matrix $(a_{kl}) = A$ with integral entries, relations (3) can be expressed by the single formula

$$B'C - C'B = A. \tag{4}$$

As was shown in Section 6, the matrix

$$\begin{pmatrix} C \\ \bar{C} \end{pmatrix} = P \tag{5}$$

is invertible. Setting $P^{-1} = (F_1 \quad F_2)$ with submatrices F_1 and F_2 each consisting of n columns we derive

$$E = P^{-1}P = F_1 C + F_2 \bar{C}. \tag{6}$$

Conversely, this formula determines F_1 and F_2 uniquely. Complex conjugation yields

$$E = \bar{F}_2 C + \bar{F}_1 \bar{C}.$$

Therefore in view of the uniqueness of F_1 and F_2, the relation $F_1 = \bar{F}_2$ holds. Dropping the indices, we can therefore write $P^{-1} = (F \quad \bar{F})$. Since we also have $PP^{-1} = E$, it follows that

$$CF = E, \qquad C\bar{F} = 0. \tag{7}$$

We now put

$$-iF'A\bar{F} = H. \tag{8}$$

Since A is real and skew symmetric, the matrices iA and H are Hermitian. In view of (4) and (7) we conclude that

$$-iH = B\bar{F}, \qquad \bar{F}'A\bar{F} = 0, \qquad F'AF = 0, \tag{9}$$

$$(P^{-1})'A\bar{P}^{-1} = \begin{pmatrix} F' \\ \bar{F}' \end{pmatrix} A(\bar{F} \quad F) = i\begin{pmatrix} H & 0 \\ 0 & -\bar{H} \end{pmatrix}. \tag{10}$$

Relation (10) implies that the determinant $|A|$ is zero if and only if the determinant $|H|$ is zero.

Let H be any Hermitian matrix with n rows. According to the definition given in Section 3 of Chapter 4 (Volume II), this matrix is said to be positive, if the real number $z'H\bar{z}$ is strictly positive for all complex rows $z \neq 0$. We denote this by $H > 0$. More generally, we call H *nonnegative*, and express this by writing $H \geqslant 0$, if $z'H\bar{z} \geqslant 0$ holds for all z. Concerning the Hermitian matrix H introduced in (8), we have the following theorem which is due to Frobenius.

Theorem 1: The Hermitian matrix H is nonnegative. If r is its rank, then there exists an invertible linear transformation $z = Q\zeta$ and a quadratic polynomial $v(z)$ such that the Jacobian function $f(z)e^{-v(z)}$ does not depend on the $n - r$ variables ζ_l, $l = r + 1, \ldots, n$.

Proof: The formula

$$x = Fz + \bar{F}\bar{z} = P^{-1}\begin{pmatrix} z \\ \bar{z} \end{pmatrix}$$

defines a real column whose elements will be denoted by $x_1, \ldots, x_{2n}$. It follows that

$$z = Cx = x_1\omega_1 + \cdots + x_{2n}\omega_{2n}. \tag{11}$$

Hence $x_1, \ldots, x_{2n}$ are the real coordinates of z with respect to the Cartesian coordinate system defined by the fundamental periods $\omega_1, \ldots, \omega_{2n}$. Let e_l be the lth column of the unit matrix of degree $2n$. Then we have

$$z + \omega_l = C(x + e_l), \qquad (l = 1, \ldots, 2n). \tag{12}$$

Our goal is to determine a function $\phi(z)$ of $z_1, \ldots, z_n$ such that the $2n$ expressions

$$S_l(z) = \phi(z + \omega_l) - \phi(z) - s_l(z), \qquad (l = 1, \ldots, 2n) \tag{13}$$

turn out to be purely imaginary. In general this function will also depend on $\bar{z}_1, \ldots, \bar{z}_n$ and therefore will not be a regular function. We make the Ansatz

$$\phi(z) = \pi i z' B x + q' x \tag{14}$$

where q is a real column consisting of $2n$ unknown elements $q_1, \ldots, q_{2n}$. In view of (2), (12), (13), and (14) we derive

$$\begin{aligned} S_l(z) &= \pi i(z + \omega_l)' B(x + e_l) + q'(x + e_l) - \pi i z' B x - q' x - 2\pi i b_l' z - \beta_l \\ &= \pi i(\omega_l' B x - b_l' z + b_l' \omega_l) + q_l - \beta_l. \end{aligned}$$

We denote the real part of a complex number or matrix α by $\rho(\alpha)$ and choose

$$q_l = \rho(\beta_l - \pi i b_l' \omega_l), \qquad (l = 1, \ldots, 2n).$$

Substituting (11) in the above we then conclude that

$$\rho(S_l(z)) = \rho(\pi i(\omega_l' B - b_l' C)x). \tag{15}$$

Since the matrix A is real, relation (4) implies that both rows $b_l' C$ and $\omega_l' B$ have the same imaginary part. In view of (15), the function $S_l(z)$ thus has the required property $\rho(S_l(z)) = 0$. According to (14), we have

$$\phi(z) = \pi i z' B(Fz + \bar{F}\bar{z}) + q'(Fz + \bar{F}\bar{z}).$$

Replacing $\bar{F}\bar{z}$ by Fz in the last term, we only change the imaginary part of $\phi(z)$ thus preserving the property that $S_l(z)$ is purely imaginary. Changing notation and using (9), we finally define

$$\phi(z) = \pi i z' B(Fz + \bar{F}\bar{z}) + 2q'Fz \tag{16}$$
$$= \pi i z' BFz + 2q'Fz + \pi z' H\bar{z}$$
$$= v(z) + \pi z' H\bar{z}.$$

Here the expression

$$v(z) = \pi i z' BFz + 2q'Fz$$

is a quadratic polynomial in $z_1, \ldots, z_n$ hence regular everywhere.

With the Jacobian function $f(z)$, we now form the function

$$F(z) = f(z)e^{-\phi(z)} \tag{17}$$

which is continuous everywhere in $\mathfrak{Z}$. In view of (13), we obtain

$$F(z + \omega_l) = f(z + \omega_l)e^{-\phi(z+\omega_l)}$$
$$= f(z)e^{s_l(z)-\phi(z+\omega_l)} = F(z)e^{-S_l(z)}.$$

Hence, the absolute values satisfy the relations

$$|F(z + \omega_l)| = |F(z)|, \qquad (l = 1, \ldots, 2n),$$

from which we conclude that the real-valued continuous function $|F(z)|$ has as period each element of the lattice, Ω, generated by $\omega_1, \ldots, \omega_{2n}$. Since that function is bounded on the period parallelotope, it follows that

$$|F(z)| \leqslant M$$

holds for all z with a suitable positive constant M. In view of (16) and (17), we therefore obtain

$$|f(z)e^{-v(z)}| \leqslant Me^{\pi z' H \bar{z}}$$

By an invertible linear transformation $z = Q\zeta$ we can now transform the Hermitian form $z'H\bar{z}$ into a diagonal form $h_1\zeta_1\bar{\zeta}_1 + \cdots + h_n\zeta_n\bar{\zeta}_n$, where we may furthermore assume $h_1 \geqslant h_2 \geqslant \cdots \geqslant h_n$. The next step is to show $H \geqslant 0$. If this is false then $h_n < 0$. The entire function

$$g(\zeta) = f(Q\zeta)e^{-v(Q\zeta)}$$

satisfies the inequality

$$|g(\zeta)| \leqslant Me^{\pi(h_1\zeta_1\bar{\zeta}_1 + \cdots + h_n\zeta_n\bar{\zeta}_n)}.$$

Hence, it is bounded for all ζ_n and arbitrary, but fixed, values $\zeta_1, \ldots, \zeta_{n-1}$. According to Liouville's theorem, the function $g(\zeta)$ is then independent of ζ_n. Letting $\zeta_n \to \infty$ the assumption $h_n < 0$ implies $g(\zeta) = 0$ for arbitrary fixed values $\zeta_1, \ldots, \zeta_{n-1}$. It follows that $f(z) = 0$ identically in z, contradicting our original assumption that $f \neq 0$. Hence the assertion $H \geqslant 0$ is proved. Finally, let r denote the rank of H. Then the coefficients $h_1, \ldots, h_r$ are positive, whereas the others vanish. Applying Liouville's theorem once again, one sees that $g(\zeta)$ is independent of $\zeta_{r+1}, \ldots, \zeta_n$. Now Theorem 1 is completely proved.

In view of (4) and (8) the Hermitian matrix H is uniquely determined by B and C. Replacing the basis $\omega_1, \ldots, \omega_{2n}$ of the period lattice Ω by another basis $\omega_1^*, \ldots, \omega_{2n}^*$ is carried out by means of a unimodular transformation. Let U be the matrix of such a transformation. Then C is replaced by $CU = C^*$. If the change of basis of the lattice Ω replaces B by B^*, we also derive the equation $B^* = BU$ from relation (4) of the preceding section combined with relation (2). In view of (4) we have $A^* = U'AU$. Relation (6) yields $F^* = U^{-1}F$. From equation (8) we then conclude $H^* = H$, thus proving that H is independent of the choice of basis of the period lattice. Finally we remark that multiplying any given Jacobian function $f(z)$ by the factor $e^{-v(z)}$ results in an equivalent linear exponent system but leaves H and A unchanged.

We now consider all Jacobian functions corresponding to a fixed linear exponent system. In accordance with the notation introduced in Section 12 of Chapter 4 (Volume II), we call them *equivariant*. Together with the function 0, they constitute a complex vector space, $\mathfrak{L}$, which will be investigated in more detail in the sequel. But first, we will discuss the case in which the rank r of H is smaller than n. Then in view of Theorem 1, multiplication by $e^{-v(z)}$ and application of a linear substitution $z = Q\zeta$, simultaneously transform all functions $f(z)$ in $\mathfrak{L}$ into functions depending only on r variables $\zeta_1, \ldots, \zeta_r$. In that case $\mathfrak{L}$ and the function $f(z)$ are said to be *degenerate*. There is the following connection with the degenerate Abelian functions introduced in Section 6. If $f \neq 0$ and h are equivariant degenerate Jacobian functions, then $q = hf^{-1}$ is a degenerate Abelian function contained in the field K corresponding to the lattice Ω. Conversely, let q be a given degenerate function contained in K. According to the theorem of Weierstrass and Cousin, we may set $q = h_0 f_0^{-1}$ where $f_0 \neq 0$ and h_0 are entire functions which are locally relatively prime everywhere. Applying a suitable invertible linear substitution $z = Q\zeta$, we may furthermore assume that $q(z)$ is independent of ζ_n. Denoting differentiation with respect to ζ_n by a prime, we then have $q' = 0$, and hence

$$f_0 h_0' = h_0 f_0'.$$

The relation $(f_0, h_0) \sim 1$ implies $f_0 \mid f_0'$. Therefore, we obtain

$$(18) \qquad f_0' = s f_0, \qquad h_0' = s h_0$$

where s is an entire function in ζ. Integration of s with respect to ζ_n yields an entire function t satisfying $t' = s$. In view of (18) both functions $f = f_0 e^{-t}$ and $h = h_0 e^{-t}$ are then independent of ζ_n. Furthermore, these functions f and h, too, are locally relatively prime everywhere on $\mathfrak{Z}$. Using the results of both preceding sections, a suitable change of t enables us to put $q = hf^{-1}$ where $f \neq 0$ and h are now equivariant Jacobian functions whose exponent system is linear in ζ and does not depend on ζ_n. In view of (1) and (2), the substitution $z = Q\zeta$ replaces B by $Q'B$. Consequently, each element of the last row of $Q'B$ vanishes, and the rank of B is at most $n - 1$. In view of (9) the rank r of H then satisfies $r < n$.

We now turn to the nondegenerate case. In view of (10), we have $|H| \neq 0$ if and only if $|A| \neq 0$. Furthermore, relation (10) implies

$$i\bar{P}A^{-1}P' = \begin{pmatrix} H^{-1} & 0 \\ 0 & -\bar{H}^{-1} \end{pmatrix},$$

which, according to (5), is equivalent to the formulas

$$i\bar{C}A^{-1}C' = H^{-1}, \qquad CA^{-1}C' = 0.$$

The positivity of H implies the same property for H^{-1} since the Hermitian form $z'H\bar{z}$ is transformed into $\zeta'H^{-1}\bar{\zeta}$ by the invertible linear transformation $z = \bar{H}^{-1}\zeta$. Hence, we obtain the following important result.

Theorem 2: If there exists a nondegenerate Jacobian function corresponding to the period matrix C, then there is an invertible skew symmetric integral matrix A such that

$$i\bar{C}A^{-1}C' > 0, \qquad CA^{-1}C' = 0. \tag{19}$$

Conditions (19) are called *period relations*. A Hermitian matrix of degree n is positive if and only if, for each $k = 1, \ldots, n$, its principal subdeterminant of degree k formed by all elements contained in the first k rows and columns, is positive. Hence, the first condition in (19) yields n inequalities. The second condition yields $n(n - 1)/2$ equations. Later we will see that the converse of Theorem 2 is also valid.

In the sequel every invertible skew symmetric integral matrix A satisfying conditions (19) will be called a *principal matrix* corresponding to C. If A is such a principal matrix and k any natural number, then the matrix kA is also a principal matrix corresponding to C. To a given period matrix there can, indeed, correspond different principal matrices which are not multiples of each other in this trivial sense. But a more detailed investigation of this question is beyond the scope of this text.

10. The reduced exponent system

Our next aim is to transform the linear exponent system, $s_l(z)$, $(l = 1, \ldots, 2n)$, considered in the preceding section, into a simple normal form by means of suitable substitutions. Three operations are involved in this normalization. First, we can replace the period matrix C by any other period matrix CU with arbitrary unimodular U. Next, we can apply a linear transformation $z = Q\zeta + b$ to the variable column z, where Q is any complex invertible matrix and b is any column of complex constants. Finally, by means of an arbitrary quadratic polynomial $\psi(z)$, we can replace the linear exponent system $s_\omega(z)$ by an equivalent, likewise linear, exponent system $s_\omega(z) + \psi(z + \omega) - \psi(z)$. We will exclude the degenerate case and therefore assume $|A| \neq 0$.

In passing from C to CU the matrix H remains unchanged, whereas B and A are replaced by BU and $U'AU$, respectively. The desired normal form of A is given by the following theorem which is also due to Frobenius.

Theorem 1: Let A be an integral skew symmetric matrix of degree $2n$ whose determinant is $\neq 0$. Then there exists a diagonal matrix T of degree n with integral diagonal entries $t_1, \ldots, t_n$ and a unimodular matrix U such that the following relations hold:

$$t_1 \,|\, t_2 \,|\, \cdots \,|\, t_n, \qquad U'AU = \begin{pmatrix} 0 & T \\ -T & 0 \end{pmatrix}.$$

Proof: Let $x'Ay$ be the bilinear form with matrix A. We use certain unimodular transformations $x \to Ux$, $y \to Uy$ which will be called elementary of first and second kind. The transformations of first kind are defined by matrices $U = (u_{kl})$ whose entries satisfy the following conditions for fixed g, h:

$$u_{gh} = u_{hg} = 1, \qquad u_{kk} = 1, \qquad k \neq g, \qquad k \neq h, \qquad u_{kl} = 0 \text{ otherwise.}$$

Analogously, the transformations of second kind are defined by $u_{kk} = 1$, $u_{gh} = q$, q an integer, and $u_{kl} = 0$ otherwise. Transition from A to $U'AU$ means, in the first case, interchanging the gth and hth rows as well as the gth and hth columns. In the second case, this means addition of q times the gth row to hth row and simultaneous addition of q times the gth column to hth column. For our purposes we need not investigate whether the whole unimodular group is generated by elementary transformations.

We use induction on n. The case $n = 1$ is trivial. Assume $n > 1$ and that the assertion is true for $n - 1$ in place of n. Since A is skew-symmetric and $|A| \neq 0$, there exists an element a_{kl} of A which is positive and as small as possible. By application of elementary transformations of first kind, we can assume it to be $a_{1,n+1}$. We now consider all matrices $U'AU = B = (b_{kl})$, with arbitrary unimodular U, such that $0 < b_{1,n+1} \leqslant |b_{kl}|$ for all indices k, l. Among these matrices, let U be chosen so that $b_{1,n+1} = t_1$ is as small as possible. Then t_1 must be a divisor of each element of the first, as well as of the $(n+1)$th row of B. Otherwise, an application of a suitable elementary transformation of second kind would yield a contradiction of the minimal property of t_1. Therefore, elementary transformations of second kind can be used to annihilate all elements of the indicated rows of B except the elements $b_{1,n+1} = t_1$ and $b_{n+1,1} = -t_1$. Hereby the difference $x'By - t_1(x_1y_{n+1} - x_{n+1}y_1)$ becomes a skew-symmetric bilinear form in two sets of $n-1$ variables with integral coefficients, whose determinant is given by $t_1^{-2}\,|B| = t_1^{-2}\,|A| \neq 0$. Application of the induction hypothesis to this bilinear form yields Theorem 1 except for the assertion that t_1 is a divisor of t_2. If t_1 is not a divisor of t_2, apply both elementary transformations of second kind with $g = 1$, $h = 2$ and arbitrary q, as well as $g = n+1$, $h = n+2$, and $q = 1$. Since t_2 is then replaced by $t_2 + qt_1$ we obtain a contradiction of the minimality of t_1. Now the proof of Theorem 1 is complete.

Furthermore, one can show that T, hence the normalized form of A, is uniquely determined. This is done by well known arguments from the theory of elementary divisors. One verifies that, for each $k = 1, \ldots, n$, the greatest common divisor of all subdeterminants of degree $2k$ of A is given by $(t_1 \cdots t_k)^2$.

Now let A be given in normal form; hence

$$A = \begin{pmatrix} 0 & T \\ -T & 0 \end{pmatrix}. \tag{1}$$

Then we consider the decomposition $C = (R \quad S)$ of C into two square submatrices of degree n and use the period relations. We obtain

$$H^{-1} = i(\bar{R} \quad \bar{S})\begin{pmatrix} 0 & -T^{-1} \\ T^{-1} & 0 \end{pmatrix}\begin{pmatrix} R' \\ S' \end{pmatrix} > 0,$$

$$(R \quad S)\begin{pmatrix} 0 & -T^{-1} \\ T^{-1} & 0 \end{pmatrix}\begin{pmatrix} R' \\ S' \end{pmatrix} = 0.$$

Hence,

$$H^{-1} = i(\bar{S}T^{-1}R' - \bar{R}T^{-1}S') > 0, \qquad ST^{-1}R' = RT^{-1}S'. \tag{2}$$

If z is any solution of the equation $\bar{R}'z = 0$, it follows that $z'H^{-1}\bar{z} = 0$, hence, $z = 0$. Therefore we obtain $|R| \neq 0$. As a second step, we apply the invertible homogeneous linear transformation $z = Q\zeta$ with $Q = RT^{-1}$ and define

$$Q^{-1}S = TR^{-1}S = W.$$

Then the matrices B and C must be replaced by $Q'B$ and $Q^{-1}C = (T \quad W)$, whereas A remains unchanged, and H becomes $Q'H\bar{Q}$. Changing notation, we again use B, C, H to designate these new matrices. Then the period matrix becomes

$$C = (T \quad W)$$

with $R = T$, $S = W$ and relation (2) yields

$$H^{-1} = i(\overline{W} - W') > 0, \qquad W = W'.$$

Therefore, the period relations now simply say that the matrix W is symmetric, and since $H^{-1} = 2Y$, the imaginary part, Y, of $W = X + iY$ is positive. Therefore, according to the notation introduced in Section 5 of Chapter 4 (Volume II), the matrix W is contained in the upper half-plane of degree n. Now we have the desired normal form of the period matrix.

Finally, as the third step, we must derive a normal form of B. Decomposing the matrix $B = (K \quad L)$ into two square submatrices of degree n and using equation (4) of Section 9 as well as relation (1), we obtain

$$B'C = \begin{pmatrix} K' \\ L' \end{pmatrix}(T \quad W) = \begin{pmatrix} K'T \\ L'T \end{pmatrix}\begin{pmatrix} K'W \\ L'W \end{pmatrix},$$

$$K'T = TK, \qquad K'W - TL = T, \qquad L'W = WL.$$

Hence, WL and the matrix $KT^{-1} = V$ are symmetric, and the relation

$$VW = L + E$$

holds. We now choose

$$\psi(z) = -\pi i z' V z$$

and write relations (1) of Section 9 in terms of matrices

$$s = 2\pi i B'z + \beta,$$

where s and β are the columns with components s_l and β_l, $l = 1, \ldots, 2n$, respectively. We obtain

$$\psi(z + \omega_l) - \psi(z) = -2\pi i \omega_l' V z + \psi(\omega_l), \qquad (l = 1, \ldots, 2n).$$

Replacing the exponent system $s_l(z)$ by $s_l(z) + \psi(z + \omega_l) - \psi(z)$, $l = 1, \ldots, 2n$, the matrix B is replaced by

$$B - VC = B - KT^{-1}(T \quad W) = (0 \quad L - VW) = (0 \quad -E).$$

We can now assume that B is in this normal form. Subsequently, modifying $\psi(z)$ by the addition of a linear form $a'z$, the matrix B remains unchanged whereas β is replaced by $\beta + C'a$. Obviously, there is a unique way to choose a so that the first n elements of the new column β vanish. We then obtain

$$s_l(z) = 0, \qquad s_{n+l}(z) = -2\pi i z_l + \beta_{n+l}, \qquad (l = 1, \ldots, n).$$

Finally, applying a translation $z \to z + b$ with constant b to z, we can arbitrarily fix the other n constants β_{n+l}, $l = 1, \ldots, n$. Rather than setting them equal to zero it is expedient for what follows to choose

$$\beta_{n+l} = -\pi i w_{ll}, \qquad (l = 1, \ldots, n),$$

where w_{ll} is the lth diagonal element of $W = (w_{kl})$.

Thus we have obtained the following result.

Theorem 2: Let $f(z)$ be a nondegenerate Jacobian function. Then there exist an entire function $v(z)$, an invertible linear transformation $z = Q\zeta + b$, a diagonal matrix T with positive integral diagonal elements $t_1, \ldots, t_n$ and a symmetric matrix W contained in the upper half-plane of degree n such that $C = (T \quad W)$ is a period matrix with respect to the variables $\zeta_1, \ldots, \zeta_n$. In addition, the exponent system corresponding to the Jacobian function $F(\zeta) = f(z)e^{v(z)}$ assumes the reduced form

$$(3) \qquad s_l(\zeta) = 0, \qquad s_{n+l}(\zeta) = -\pi i(2\zeta_l + w_{ll}), \qquad (l = 1, \ldots, n).$$

Furthermore, we have to clarify the extent to which $v(z)$, Q, b, and C in Theorem 2 are uniquely determined by the function $f(z)$. If we require, as above, the matrix T to satisfy the conditions $t_1 \mid t_2 \mid \cdots \mid t_n$, then T, hence A in (1), is uniquely determined. Since our previous choice of U was not unique, we subsequently may pass from C to CU, where U is any unimodular solution of the matrix equation

$$(4) \qquad U'AU = A = \begin{pmatrix} 0 & T \\ -T & 0 \end{pmatrix}.$$

Putting

$$CU = (R \quad S) = RT^{-1}(T \quad W^*), \qquad U = \begin{pmatrix} M_1 & M_2 \\ M_3 & M_4 \end{pmatrix}$$

with square submatrices $M_1, \ldots, M_4$ of degree n, we obtain

$$R = TM_1 + WM_3, \qquad S = TM_2 + WM_4,$$

$$W^* = TR^{-1}S = (WM_3T^{-1} + TM_1T^{-1})^{-1}(WM_4 + TM_2).$$

If, in passing to the normal form of A, the matrix U is chosen fixed, then Q and b are uniquely determined, whereas the function $v(z)$ is uniquely determined up to an arbitrary additive constant. This is clear from the considerations leading to Theorem 2. Furthermore, as a consequence of (4), if we put

$$\begin{pmatrix} 0 & T \\ E & 0 \end{pmatrix} = D, \qquad D^{-1}U'D = M, \qquad J = \begin{pmatrix} 0 & E \\ -E & 0 \end{pmatrix},$$

then we obtain

$$A = DJ^{-1}D', \qquad M'JM = J, \quad M = \begin{pmatrix} M_4' & M_2'T \\ T^{-1}M_3' & T^{-1}M_1'T \end{pmatrix}.$$

Hence, M is symplectic and the matrix

$$W^* = (M_4'W + M_2'T)(T^{-1}M_3'W + T^{-1}M_1'T)^{-1} \tag{5}$$

arises from W by the fractional linear transformation corresponding to M. Therefore, the matrix W^*, likewise, lies in the upper half-plane of degree n. If U is any integral solution of (4) it follows by forming the determinant that U is unimodular. Thus, the totality of integral solutions U of (4) constitutes a multiplicative group which, in the special case $T = E$, is the modular group of degree n. This group is called *the modular group of level* T. Correspondingly, relation (5) is called a *modular substitution of level* T. Obviously, equation (4) remains valid if T is replaced by qT with any scalar factor q. Hence the modular group of level T only depends on the elementary divisors $t_2/t_1, \ldots, t_n/t_{n-1}$. Thus we have attained the following result concerning the normalized form $C = (T \quad W)$ of the period matrix.

Theorem 3: Let a matrix C_0 with n rows and $2n$ columns satisfy the period relations

$$i\bar{C}_0A^{-1}C_0' > 0, \qquad C_0A^{-1}C_0' = 0 \tag{6}$$

with an invertible skew symmetric integral matrix A. Determining T and U according to Theorem 1 and setting

$$C_0U = (R \quad S) = QC, \qquad C = (T \quad W), \qquad W = TR^{-1}S,$$

the matrix T is then uniquely determined, whereas the matrix W is uniquely determined up to an arbitrary modular substitution of level T.

This result can be compared with Theorem 6 in Section 5 of Chapter 4 (Volume II) by putting, in particular, $T = E$, $A = J$ and by considering W in place of the former period matrix Z. In our previous discussions the modular group of degree n arose when we passed to the period matrix Z^* corresponding to an arbitrary canonical decomposition of the Riemann surface. Now, the period matrix W is subject to modular substitutions of level T if the principal matrix A is transformed into the normal form $U'AU$ in all possible ways.

Finally we notice that our preceding investigation involved, in addition to the period lattice Ω, a nondegenerate Jacobian function $f(z)$ with divisor $\mathfrak{d}$. After fixing a lattice basis this function uniquely determined the principal matrix A occurring in the period relations. Keeping the lattice basis fixed, but selecting any other nondegenerate Jacobian function, not necessarily equivariant with $f(z)$, the principal matrix will, in general, change. If that change is not merely by a scalar factor, the corresponding modular group of level T changes too.

We already pointed out the arithmetical problem of determining all principal matrices corresponding to a given period matrix at the end of the preceding section.

11. Existence proofs

We now turn to the second important part of the theory. We will prove that the period relations are not only necessary but also sufficient for the existence of nondegenerate Jacobian and Abelian functions corresponding to a given lattice Ω. In the process the desired functions will be expressed explicitly by means of the theta functions $\Theta(Z, s)$ introduced in Section 9 of Chapter 4 (Volume II). We will again use the abbreviation $e^{\pi i a} = \varepsilon(a)$.

In view of Theorem 2 in Section 10, a more detailed investigation of Jacobian functions can be restricted to the case where the entire function $f(z)$ satisfies the conditions

$$(1) \qquad f(z + Tg) = f(z), \qquad f(z + Wh) = \varepsilon(-h'Wh - 2h'z)f(z)$$

with arbitrary integral columns g and h. This is seen by induction from relations (3) in Section 10 which correspond to the generators $\omega_1, \ldots, \omega_{2n}$ of the period lattice. Two integral columns r and s are said to be *congruent modulo* T if the column $T^{-1}(r - s)$ is also integral; in that case, we write $r \equiv s(\text{mod } T)$. Furthermore, if the symbol $r(\text{mod } T)$ appears below a summation sign, the letter r is meant to range over a complete residue system modulo T.

We now define

$$(2) \qquad \phi_r(z) = \sum_{s(\text{mod } T)} \varepsilon(-2r'T^{-1}(z + s))f(z - WT^{-1}r + s)$$

and remark that, in view of (1), the general term of this sum remains unchanged if s is replaced by $s + Tg$. Hence, the definition of $\phi_r(z)$ does not depend on the choice of residues modulo T. Furthermore, if s ranges over a complete system of residues modulo T, so does $s + g$ for every fixed integral column g. Therefore, we obtain

$$\phi_r(z + g) = \phi_r(z).$$

Relation (1) also yields

$$\phi_r(z + Wh) = \sum_{s(\mathrm{mod}\ T)} \varepsilon(-2r'T^{-1}(z + s + Wh))f(z - WT^{-1}r + s + Wh)$$

$$= \sum_{s(\mathrm{mod}\ T)} \varepsilon(-2r'T^{-1}(z + s))\varepsilon(-2h'WT^{-1}r)$$

$$\cdot\, \varepsilon(-h'Wh - 2h'(z - WT^{-1}r + s))f(z - WT^{-1}r + s)$$

$$= \varepsilon(-h'Wh - 2h'z)\phi_r(z).$$

Consequently, the entire function $\phi_r(z)$ satisfies the assumptions of Theorem 2 in Section 9 of Chapter 4 (Volume II) if we replace the symbols s and Z by z and W. Hence the function $\phi_r(z)$ only differs from $\Theta(W, z)$ by a factor independent of z. We now multiply equation (2) by $\varepsilon(2r'T^{-1}z)$, replace z by $z + WT^{-1}r$, and sum over $r(\mathrm{mod}\ T)$. The determinant

$$|T| = t_1 \cdots t_n = t$$

is equal to the number of different residue classes modulo T. Hence, dividing by t, we obtain the explicit formula

$$f(z) = \sum_{r(\mathrm{mod}\ T)} \rho_r \varepsilon(2r'T^{-1}z)\Theta(W, z + WT^{-1}r) \tag{3}$$

with certain coefficients ρ_r not depending on z. Conversely, one easily verifies that every function $f(z)$ defined by (3) with arbitrary constants ρ_r is an entire function and satisfies conditions (1).

Finally, we denote the quadratic form $x'Wx$ by $W[x]$ and observe that the theta function $\Theta(W, z)$ is defined by

$$\Theta(W, x) = \sum_g \varepsilon(W[g] + 2g'z).$$

Then we obtain

$$\varepsilon(2r'T^{-1}z)\Theta(W, z + WT^{-1}r)$$
$$= \varepsilon(-W[T^{-1}z]) \sum_g \varepsilon(W[g + T^{-1}r] + 2(g + T^{-1}r)'z)$$

Putting

$$\sum_g \varepsilon(W[g + T^{-1}r] + 2(g + T^{-1}r)'z) = \Theta_r(T, W, z), \tag{4}$$

these general theta functions $\Theta_r(T, W, z)$ are connected with the basic theta function $\Theta(W, z)$ by the formula

$$\Theta_r(T, W, z) = \varepsilon(W[T^{-1}r] + 2r'T^{-1}z)\Theta(W, z + WT^{-1}r).$$

Relation (3) yields the representation

$$f(z) = \sum_{r(\text{mod } T)} \gamma_r \Theta_r(T, W, z) \tag{5}$$

with constant coefficients γ_r. Each of the theta functions $\Theta_r(T, W, z)$ defined by (4) is a Jacobian function corresponding to the given reduced linear exponent system. Since the equations

$$(g + T^{-1}r)'z = (Tg + r)'T^{-1}z, \qquad Tg + r \equiv r(\text{mod } T)$$

hold, the function given by (4) is a Fourier series with respect to the variables $T^{-1}z$. Here the exponents of terms which actually occur are determined by the residue class r modulo T. Hence the t functions $\Theta_r(T, W, z)$ are linearly independent, and we have established the following theorem.

Theorem 1: If the period relations are satisfied, then the Jacobian functions corresponding to a given exponent system form a complex vector space of dimension $t = |A|^{1/2}$ which, for a reduced exponent system, has the functions $\Theta_r(T, W, z)$ as a basis.

Finally, we will construct Abelian functions by means of Jacobian functions.

Theorem 2. Let C be a matrix with $2n$ columns which generate a lattice Ω in complex n-space. There exists an Abelian function whose period group coincides with Ω if and only if the period relations

$$i\bar{C}A^{-1}C' > 0, \qquad CA^{-1}C' = 0$$

are satisfied by some invertible skew symmetric integral matrix A.

Proof: If the function $\phi(z)$ has the desired property it cannot be degenerate since otherwise its period group would not be discrete. Therefore, the function $\phi(z)$ can be represented as a quotient of two equivariant Jacobian functions which are also nondegenerate. Theorem 2 of Section 9 then implies the period relations.

Conversely, using the period relations, we must construct an Abelian function $\phi(z)$ whose period lattice is precisely Ω. We are confronted with the following difficulty. In view of Theorem 1, we can easily construct two nonidentically vanishing Jacobian functions corresponding to the reduced exponent system (3) of Section 10 whose quotient is an Abelian function.

However, this function could have other periods besides those contained in Ω. For instance, when $A = J$ the equation $t = 1$ implies that the Abelian function so constructed is constant. To overcome this difficulty we start with some auxiliary considerations. Let $f(z)$ be a nonidentically vanishing Jacobian function. We will determine a constant column q such that the functions $f(z)$ and $f(z + q)$ are locally relatively prime everywhere. Since the quotient $f(z + \omega)/f(z)$ is a unit for every element, ω, of Ω, it suffices to satisfy the condition in question on the period parallelotope $\mathfrak{P}$ instead of on the whole space $\mathfrak{Z}$. Here $\mathfrak{P}$ is considered to be closed.

The function $f(z)$ is represented by an everywhere convergent power series. According to the Preparation theorem, at every point a of $\mathfrak{P}$ there exists an invertible linear transformation $z - a = x = Ly$ such that

$$f(z) = f_a(x) = (y_1^k + \alpha_1 y_1^{k-1} + \cdots + \alpha_k)u(y), \tag{6}$$

where k is the order of the local part $f_a(x)$ and $u(y)$ is a local unit. The coefficients $\alpha_1, \ldots, \alpha_k$ are power series in the $n - 1$ remaining variables $y_2, \ldots, y_n$ and have no constant terms. We select a positive number ρ so small that the power series $\alpha_1, \ldots, \alpha_k$ and u are all convergent for $|y| < 2\rho$ and $u(y) \neq 0$ there. Then let $\sigma > 0$ be determined so that the condition $|z - a| < \sigma$ implies $|y| < \rho$. Here σ will depend on a. Finally, let finitely many points, a, be selected so that $\mathfrak{P}$ is entirely covered by the corresponding n-disks $|z - a| < \sigma$. Then the product

$$g(x) = \prod_a f_a(x) = \prod_a f(x + a)$$

is an entire function of x not identically zero. We now fix any point q satisfying $|q| < \min_a \sigma$ and $g(q) \neq 0$. We will show that $f(z)$ and $f(z + q)$ are then locally relatively prime everywhere on $\mathfrak{P}$. In view of (6), it suffices to demonstrate this assertion for the functions

$$P(y) = y_1^k + \alpha_1 y_1^{k-1} + \cdots + \alpha_k \tag{7}$$

and $P(y + r)$ on the n-disk $|y| < \rho$. Here $r = L^{-1}q$, and we can restrict ourselves to the case $k > 0$.

The condition $f_a(q) \neq 0$ is equivalent to $P(r) \neq 0$. Consequently, the power series $P(y + r)$ has order 0, whereas $P(y)$ has positive order k. Considered as functions of y_1, both expressions $P(y)$ and $P(y + r)$ are polynomials of degree k with coefficients contained in the ring Σ of convergent power series in the remaining $n - 1$ variables $y_2, \ldots, y_n$. Obviously both polynomials are primitive. Supposing they are not relatively prime with respect to y_1, the argument used in the proof of Theorem 2 of Section 2 yields the existence of a primitive polynomial $Q(y)$ which divides $P(y)$ and $P(y + r)$, which actually contains y_1, and whose coefficients are in Σ. Because $P(r) \neq 0$ the polynomial $Q(y)$ is a local unit and the quotient $P(y)/Q(y)$ has order k as power series; but as a polynomial in y_1 its degree is smaller than k. Now as in the proof of Theorem 2 of Section 2 we obtain a contradiction of (7). Thus it is proved that the resultant of $P(y)$ and $P(y+r)$ with respect to y_1 is not zero. Now the rest of the demonstration is easily completed by recapitulating the corresponding argument in the proof of Theorem 1 of Section 3.

Now we define

$$f(z) = \sum_{r(\mathrm{mod}\ T)} \gamma_r \Theta_r(T, W, z) \tag{8}$$

with t constants γ_r each of which is assumed to be unequal to 0. With q the same as above we form the quotient

$$\phi(z) = \frac{f(z+q)f(z-q)}{f(z)f(z)}.$$

Then $\phi(z)$ is a meromorphic function of z which, in view of (1), possesses all elements of Ω as periods. However, it could have additional periods. Let ω be one of them. Since we may assume that $f(z)$ and $f(z+q)$ are relatively prime the relation

$$(f(z))^2 f(z+\omega+q)f(z+\omega-q) = (f(z+\omega))^2 f(z+q)f(z-q)$$

implies the following divisibility properties:

$$(f(z))^2 \mid (f(z+\omega))^2 \mid (f(z))^2.$$

Hence the quotient $f(z+\omega)/f(z)$ is a global unit and we can write

$$f(z+\omega) = e^{-2\pi i p(z)} f(z) \tag{9}$$

with an entire function $p(z)$. In view of (1), we obtain

$$p(z+Tg) - p(z) = \alpha_g, \qquad p(z+Wh) - p(z) = h'\omega - \beta_h$$

where α_g and β_h are certain integers which may depend on the arbitrary integral columns g and h. The partial derivatives of $p(z)$ turn out to be entire Abelian functions, hence constants. It follows that

$$p(z) = c'z + \gamma \tag{10}$$

with a constant column c and a constant complex number γ. Since all conditions

$$c'Tg = \alpha_g, \qquad c'Wh = h'\omega - \beta_h$$

must be satisfied, we can, by a suitable choice of g and h, establish the existence of integral columns a and b such that

$$c = T^{-1}a, \qquad \omega = b + WT^{-1}a. \tag{11}$$

Since the function $\phi(z)$ has all columns $Tg + Wh$ as periods, we may replace ω by $\omega + Tg + Wh$. Hence we can restrict the columns a and b in (11) to a fixed set of representatives of residue classes modulo T. Putting $\rho = \varepsilon(-2\gamma)$, the relations (8), (9), (10), and (11) then imply

$$\sum_{r(\text{mod } T)} \gamma_r \Theta_r(T, W, z + \omega) = \rho \sum_{r(\text{mod } T)} \gamma_r \varepsilon(-2a'T^{-1}z)\Theta_r(T, W, z),$$

$$\sum_{r(\text{mod } T)} \gamma_r \sum_g \varepsilon(W[g + T^{-1}r] + 2(g + T^{-1}r)'(z + \omega))$$

$$= \rho \sum_{r(\text{mod } T)} \gamma_r \sum_g \varepsilon(W[g + T^{-1}r] + 2(g + T^{-1}r - T^{-1}a)'z).$$

We define γ_s for all integral columns s by $\gamma_s = \gamma_r$ whenever $s \equiv r(\text{mod } T)$ and replace r by $r + a$ on the right-hand side. We obtain

$$\gamma_r \varepsilon(2(g + T^{-1}r)'\omega) = \rho \gamma_{r+a} \varepsilon(W[T^{-1}a] + 2(T^{-1}a)'W(g + T^{-1}r)),$$

$$\gamma_r \varepsilon(2r'T^{-1}b) = \rho \gamma_{r+a} \varepsilon(W[T^{-1}a]).$$

For simplicity we write

$$\eta = \rho \varepsilon(W[T^{-1}a])$$

and obtain

$$\gamma_r \varepsilon(2r'T^{-1}b) = \gamma_{r+a}\eta. \tag{12}$$

Multiplication over the t different residue classes $r(\text{mod } T)$ yields $\eta^{2t} = 1$. Supposing $a \equiv 0(\text{mod } T)$ implies $\gamma_r = \gamma_{r+a}$. Hence $\varepsilon(2r'T^{-1}b)$ is independent of r in that case. We conclude $b \equiv 0(\text{mod } T)$, thus showing that ω is an element of Ω, whereas we assumed that it is not. Therefore, we have $a \not\equiv 0(\text{mod } T)$. We now choose $\gamma_0 = 1$ and $\gamma_r^{2t} \neq 1$, for instance $\gamma_r = 2$, whenever $r \not\equiv 0(\text{mod } T)$. Then (12) leads to a contradiction, and Theorem 1 is proved.

If a period matrix satisfies the period relations with a suitable invertible skew symmetric integral matrix A, it is traditionally called a *Riemann matrix*. The reason is that Riemann was the first to recognize that the period relations are necessary and sufficient for the existence of nondegenerate Abelian functions. However, his formulation was incomplete and he did not supply a proof. Later, Weierstrass also failed to establish a complete proof despite his many efforts in this direction. Complete proofs were finally attained by Appell for the case $n = 2$ and by Poincaré for arbitrary n.

Given a period matrix C, let K be the corresponding field of Abelian functions. If C is not a Riemann matrix, all functions in K are degenerate in view of Theorem 2 of the preceding section. According to the notation introduced in Section 6, the field K itself is then degenerate. Conversely, if C is a Riemann matrix, Theorem 2 of the present section implies the existence of a nondegenerate function in K. Hence, K is nondegenerate if and only if C is a Riemann matrix.

From Theorem 2 we draw still another simple conclusion.

Theorem 3: Let K be a nondegenerate field of Abelian functions corresponding to a period lattice Ω. If a and b are any two points of $\mathfrak{Z}$ whose difference $a - b$ is not contained in Ω, then there exists a function $f(z)$ in K which is regular at a and b and satisfies the condition $f(a) \neq f(b)$.

Proof: In view of Theorem 2 there exists a function $\phi(z)$ in K whose period group coincides with Ω. Since, by assumption, $b - a$ is not a period, the difference $\phi(z + a) - \phi(z + b) = \psi(z)$ is not identically 0. Let c be a point where the functions $\phi(z + a)$ and $\phi(z + b)$ are both regular and $\psi(c) \neq 0$. Then the function $f(z) = \phi(z + c)$ obviously has the desired properties.

Theorem 3 can be formulated in a different way by considering the period torus $\mathfrak{T} = \mathfrak{Z}/\Omega$ in place of $\mathfrak{Z}$. The points of $\mathfrak{T}$ are the congruence classes of points of $\mathfrak{Z}$ modulo Ω. Again, let K be the field of Abelian functions corresponding to Ω. Then the points of $\mathfrak{T}$ are separated by functions in K. With this result, Theorem 3 of Section 6 of Chapter 3 (Volume II) is, in a certain sense, extended to the present case. The present result is even more satisfactory since there are no exceptional points to be excluded. But, in this connection we emphasize that in dealing with Abelian functions we must assume that the period relations are satisfied; whereas, in our previous discussions of automorphic functions of one variable, there were no additional conditions.

By means of the nondegenerate function $\phi(z)$ in K constructed above, we are now able to build up the whole field K in several steps outlined in the sequel.

Theorem 4: Corresponding to every Riemann period matrix there exist n Abelian functions which are not related by an algebraic equation with constant coefficients.

Proof: Let $f(z)$ be a nondegenerate Abelain function possessing each column of the given Riemann matrix as a period. Let $z = z^{(k)}$ $(k = 1, \ldots, n)$ be n independently varying points and consider the n partial derivatives

$$F_l(z) = \frac{\partial f(z)}{\partial z_l}, \qquad (l = 1, \ldots, n).$$

Then form the determinant

$$\Delta = \Delta(z^{(1)}, \ldots, z^{(n)}) = |F_l(z^{(k)})|$$

which, as a function of $z^{(1)}, \ldots, z^{(n)}$, is obviously a meromorphic function of n^2 independent complex variables. Suppose Δ is identically zero. Then there exists a smallest natural number $q \leqslant n$ such that each $q \times q$ subdeterminant of Δ vanishes identically. Since $f(z)$ is not constant we have $q > 1$. We can choose the notation so that the subdeterminant Δ_q formed by the first $q - 1$ rows and columns of Δ does not vanish identically in $z^{(1)}, \ldots, z^{(q-1)}$. We expand the subdeterminant formed by the first q rows and columns of Δ with respect to the qth row and put $z^{(q)} = z$. Then, we obtain an equation

$$\Delta_1 \frac{\partial f(z)}{\partial z_1} + \cdots + \Delta_q \frac{\partial f(z)}{\partial z_q} = 0 \tag{13}$$

where the coefficients $\Delta_1, \ldots, \Delta_q$ are meromorphic functions in $z^{(1)}, \ldots, z^{(q-1)}$. In place of the $n(q - 1)$ complex variables occurring in $z^{(1)}, \ldots, z^{(q-1)}$ we insert complex values for which $\Delta_1, \ldots, \Delta_q$ are regular at these points and $\Delta_q \neq 0$. Now, let L stand for a constant invertible matrix of degree n whose first column has components $\Delta_1, \ldots, \Delta_q, 0, \ldots, 0$. Then in view of (13), $f(L\zeta)$ as a function of $\zeta_1, \ldots, \zeta_n$ is a meromorphic function whose derivative with respect to ζ_1 vanishes identically. This is a contradiction since $f(z)$ is nondengenerate. Hence Δ is not identically zero.

We replace $z^{(k)}$ by $z + z^{(k)}$ $(k = 1, \ldots, n)$ where z varies independently of $z^{(1)}, \ldots, z^{(n)}$. Then the variables $z^{(k)}$ can assume values $c^{(k)}$ $(k = 1, \ldots, n)$ such that the determinant

$$\Delta(z + c^{(1)}, \ldots, z + c^{(n)}) = |F_l(z + c^{(k)})| \tag{14}$$

does not vanish identically in z. For every fixed point c the function $f(z + c)$ is also an element of the field K of all Abelian functions corresponding to the given period lattice. In particular, the n functions

$$f_k(z) = f(z + c^{(k)}), \qquad (k = 1, \ldots, n)$$

are contained in K and their Jacobian determinant is given by (14), which by construction is not identically zero. Hence the functions $f_1, \ldots, f_n$ are analytically independent.

The following argument for the proof of the algebraic independence of these n functions now suggests itself. Let $P(u_1, \ldots, u_n)$ be a polynomial of smallest possible total degree such that the equation

$$P(f_1, \ldots, f_n) = 0$$

is satisfied identically in z. By partial differentiation with respect to $z_1, \ldots, z_n$ we obtain the equations

$$\frac{\partial P}{\partial f_1} F_l(z + c^{(1)}) + \cdots + \frac{\partial P}{\partial f_n} F_l(z + c^{(n)}) = 0, \qquad (l = 1, \ldots, n)$$

identically in z. Since the determinant (14) is not identically zero, the derivatives $\partial P/\partial f_k$ $(k = 1, \ldots, n)$ must vanish identically in z. This contradicts the minimality of the degree of P.

In view of Theorem 4, every nondegenerate field of Abelian functions has transcendence degree at least n.

Theorem 5: Every nondegenerate field of Abelian functions is an algebraic function field of transcendence degree n.

Proof: In establishing Theorem 4, we construct n functions $f_k(z)$ $(k = 1, \ldots, n)$ in K whose Jacobian determinant $|\partial f_k/\partial z_l|$ is not identically zero. Therefore, these functions are algebraically independent. Assume that these functions can be simultaneously transformed by an invertible linear substitution $z = L\zeta$ into functions of the $n - 1$ variables $\zeta_2, \ldots, \zeta_n$. Then these functions are independent of ζ_1. This yields n equations

$$a_1 \frac{\partial f_k}{\partial z_1} + \cdots + a_n \frac{\partial f_k}{\partial z_n} = 0, \qquad (k = 1, \ldots, n)$$

where the constant coefficients $a_1, \ldots, a_n$ are not all zero. Hence we have a contradiction.

The n divisors (f_k) can be written as quotients

$$(f_k) = \frac{\mathfrak{g}_k}{\mathfrak{d}}, \qquad (k = 1, \ldots, n)$$

where $\mathfrak{g}_k, \mathfrak{d} \neq (0)$ are periodic integral divisors satisfying $(\mathfrak{g}_1, \ldots, \mathfrak{g}_n, \mathfrak{d}) \sim 1$. In view of Theorem 1 in Section 8, we obtain representations

$$f_k(z) = \frac{g_k(z)}{h(z)}, \qquad (k = 1, \ldots, n) \tag{15}$$

where $g_1, \ldots, g_n, h \neq 0$ are equivariant Jacobian functions with a common linear exponent system. To a fixed period matrix C there correspond exponents

$$s_l(z) = 2\pi i \sum_{k=1}^{n} b_{kl} z_k + \beta_l, \qquad (l = 1, \ldots, 2n), \tag{16}$$

and as in equation (4) of Section 9, the matrix $(b_{kl}) = B$ satisfies a relation

$$B'C - C'B = A, \tag{17}$$

where A is integral and skew symmetric. Theorem 1 in Section 9 now implies that the determinant $|A|$ is not 0. Otherwise, by a suitable invertible linear transformation we could remove one variable from the functions $g_1, \ldots, g_n$, h and hence from each function $f_k(z)$. This would contradict the result obtained at the beginning of the proof.

Now, let $f_0(z)$ be any function in K. It can also be represented as a quotient

$$f_0(z) = \frac{g_0(z)}{h_0(z)},$$

where $g_0, h_0 \neq 0$ are equivariant Jacobian functions with a linear exponent system

$$s_l^*(z) = 2\pi i \sum_{k=1}^{n} b_{kl}^* z_k + \beta_l^*, \qquad (l = 1, \ldots, 2n).$$

We put $(b_{kl}^*) = B_0$ and define

$$A_0 = B_0'C - C'B_0. \tag{18}$$

Then, the matrix A_0 is integral and skew symmetric, but may not be invertible. For given natural numbers q and r, we denote the power products $f_0^\rho f_1^{\sigma_1} \cdots f_n^{\sigma_n}$ $(\rho = 0, \ldots, q)$, $(\sigma_1, \ldots, \sigma_n = 0, \ldots, r)$ in a fixed order by $F_1, \ldots, F_p$ where the relation

$$p = (q+1)(r+1)^n > (q+1)r^n \tag{19}$$

holds. Furthermore, we set

$$h_0^q h^{nr} = G, \qquad GF_k = G_k, \qquad (k = 1, \ldots, p).$$

Then the functions $G_1, \ldots, G_p$ are obviously equivariant Jacobian functions with exponent system $qs_l^*(z) + nrs_l(z)$, $(l = 1, \ldots, 2n)$. In view of (17) and (18), the matrices B and A corresponding to this exponent system are given by $qB_0 + nrB$ and

$$A^* = qA_0 + nrA.$$

We now use the fact that the sum of a positive and a nonnegative Hermitian matrix is positive, and we recall relation (10) in Section 9 as well as Theorem 1 of that section. It follows that $|A^*| > 0$. On the other hand, we have

$$|A^*| = |qA_0 + nrA| = r^{2n}\left|\frac{q}{r}A_0 + nA\right|.$$

We put

$$t^* = |A^*|^{1/2} = r^n \left|\frac{q}{r}A_0 + nA\right|^{1/2}, \tag{20}$$

where we take the positive square root. According to Theorem 1, there are no more than t^* linearly independent Jacobian functions corresponding to the exponent system $qs_l^* + nrs_l$. Therefore, if the inequality $p > t^*$ holds, the functions $G_1, \ldots, G_p$ must satisfy, identically in z, a homogeneous linear equation

(21) $$c_1G_1 + \cdots + c_pG_p = 0$$

with constant coefficients $c_1, \ldots, c_p$, not all of which are zero. In view of (19) and (20), the inequality $p > t^*$ is evidently fulfilled whenever

(22) $$q + 1 > \left| \frac{q}{r} A_0 + nA \right|^{1/2}$$

holds. We now choose

(23) $$q = |nA|^{1/2} = n^n \, |A|^{1/2}.$$

Then, for sufficiently large r, relation (22) is satisfied. Dividing (21) by G yields an algebraic equation

(24) $$P(f_0, f_1, \ldots, f_n) = 0,$$

where $P(u_0, u_1, \ldots, u_n)$ is a polynomial with constant coefficients, not all of which vanish. Its degree with respect to the indeterminate u_0 is at most q and with respect to any of the other indeterminates is at most r.

Finally, we notice that the natural number q defined by (23) does not depend on the choice of f_0, but is uniquely determined by $f_1, \ldots, f_n$. Further more, we can assume that equation (24) is irreducible. The function f_0 actually occurs in this equation since $f_1, \ldots, f_n$ are algebraically independent. We select f_0 so that the degree of P with respect to f_0 is as large as possible. Then the field K is obtained by adjoining f_0 to the field of all rational functions of $f_1, \ldots, f_n$ by means of equation (24). Thus Theorem 5 is completely proved.

We turn again to a degenerate Abelian function field in order to clarify this case.

Theorem 6: To every degenerate field K of Abelian functions of n variables $z_1, \ldots, z_n$ there corresponds a nonnegative integer $r < n$ such that K is transformed by a suitable invertible linear substitution $z = L\zeta$ into a nondegenerate field of Abelian functions of the r variables $\zeta_1, \ldots, \zeta_r$.

Proof: For every function $f(z)$ of K let the column $\mathfrak{f}$ be formed by the n partial derivatives $\partial f(z)/\partial z_k$ $(k = 1, \ldots, n)$. Select functions $f_1, \ldots, f_p$ in K so that there is at least one point z where the matrix

$$M = (\mathfrak{f}_1, \ldots, \mathfrak{f}_p)$$

has rank p, and let p be the maximal number with this property. Evidently, we then have $0 \leqslant p \leqslant n$. The case $p = 0$ occurs if and only if K consists only of constants, and in that case the statement of our theorem is true with $r = 0$ and $L = E$. Let us now assume that $p > 0$. We now consider the set of all constant columns c satisfying the equation $M'c = 0$ identically in z. Obviously, because of the maximality of p, for every such solution and every function f in K, the equation $c'f = 0$ also holds identically in z. The solutions c form a complex vector space. Let the columns $c_1, \ldots, c_s$ be a basis of that space. Put $n - s = r$. Then we have $p \leqslant r \leqslant n$. Let L be an invertible constant matrix whose last s columns are $c_1, \ldots, c_s$ and make the substitution $z = L\zeta$. The transformed Abelian functions $f(L\zeta)$ are then independent of $\zeta_{r+1}, \ldots, \zeta_n$.

Now, with a change of notation, we replace ζ by z. Therefore, we can assume that each function $f(z)$ in K depends only on the variables $z_1, \ldots, z_r$. Hence the last s rows of M consist entirely of zeros. Recalling the meaning of s, the first r components of each constant solution of $M'c = 0$ must then vanish. Therefore, it is impossible to transform the p functions $f_1, \ldots, f_p$ simultaneously by an invertible linear substitution into functions which depend on fewer than the r variables $z_1, \ldots, z_r$. These p functions are now treated in exactly the same way as $f_1, \ldots, f_n$ are treated in the preceding proof. We obtain formulas corresponding to (15), (16), and (17); however, in (15) we need only consider the indices $k = 1, \ldots, p$. Now, the Jacobian functions $g_1, \ldots, g_p, h$ as well as the exponents $s_l(z)$ $(l = 1, \ldots, 2n)$ only depend on the variables $z_1, \ldots, z_r$. Hence the last s rows of the matrix B consist entirely of zeros. Let C^* be the matrix formed by the first r rows of the period matrix C. Then the $2n$ columns of C^* generate an additive group Ω^* which, upon decomposition into real and imaginary parts, can be considered as a vector group in real $2r$-space. We suppose Ω^* is not discrete and recall the arguments of Section 6, in particular, Theorem 2 of that section. Then we conclude that another variable can be eliminated simultaneously from all functions f_k $(k = 1, \ldots, p)$ by a suitable invertible linear substitution. This yields a contradiction to the definition of s. Hence Ω^* is a lattice with a basis of m elements that we consider as columns of a matrix C_0 with r rows. It follows that $m \leqslant 2r$. Since the matrix P in relation (5) of Section 9 is invertible, the matrix formed analogously from C^* and $\bar{C}^*$ must have rank $2r$. Therefore, there exist $2r$ columns of C^* which are linearly independent over the reals. Thus we obtain $m \geqslant 2r$; hence, $m = 2r$. The definition of C^* and C_0 implies the existence of a unimodular matrix U satisfying

$$C^*U = (C_0 \quad 0).$$

Moreover, we obtain

$$BU = \begin{pmatrix} B_0 & 0 \\ 0 & 0 \end{pmatrix},$$

where B_0 is formed by r rows and $2r$ columns. Putting

$$A_0 = B_0'C_0 - C_0'B_0$$

we derive

$$U'AU = (BU)'(CU) - (CU)'(BU) = \begin{pmatrix} A_0 & 0 \\ 0 & 0 \end{pmatrix}.$$

Here C_0 is the period matrix of the lattice Ω^* to which the field K of Abelian functions of $z_1, \ldots, z_r$ corresponds. Analogously, the matrices B_0 and A_0 assume the role of B and A. Since it is impossible to eliminate another variable simultaneously from $f_1, \ldots, f_p$ by an invertible linear transformation, Theorem 1 in Section 9 implies $|A_0| \neq 0$. Therefore, the matrix C_0 turns out to be a Riemann matrix, and K is a nondegenerate field of functions depending on the variables $z_1, \ldots, z_r$. Hence $r < n$ and Theorem 6 is proved. Moreover, in view of Theorem 5, the proof yields the fact that $p = r$.

As is well known in the case of elliptic functions, that is, the case $n = 1$, there always exist nonconstant periodic meromorphic functions corresponding to a given lattice of real dimension 2 in the complex plane. Hence, in this case, the field K is never degenerate, and as a matter of fact, the period relations are indeed satisfied with $A = \pm J$. On the other hand, for $n > 1$ there do exist degenerate fields of Abelian functions. We will give an example in case $n = 2$ where K contains only constants.

As period matrix we take

$$C = (E \quad iV), \qquad E = \begin{pmatrix} 1 & 0 \\ 0 & 1 \end{pmatrix}, \qquad V = \begin{pmatrix} \alpha & \beta \\ \gamma & \delta \end{pmatrix}$$

where $\alpha, \beta, \gamma, \delta$ are real numbers linearly independent over the rationals; furthermore, we assume $\alpha\delta - \beta\gamma$ is irrational. For instance, we can choose $\alpha = 1, \beta = \sqrt{2}, \gamma = \sqrt{3}, \delta = \sqrt{5}$. The columns of C generate a lattice Ω of rank 4 because they are linearly independent over the reals. First, we show that C is not a Riemann matrix. Supposing C is a Riemann matrix implies the equation $CA^{-1}C' = 0$ with a rational invertible skew symmetric matrix

$$A^{-1} = \begin{pmatrix} A_1 & -F' \\ F & A_2 \end{pmatrix}, \qquad A_1 = a_1\begin{pmatrix} 0 & 1 \\ -1 & 0 \end{pmatrix},$$

$$A_2 = a_2\begin{pmatrix} 0 & 1 \\ -1 & 0 \end{pmatrix}, \qquad F = \begin{pmatrix} p & q \\ r & s \end{pmatrix}.$$

The equation

$$0 = CA^{-1}C' = (E \quad iV)\begin{pmatrix} A_1 & -F' \\ F & A_2 \end{pmatrix}\begin{pmatrix} E \\ iV' \end{pmatrix} = A_1 + i(VF - F'V') - VA_2V'$$

yields the symmetry of VF as well as the relation

$$A_1 = VA_2V'.$$

Hence, we obtain

$$\alpha q + \beta s = \gamma p + \delta r, \qquad a_1 = a_2(\alpha\delta - \beta\gamma).$$

In view of the assumptions on α, β, γ, δ, we conclude $A^{-1} = 0$, which is a contradiction.

Since C is not a Riemann matrix, the field K is degenerate and we may apply Theorem 6. A suitable invertible linear substitution $z = L\zeta$ transforms all elements $f(z)$ of K into function of r variables, where r can assume the values $r = 0$ or $r = 1$. We suppose $r = 1$. Then, according to the notation in the proof of Theorem 6, the matrix C^* is the first row of $L^{-1}C$ whose components are of the form ρ, σ, $i(\rho\alpha + \sigma\gamma)$, $i(\rho\beta + \sigma\delta)$ with certain complex numbers ρ and σ. These four components generate a lattice Ω^* of rank 2 in the complex plane. Let λ, μ be a basis of Ω^*. Then we have representations

$$\text{(25)} \qquad \begin{aligned} \rho &= p_1\lambda + q_1\mu, & \sigma &= p_2\lambda + q_2\mu, \\ i(\rho\alpha + \sigma\gamma) &= p_3\lambda + q_3\mu, & i(\rho\beta + \sigma\delta) &= p_4\lambda + q_4\mu \end{aligned}$$

with integral coefficients $p_1, \ldots, q_4$. Elimination of ρ and σ yields

$$\{(p_1\alpha + p_2\gamma)i - p_3\}\lambda + \{(q_1\alpha + q_2\gamma)i - q_3\}\mu = 0,$$
$$\{(p_1\beta + p_2\delta)i - p_4\}\lambda + \{(q_1\beta + q_2\delta)i - q_4\}\mu = 0.$$

Thus we are led to two homogeneous linear equations admitting a nontrivial solution λ, μ. Hence, the determinant must vanish and we obtain

$$\begin{aligned} 0 &= \{(p_1\alpha + p_2\gamma)i - p_3\}\{(q_1\beta + q_2\delta)i - q_4\} \\ &\quad - \{(q_1\alpha + q_2\gamma)i - q_3\}\{(p_1\beta + p_2\delta)i - p_4\} \\ &= -(\alpha\delta - \beta\gamma)(p_1q_2 - p_2q_1) + (p_3q_4 - p_4q_3) \\ &\quad - \alpha i(p_1q_4 - p_4q_1) + \beta i(p_1q_3 - p_3q_1) \\ &\quad - \gamma i(p_2q_4 - p_4q_2) + \delta i(p_2q_3 - p_3q_2). \end{aligned}$$

In view of the assumptions about α, β, γ, δ each 2×2 subdeterminant of the matrix formed from the rows $(p_1 \quad p_2 \quad p_3 \quad p_4)$ and $(q_1 \quad q_2 \quad q_3 \quad q_4)$ must vanish. Relations (25) then imply that Ω^* is at most one-dimensional. Assuming $r = 1$ leads to a contradiction. Therefore, we have $r = 0$ and the field of Abelian functions corresponding to Ω contains only the constant functions.

12. Picard varieties

Let K be a nondegenerate field of Abelian functions. Theorem 5 of the previous section then shows that K is an algebraic function field of transcendence degree n. Accordingly, there are $n + 1$ functions $f_1, \ldots, f_n, f_{n+1}$ in K such that $f_1, \ldots, f_n$ are algebraically independent, and f_{n+1} is connected to $f_1, \ldots, f_n$ by means of an irreducible algebraic equation, $P(f_1, \ldots, f_{n+1}) = 0$. Furthermore, every function in K may be expressed as a rational combination of $f_1, \ldots, f_{n+1}$ with constant coefficients. On the other hand, the equation $P(x_1, \ldots, x_{n+1}) = 0$ defines an irreducible algebraic variety, $\mathfrak{M}$, in $n+1$ dimensional complex space with coordinates $x_1, \ldots, x_{n+1}$. The points of $\mathfrak{M}$ are precisely the totality of solutions of the equation. In case $n = 1$ we can take, in particular,

$$f_1 = \wp(z), \qquad f_2 = \frac{d\wp(z)}{dz}, \qquad P(x_1, x_2) = x_2^2 - 4x_1^3 + g_2x_1 + g_3,$$

and conversely the variable z can be expressed as an elliptic integral of first kind. A one to one analytic correspondence is thus established between the points of $\mathfrak{M}$ and the period parallelogram $\mathfrak{P}$ provided we complete the curve $\mathfrak{M}$ by adding its point at infinity. There now arises the problem of generalizing this simple fact to the case of arbitrary values of n. However, in doing this we have to overcome some difficulties.

Making the Ansatz

$$x_k = f_k(z) \qquad (k = 1, \ldots, n + 1),$$

we clearly obtain a mapping into $\mathfrak{M}$ of those points on the period torus $\mathfrak{T} = \mathfrak{Z}/\Omega$ where $f_1, \ldots, f_{n+1}$ are simultaneously regular. However, at the singular points a peculiarity occurs because, for $n > 1$, the possible points of indeterminacy must be considered. For this reason we introduce the homogeneous coordinates $x_0, x_1, \ldots, x_{n+1}$ where the old coordinates x_k $(k = 1, \ldots, n + 1)$ are now replaced by the quotients x_k/x_0. The irreducible equation of $\mathfrak{M}$ then becomes homogeneous, and we denote this form by $P(x_0, x_1, \ldots, x_{n+1}) = 0$. In this connection we should bear in mind that all nontrivial solutions, that is those for which not all x_k $(k = 0, \ldots, n + 1)$ are zero, are admissible. We also remark that x_k and tx_k with arbitrary $t \neq 0$ $(k = 0, \ldots, n + 1)$ yield the same point in the $n + 1$ dimensional projective space $\mathfrak{Q}_{n+1}$. This introduction of homogeneous coordinates corresponds to the representation of the

$$(1) \qquad f_k(z) = \frac{p_k(z)}{p_0(z)} \qquad (k = 1, \ldots, n + 1), \qquad (p_0, \ldots, p_{n+1}) \sim 1,$$

as quotients of equivariant Jacobian functions $p_0, \ldots, p_{n+1}$. Then if we put $x_k = p_k(z)$, the equation $P(x_0, \ldots, x_{n+1}) = 0$ is obviously satisfied identically in z. Passing from z to a congruent point $z + \omega$, all the $p_k(z)$ are multiplied by the same nonzero factor $e^{s\omega(z)}$; therefore, we need only investigate the mapping defined by $x_k = p_k(z)$ on the period parallelotope $\mathfrak{P}$. By introducing these homogeneous coordinates we eliminate the difficulty only at those singular points where not all of the $n + 2$ Jacobian functions $p_0, \ldots, p_{n+1}$ vanish simultaneously. In general we cannot prove that this situation is impossible. Here the singular points are given by the zeros of $p_0(z)$ in (1). For this reason we increase the dimension of the projective space in which $\mathfrak{M}$ is embedded. Let $m \geqslant n + 1$. We select m functions $\phi_1, \ldots, \phi_m$ from K which generate this field and we can again set

$$\phi_k = \frac{g_k(z)}{g_0(z)} \qquad (k = 1, \ldots, m), \qquad (g_0, g_1, \ldots, g_m) \sim 1,$$

where $g_0, \ldots, g_m$ are $m + 1$ equivariant Jacobian functions. We will find that, for sufficiently large m and a suitable choice of $\phi_1, \ldots, \phi_m$, the functions $g_0, \ldots, g_m$ are nowhere simultaneously zero.

Let us consider the Abelian functions $\phi_1, \ldots, \phi_m$ together with all the algebraic relations, again in homogeneous form, among them. Thus we obtain infinitely many homogeneous polynomials $P(x_0, \ldots, x_m)$, and the algebraic variety $\mathfrak{N}$ which is to be studied is now defined in projective space $\mathfrak{Q}_m$ by the simultaneous nontrivial solutions of the equations $P = 0$. By the Hilbert theorem, which is known from algebra, we see that $\mathfrak{N}$ is actually defined by finitely many of these equations. If we put $x_k = g_k(z)$, $(k = 0, \ldots, m)$, we again obtain a mapping from $\mathfrak{T}$ to $\mathfrak{N}$. We will now determine the functions $\phi_1, \ldots, \phi_m$ so that $\mathfrak{N}$ has the following three important properties. First, the mapping from $\mathfrak{T}$ to $\mathfrak{N}$ will be one to one and onto; thus each point on $\mathfrak{N}$ will correspond to exactly one point of $\mathfrak{T}$. Next $\mathfrak{N}$ will have no singularities. This means that for each point $\mathfrak{p}$ on $\mathfrak{N}$ there exist n suitable inhomogeneous coordinates $u_k = x_{l_k}/x_{l_0}$ $(k = 1, \ldots, n)$ with $x_{l_0} \neq 0$ upon which the remaining $m - n$ inhomogeneous coordinates x_l/x_{l_0} $(l \neq l_0, l_1, \ldots, l_n)$ depend regularly in a whole neighborhood of $\mathfrak{p}$. Finally, in this neighborhood, $z_1, \ldots, z_n$ will also be regular functions of $u_1, \ldots, u_n$.

We reduce the problem of establishing the second and third properties to investigating a certain functional matrix. Let ζ be a point in $\mathfrak{Z}$ which is not a simultaneous zero of $g_0(z), \ldots, g_m(z)$ and let $\mathfrak{p}$ be its image on $\mathfrak{N}$. If $g_l(\zeta) \neq 0$ for $l = l_0$, then we consider the m inhomogeneous coordinates

$$\frac{x_k}{x_{l_0}} = \frac{g_k(z)}{g_{l_0}(z)} \qquad (k \neq l_0) \tag{2}$$

in a neighborhood of $z = \zeta$. If the matrix

$$F = \left(\frac{\partial}{\partial z_l}\left\{\frac{g_k(z)}{g_{l_0}(z)}\right\}\right) \tag{3}$$

formed from the partial derivatives with respect to $z_1, \ldots, z_n$ has rank n at $z = \zeta$, then the inverse of the mapping (2) exists and is regular in a neighborhood of $\mathfrak{p}$ on $\mathfrak{R}$. Now we have

$$\frac{\partial}{\partial z_l}\left(\frac{g_k}{g_{l_0}}\right) = \frac{\dfrac{\partial g_k}{\partial z_l} - \dfrac{g_k}{g_{l_0}}\dfrac{\partial g_{l_0}}{\partial z_l}}{g_l}. \tag{4}$$

In order to eliminate the special role of l_0, we consider the matrix M formed from the $m + 1$ rows $(g_k\ (\partial g_k/\partial z_1) \cdots (\partial g_k/\partial z_n))$, where $k = 0, 1, \ldots, m$. If we multiply the l_0th row of M by $-g_k/g_{l_0}$ $(k \neq l_0)$ and add the result to the kth row, then, omitting the first column and l_0th row, we see that (4) gives rise to the matrix $g_{l_0}F$. On the other hand, the cancelled first column has g_{l_0} in the l_0th place and zero everywhere else. Thus F has rank n at $z = \zeta$ if and only if M has rank $n + 1$ there. We should observe that in order to define M the assumption $g_{l_0}(\zeta) \neq 0$ is unnecessary. But if $g_k(\zeta) = 0$ for $k = 0, \ldots, m$, then the entire first column of M vanishes at $z = \zeta$, and its rank there can be at most n. Consequently, from now on we will assume the equivalent hypothesis, namely, that M has rank $n + 1$ everywhere.

We suppose the exponent system of the Jacobian functions appearing in (1) to be in reduced form and again put

$$A = \begin{pmatrix} 0 & T \\ -T & 0 \end{pmatrix}, \qquad C = (T \quad W),$$

where T is a diagonal matrix whose diagonal entries are $t_1, \ldots, t_n$. The corresponding Jacobian functions $p(z)$ thus satisfy the functional equations

$$p(z + Tr) = p(z), \qquad p(z + Ws) = \varepsilon(-s'Ws - 2s'z)p(z) \tag{5}$$

for arbitrary integral columns r and s. Now in order to satisfy the conditions placed on $g_0, \ldots, g_m$, we take these $m + 1$ desired functions as solutions of the modified functional equations

$$g(z + Tr) = g(z), \qquad g(z + Ws) = \varepsilon(-3s'Ws - 6s'z)g(z), \tag{6}$$

which are clearly satisfied by $(p(z))^3$. As a result of (1) and (5) in Section 11, the totality of entire solutions of (6) is given by the functions

$$g(z) = \sum_{r(\mathrm{mod}\ 3T)} \gamma_r \Theta_r(3T, 3W, 3z) \tag{7}$$

with arbitrary constants γ_r. We put

$$|3T| = 3^n t_1 \cdots t_n = m + 1 \tag{8}$$

and denote the $m + 1$ distinct functions $\Theta_r(3T, 3W, 3z)$ in a fixed order by $g_0(z), \ldots, g_m(z)$. We must now show that they possess all the required properties. Since each of the $n + 2$ functions $p_0^2 p_k$ $(k = 0, \ldots, n + 1)$ are equivariant with $g(z)$ they are also of the form given by (7). Therefore, the $n + 1$ quotients $f_k = (p_0^2 p_k)/p_0^3$ $(k = 1, \ldots, n + 1)$ may be expressed as fractions whose numerators and denominators are linear functions of the m quotients $\phi_k = g_k/g_0$ $(k = 1, \ldots, m.)$. Because of the assumption about $f_1, \ldots, f_{n+1}$ which we made at the outset, $\phi_1(z), \ldots, \phi_m(z)$ also generate the field K. Moreover, it follows from (8) that the number $m \geqslant 3^n - 1 \geqslant n + 1$.

Now we begin by showing that two distinct points of $\mathfrak{T}$ can never be mapped to the same point of $\mathfrak{N}$. Let $p(z)$ be a nonidentically zero Jacobian function satisfying (5). Given two arbitrary columns x and y, we form

$$g(z) = p(z + x)p(z + y)p(z - x - y), \tag{9}$$

which obviously satisfies (6). Thus we have the representation

$$g(z) = \sum_{k=0}^{m} \gamma_k g_k(z) \tag{10}$$

where now, however, the coefficients $\gamma_0, \ldots, \gamma_m$ depend on x and y. Now, if the points η and ζ in $\mathfrak{Z}$ are mapped to the same point in $\mathfrak{N}$, then

$$\alpha g_k(\eta) = \beta g_k(\zeta) \qquad (k = 0, \ldots, m) \tag{11}$$

where α and β are constants independent of x and y and not both zero. From (9), (10), and (11) it follows that

$$\alpha g(\eta) = \beta g(\zeta),$$

$$\alpha p(\eta + y)p(\eta - x - y) = \beta \frac{p(\zeta + x)}{p(\eta + x)} p(\zeta + y)p(\zeta - x - y) \tag{12}$$

holds identically in the variables x and y. First, it follows that $\alpha\beta \neq 0$. Furthermore, since the left-hand side of (12) is an entire function of x for any fixed y, the same is also true of the right-hand side. Now, if the meromorphic function

$$\frac{p(\zeta + x)}{(\eta + x)} = \psi(x)$$

of the variable x is not regular at any point $x = \xi$ of $\mathfrak{Z}$, then $p(\zeta + y)p(\zeta - \xi - y)$ must vanish identically in y. This, however, is not true. Similarly, it follows that the function ψ^{-1} is regular everywhere; hence,

$\psi(x)$ is a unit when viewed as a function of x. Furthermore, if we put $\zeta - \eta = \omega$, $x + \eta = z$, then the functional equation

$$p(z + \omega) = e^{w(z)}p(z)$$

holds for a suitable entire function $w(z)$. Here $p(z)$ is an arbitrary Jacobian function which satisfies (5). However, it was shown in the sequel to functional equation (9) in Section 11 that this is impossible unless ω is a period.

Furthermore, it follows from the previous considerations that the $m + 1$ functions $g_0(z), \ldots, g_m(z)$ have no common zero. Namely, if $z = \zeta$ is such a zero, then (11) would be satisfied with $\alpha = 0$, $\beta = 1$ and η arbitrary. Expression (12) now furnishes a contradiction. This result implies that $g_0, \ldots, g_m$ are relatively prime.

We will now show that the matrix M formed from the $m + 1$ rows

$$\left(g_k \frac{\partial g_k}{\partial z_1} \cdots \frac{\partial g_k}{\partial z_n}\right) \qquad (k = 0, \ldots, m)$$

has rank $n + 1$ everywhere. If the rank of M is less than $n + 1$ at a point $z = \zeta$, then we have the relations

$$(13) \quad c_0 g_k(\zeta) + c_1 \left(\frac{\partial g_k(z)}{\partial z_1}\right)_{z=\zeta} + \cdots + c_n \left(\frac{\partial g_k(z)}{\partial z_n}\right)_{z=\zeta} = 0 \qquad (k = 0, \ldots, m)$$

with certain constants $c_0, \ldots, c_n$ which are not all zero. From (10) and (13) we also obtain the relation

$$(14) \qquad c_0 g(\zeta) + c_1 g_{z_1}(\zeta) + \cdots + c_n g_{z_n}(\zeta) = 0$$

with the same constants $c_0, \ldots, c_n$. Next, let

$$r(z) = c_1 p_{z_1}(z) + \cdots + c_n p_{z_n}(z).$$

From (9) we obtain

$$\frac{g_{z_l}(z)}{g(z)} = \frac{p_{z_l}(z + x)}{p(z + x)} + \frac{p_{z_l}(z + y)}{p(z + y)} + \frac{p_{z_l}(z - x - y)}{p(z - x - y)}.$$

Using this expression, we change (14) into the relation

$$(15) \qquad c_0 + \frac{r(\zeta + x)}{p(\zeta + x)} + \frac{r(\zeta + y)}{p(\zeta + y)} + \frac{r(\zeta - x - y)}{p(\zeta - x - y)} = 0,$$

which holds identically in x and y. By considerations analogous to those employed in an earlier proof, we conclude from the independence of x and y that the first quotient

$$\frac{r(\zeta + x)}{p(\zeta + x)} = s(\zeta + x)$$

in (15) must be an entire function of the variable x. Therefore, we have

$$s(z) = c_1 \frac{p_{z_1}(z)}{p(z)} + \cdots + c_n \frac{p_{z_n}(z)}{p(z)}, \tag{16}$$

and in view of (5), the n partial derivatives $s_{z_1}, \ldots, s_{z_n}$ are seen to be entire functions belonging to K. Therefore, they are all constants. Consequently, $s(z)$ is a linear function of $z_1, \ldots, z_n$ which has period $t_k \neq 0$ in the variable z_k $(k = 1, \ldots, n)$; therefore, it must be a constant. From the second equation in (5) together with (9), (14), and (16), there arises the contradiction that $c_0, c_1, \ldots, c_n$ must all be 0.

It remains to show that every point $\mathfrak{p}$ of $\mathfrak{N}$ is the image of a point of $\mathfrak{Z}$. With this purpose in mind, we will first show that the algebraic variety $\mathfrak{N}$ is irreducible; that is, it cannot be represented as the union of two algebraic varieties $\mathfrak{N}_1$ and $\mathfrak{N}_2$ in $\mathfrak{Q}_m$ which are proper subsets of $\mathfrak{N}$. Otherwise, for $l = 1, 2$ there is a homogeneous polynomial R_l in $x_0, \ldots, x_m$ which vanishes on all of $\mathfrak{N}_l$, but not everywhere on $\mathfrak{N}$. The polynomial $P_0 = R_1R_2$ then vanishes everywhere on $\mathfrak{N}$ and, in particular, upon applying the substitution $x_k = g_k(z)$ $(k = 0, \ldots, m)$ vanishes identically in z. It follows that upon applying this substitution at least one of the polynomials R_l vanishes identically in z. Hence, this polynomial belongs to the ideal defining $\mathfrak{N}$. Thus it vanishes on $\mathfrak{N}$, which is a contradiction.

Let the image of $\mathfrak{T}$ under the mapping $x_k = g_k(z)$ $(k = 0, \ldots, m)$ be denoted by $\mathfrak{N}^*$. Since $\mathfrak{T}$ is compact, $\mathfrak{N}^*$ is also compact because of the continuity of $g_k(z)$. Clearly, $\mathfrak{N}^*$ is a subset of $\mathfrak{N}$, and we must show that $\mathfrak{N}^* = \mathfrak{N}$. We consider an arbitrary permutation $l_0, l_1, \ldots, l_m$ of the $m + 1$ numbers $0, 1, \ldots, m$ and put

$$\frac{g_{l_r}}{g_{l_0}} = \phi_r \qquad (r = 1, \ldots, m). \tag{17}$$

If the functions $\phi_1, \ldots, \phi_n$ are algebraically independent, then for $r = n + 1, \ldots, m$, we consider the irreducible algebraic equation satisfied by ϕ_r and $\phi_1, \ldots, \phi_n$. If we make these equations homogeneous and apply all possible permutations $l_0, l_1, \ldots, l_m$ we obtain finitely many homogeneous polynomials $P_\nu(x_0, \ldots, x_m)$, $(\nu = 1, \ldots, h)$, which vanish throughout $\mathfrak{N}$. However, because of the irreducibility of P_ν, the partial derivative

$$\frac{\partial P_\nu}{\partial x_{l_r}} = Q_\nu(x_0, \ldots, x_m) \tag{18}$$

with respect to the designated variable x_{l_r} does not vanish on all of $\mathfrak{N}$. Finally, let

$$Q = Q_1Q_2 \cdots Q_n. \tag{19}$$

By a theorem of algebra, every irreducible algebraic variety $\mathfrak{N}$ in complex projective space $\mathfrak{Q}_m$ is connected. Moreover, for an arbitrarily given fixed form Q not vanishing everywhere on $\mathfrak{N}$ and two arbitrary points $\mathfrak{p}$ and $\mathfrak{q}$ on $\mathfrak{N}$, we can join $\mathfrak{p}$ and $\mathfrak{q}$ with a curve $\mathfrak{C}$ on $\mathfrak{N}$ on which Q is not 0 except possibly at $\mathfrak{p}$ and $\mathfrak{q}$.

Now let ζ be a point of $\mathfrak{Z}$ to which the point $\mathfrak{p}$ of $\mathfrak{N}^*$ corresponds. Since the form Q defined by (19) does not vanish identically in z for $x_k = g_k(z)$ $(k = 0, \ldots, m)$, we can, in particular, select the point ζ so that $Q \neq 0$ at ζ. Furthermore, the matrix M has rank $n + 1$ at the point ζ. Therefore, we can determine the permutation $l_0, \ldots, l_m$ of $0, \ldots, m$ in such a way that the n functions $\phi_1, \ldots, \phi_n$ defined by (17) are regular at $z = \zeta$ and have a nonzero functional determinant there. Thus the system of n equations

$$\phi_k = u_k \qquad (k = 1, \ldots, n)$$

is invertible in a sufficiently small neighborhood of $z = \zeta$, and from this system we obtain $z_1, \ldots, z_n$ as regular functions of the local coordinates $u_1, \ldots, u_n$ on $\mathfrak{N}$. The remaining $m - n$ coordinates

$$u_r = \frac{x_{l_r}}{x_{l_0}}$$

of points of $\mathfrak{N}$ in a sufficiently small neighborhood of the point $\mathfrak{p}$ arise from solutions of the corresponding equation $P_\nu = 0$ with respect to u_r. We use the fact that the partial derivative Q_ν defined by (18) is not zero at $z = \zeta$. Consequently, the u_r are uniquely determined in that neighborhood of $\mathfrak{p}$ and thus are given by the formula

$$u_r = \phi_r \qquad (r = n + 1, \ldots, m).$$

From this we also see that $u_{n+1}, \ldots, u_m$ are regular functions of the n independent variables $u_1, \ldots, u_n$ in a neighborhood of $\mathfrak{p}$. In addition, we have shown that in this neighborhood of $\mathfrak{p}$ every point of $\mathfrak{N}$ is also on $\mathfrak{N}^*$.

Now if $\mathfrak{q}$ is any other point of $\mathfrak{N}$, then we may join it to $\mathfrak{p}$ by a curve $\mathfrak{C}$ on $\mathfrak{N}$ on which $Q \neq 0$ everywhere except possibly at $\mathfrak{q}$. Now if the whole of $\mathfrak{C}$ including $\mathfrak{q}$ does not belong to $\mathfrak{N}^*$, then we consider the largest connected open subset $\mathfrak{C}^*$ of $\mathfrak{C}$ beginning at $\mathfrak{p}$ and lying entirely in $\mathfrak{N}^*$. Its endpoint $\mathfrak{q}^*$ cannot belong to $\mathfrak{N}^*$ since otherwise $Q \neq 0$ at $\mathfrak{q}^*$, and the previous considerations with $\mathfrak{q}^*$ in place of $\mathfrak{p}$ supply a contradiction to the maximality of $\mathfrak{C}^*$. On the other hand, $\mathfrak{q}^*$ must belong to $\mathfrak{N}^*$ since $\mathfrak{N}^*$ is indeed closed. Thus the whole curve $\mathfrak{C}$ including $\mathfrak{q}$ does in fact lie on $\mathfrak{N}^*$. Hence, we have proved that $\mathfrak{N} = \mathfrak{N}^*$.

We must still investigate the mapping of a neighborhood of $\mathfrak{q}$ on $\mathfrak{N}$ onto $\mathfrak{T}$ in case the polynomial Q vanishes at $\mathfrak{q}$. Let $z = \eta$ be the point on $\mathfrak{T}$ corresponding to $\mathfrak{q}$. If a complete neighborhood of $\mathfrak{q}$ on $\mathfrak{N}$ is not filled up by the image of any complete neighborhood $\mathfrak{U}$ of $z = \eta$ under the mapping $x_k = g_k(z)$ $(k = 0, \ldots, m)$, then there is a sequence of points $\zeta^{(k)}$ $(k = 1, 2, \ldots)$ outside $\mathfrak{U}$ converging to a point ξ of $\mathfrak{T}$ such that the corresponding image points $\mathfrak{q}_k$ converge to $\mathfrak{q}$. Then the distinct points ξ and η of $\mathfrak{T}$ have the same image point $\mathfrak{q}$, which is impossible. Thus under this mapping every neighborhood of η covers all the points of a neighborhood of $\mathfrak{q}$ exactly once. Once again, we obtain n independently varying inhomogeneous coordinates $u_1, \ldots, u_n$ at $\mathfrak{q}$, and the remaining $m - n$ coordinates are regular functions of these. We have now completely proved that $\mathfrak{N}$ actually possesses the three required properties. In this connection, $\mathfrak{N}$ has a parametric representation $x_k = g_k(z)$, $(k = 0, \ldots, m)$, where the $g_k(z)$ are the distinct members of the sequence of theta functions $\Theta_r(3T, 3W, 3z)$. If, for simplicity, we write T, W, z in place of $3T$, $3W$, $3z$, then K remains unchanged because z does not occur as an element of it. We collect the established facts in the modified notation.

Theorem 1: Let K be a nondegenerate Abelian function field. Then we can select a normal form $(T\ W)$ for the period matrix so that the quotients of the theta functions $\Theta_r(T, W, z)$ generate the field K. We denote the distinct theta functions which occur by $g_0(z), \ldots, g_m(z)$ and write out all the homogeneous algebraic relations

$$P(g_0(z), \ldots, g_m(z)) = 0$$

with constant coefficients which occur among them. Then the equations

$$P(x_0, \ldots, x_m) = 0$$

define a nonsingular irreducible algebraic variety $\mathfrak{N}$ in m dimensional projective space which may be mapped biregularly onto the period torus $\mathfrak{T}$ by the correspondence $x_k = g_k(z)$, $(k = 0, \ldots, m)$.

Upon consideration of Theorem 6 of the previous section, the first statement shows that every Abelian function can be expressed as a rational function of theta functions. This is the so called "*theta-theorem*" first stated without proof by Riemann. Subsequently, despite his many clever considerations, Weierstrass also failed to give a complete proof. Poincaré was the first to succeed in overcoming all of the essential difficulties.

Now let us consider the second statement of Theorem 1 without assuming that $g_0(z), \ldots, g_m(z)$ are the theta functions $\Theta_r(T, W, z)$. More generally, we assume that the functions $g_0(z), \ldots, g_m(z)$ represent any system of equivariant Jacobian functions satisfying the indicated conditions. Then $\mathfrak{N}$ is called a *Picard variety* corresponding to the field K. Here we need not assume that K is generated by the quotients of the nonidentically vanishing functions g_k $(k = 0, \ldots, m)$. Rather, this may be derived in the following way. Let Λ be the subfield of K that is generated by the quotients. Clearly, Λ has transcendence degree n since $\mathfrak{T}$ is of complex dimension n. Hence, K is an algebraic extension of Λ and is obtained by adjoining a single function $f(z)$ to Λ. Including the irreducible equation of $f(z)$ with respect to Λ among the system of defining equations given in the theorem yields an irreducible algebraic variety $\mathfrak{R}$ in $(m+1)$-dimensional projective space $\mathfrak{Q}_{m+1}$. Of course, $\mathfrak{R}$ may have singularities. Now, select any point ζ on $\mathfrak{T}$ such that the function $f(z)$ as well as each coefficient of the irreducible equation of $f(z)$ with respect to Λ is regular at ζ and such that the discriminant of this equation differs from 0 at ζ. Assuming K does not coincide with Λ, there exists, for $z = \zeta$, at least one solution of that equation different from $f(\zeta)$. Hence, to the point $z = \zeta$ of $\mathfrak{T}$ there corresponds only one point $\mathfrak{p}$ of $\mathfrak{N}$ but more than one point of $\mathfrak{R}$ lying over $\mathfrak{p}$. We join the point of $\mathfrak{R}$ corresponding to $f(\zeta)$ with another point of $\mathfrak{R}$ lying over $\mathfrak{p}$ by a curve such that each coefficient of the equation of $f(z)$ remains regular and its discriminant remains different from 0 along this curve. We arrive at a contradiction since the solution of the equation, considered as a function on the curve, is given by $f(z)$ throughout the curve and hence returns to the value $f(\zeta)$ at the endpoint of the curve.

In the well-known fashion, we assign a field K to each nonempty irreducible algebraic variety $\mathfrak{N}$ in projective space with the homogeneous coordinates $x_0, \ldots, x_m$. Namely, we consider all quotients P/S of two homogeneous polynomials of the same degree whose denominator S does not vanish everywhere on $\mathfrak{N}$. Two such quotients P_1/S_1 and P_2/S_2 are to be identified in K if and only if $P_1 S_2 = P_2 S_1$ everywhere on $\mathfrak{N}$. K can be proved to be an algebraic function field whose transcendence degree n is the dimension of $\mathfrak{N}$. Two such varieties $\mathfrak{N}$ and $\mathfrak{R}$ with the respective projective coordinates $x_0, \ldots, x_m$ and $y_0, \ldots, y_h$ are said to be *birationally equivalent* if there exists an isomorphism of their corresponding fields K and Λ which leaves all constants invariant. We assume the notation is so chosen that neither x_0 on $\mathfrak{N}$ nor y_0 on $\mathfrak{R}$ vanish identically. Thus the functions $x_1/x_0, \ldots, x_m/x_0$ in K are the images of certain functions $Q_1/Q_0, \ldots, Q_m/Q_0$ in Λ under an isomorphism of the type under consideration. Conversely, the functions $y_1/y_0, \ldots, y_h/y_0$ in Λ are also the images of certain functions $P_1/P_0, \ldots, P_h/P_0$ in K. Then K and Λ can be identified by the rule

$$\frac{y_k}{y_0} = \frac{P_k(x_0, \ldots, x_m)}{P_0(x_0, \ldots, x_m)}, \qquad (k = 1, \ldots, h), \tag{20}$$

which, conversely, again gives rise to

$$\frac{x_k}{x_0} = \frac{Q_k(y_0, \ldots, y_h)}{Q_0(y_0, \ldots, y_h)}, \qquad (k = 1, \ldots m). \tag{21}$$

Now let $\mathfrak{p}$ be any point of $\mathfrak{N}$ at which both polynomials $P_0(x_0, \ldots, x_m)$ and $Q_0(P_0, \ldots, P_h)$ are different from zero. By (19), there corresponds to $\mathfrak{p}$ a unique image point $\mathfrak{q}$ on $\mathfrak{R}$ at which $Q_0(y_0, \ldots, y_h) \neq 0$. Conversely by (20), $\mathfrak{p}$ is the unique image point of $\mathfrak{q}$ at which we also have $P_0(Q_0, \ldots, Q_m) \neq 0$. In the usual terminology, this mapping is a birational mapping of each of the varieties into the other. Although the mapping is one to one for the indicated points, it need not be so for the remaining points if the dimension n is greater than 1.

A particular case which is of importance for our purposes arises when the mapping between $\mathfrak{N}$ and $\mathfrak{R}$ is defined everywhere. In the sequel we assume that both varieties, $\mathfrak{N}$ and $\mathfrak{R}$, are nonsingular; that is, for each point $\mathfrak{p}$ of $\mathfrak{N}$ and a suitably chosen permutation $l_0, \ldots, l_m$ of $0, \ldots, m$, the inhomogeneous coordinates

$$t_r = \frac{x_{l_r}}{x_{l_0}} \qquad (r = 1, \ldots, m)$$

are regular functions of the independent variables $t_1, \ldots, t_n$ in a neighborhood of $\mathfrak{p}$ on $\mathfrak{N}$. The analogous statement holds for each point $\mathfrak{q}$ of $\mathfrak{R}$; that is, with a permutation $k_0, \ldots, k_h$ of $0, \ldots, h$ the coordinates

$$u_s = \frac{y_{k_s}}{y_{k_0}} \qquad (s = 1, \ldots, h)$$

depend regularly on $u_1, \ldots, u_n$. Furthermore, it is required that by means of (19) the h functions u_s $(s = 1, \ldots, h)$ on $\mathfrak{N}$ are regular functions of $t_1, \ldots, t_n$ in a neighborhood of $\mathfrak{p}$, and by (20) the m functions t_r, $(r = 1, \ldots, m)$, on $\mathfrak{R}$ depend on $u_1, \ldots, u_n$ regularly in a neighborhood of $\mathfrak{q}$. Under these conditions we call the birational transformation *regular;* it is then one to one without exceptions.

Once again let $\mathfrak{N}$ be a Picard variety corresponding to the field K of Abelian functions. It is then clear that K is precisely that field which was assigned to the irreducible algebraic variety $\mathfrak{N}$ under the definition given two paragraphs before this one.

Theorem 2: Let $\mathfrak{N}$ be a Picard variety and $\mathfrak{R}$ an irreducible algebraic variety. $\mathfrak{R}$ is a Picard variety corresponding to K if and only if there exists a birational regular transformation of $\mathfrak{N}$ onto $\mathfrak{R}$.

Proof: If $\mathfrak{R}$ is a Picard variety that corresponds to the same Abelian function field as $\mathfrak{N}$, then $\mathfrak{R}$ and $\mathfrak{N}$ are birationally equivalent. In addition to this, by Theorem 1, both $\mathfrak{R}$ and $\mathfrak{N}$ are mapped biregularly onto the period torus $\mathfrak{T}$. Consequently, the birational mapping of $\mathfrak{R}$ onto $\mathfrak{N}$ is regular throughout.

Conversely, let the irreducible algebraic variety $\mathfrak{R}$ be mapped onto $\mathfrak{N}$ by means of a birational regular transformation given by (19) and (20). If $x_k = g_k(z)$, $(k = 0, \ldots, m)$, is the parametric representation of $\mathfrak{N}$ by means of equivariant Jacobian functions, then

$$\frac{y_k}{y_0} = \frac{P_k(g_0, \ldots, g_m)}{P_0(g_0, \ldots, g_m)}, \qquad (k = 1, \ldots, h)$$

show that the inhomogeneous coordinates on $\mathfrak{R}$ belong to K. They may then be represented in the form $p_k(z)/p_0(z)$, that is, as quotients of equivariant Jacobian functions where we can demand that $(p_0, p_1, \ldots, p_h) \sim 1$. By (20) we also have

$$\frac{x_k}{x_0} = \frac{Q_k(p_0, \ldots, p_h)}{Q_0(p_0, \ldots, p_h)}, \qquad (k = 1, \ldots, m).$$

Because of the regularity of the mappings of $\mathfrak{R}$ onto $\mathfrak{N}$ and of $\mathfrak{N}$ onto $\mathfrak{T}$, it follows that the functions $p_0, \ldots, p_h$ possess the properties of the functions $g_0, \ldots, g_m$ in Theorem 1. Thus, $\mathfrak{R}$ is a Picard variety corresponding to K.

13. The addition theorem

Chapter 4 was concerned with a special case of Abelian functions, the Jacobi–Abelian functions. At the end of that chapter we proved the addition theorem in this special case, which, in turn, represents a generalization of the addition theorem for elliptic functions. The corresponding property of elliptic integrals served as the starting point in Chapter 1. We will now show that such an addition theorem also exists for arbitrary Abelian functions. Its content is as follows.

Theorem 1: Let z and w be two columns varying independently of each other and let $f_0, f_1, \ldots, f_n$ be generators of a nondegenerate Abelian function field. Then there exist $n + 1$ rational functions

$$R_k = R_k(u_0, \ldots, u_n; v_0, \ldots, v_n), \qquad (k = 0, \ldots, n)$$

of $2n + 2$ variables $u_0, \ldots, u_n$ and $v_0, \ldots, v_n$ with constant coefficients such that the equations

$$f_k(z + w) = R_k(f_0(z), \ldots, f_n(z); f_0(w), \ldots, f_n(w)), \qquad (k = 0, \ldots, n)$$

hold and none of the denominators vanish identically in z and w.

Proof: We may assume that the functions $f_1, \ldots, f_n$ are algebraically independent. Then $x_0 = f_0$ is a root of an irreducible algebraic equation

$$P_0 x_0^h + P_1 x_0^{h-1} + \cdots + P_h = 0, \tag{1}$$

where $P_0, P_1, \ldots, P_h$ are polynomials in the n variables $x_1 = f_1, \ldots, x_n = f_n$ with constant coefficients. The elements y of the field K of Abelian functions can then be uniquely represented as

$$y = S_1 x_0^{h-1} + S_2 x_0^{h-2} + \cdots + S_h, \tag{2}$$

where $S_1, S_2, \ldots, S_h$ range over all rational functions of $x_1, \ldots, x_n$. Now let $f(z)$ be any Abelian function in K. Then, for every fixed column w, the function $f(z + w)$ is also an element of K. We apply (2) to $y = f(z + w)$ and show that the numerators, as well as the denominators, of all reduced quotients $S_1, \ldots, S_h$, viewed as polynomials in $x_1, \ldots, x_n$, have total degrees not exceeding a bound which is independent of w.

For this purpose we first consider $x_1, \ldots, x_n$ as indeterminates rather than as given functions of z. We denote the field of rational functions of $x_1, \ldots, x_n$ by K_0, the h roots of equation (1) by $\xi_1, \ldots, \xi_h$ and the corresponding conjugates of y in (2) by $y^{(1)}, \ldots, y^{(h)}$. Forming the $h \times h$ square matrix

$$L = (\xi_l^{h-k}),$$

we obtain

$$(y^{(1)} y^{(2)} \cdots y^{(h)}) = (S_1 S_2 \cdots S_h)L. \tag{3}$$

Introducing the trace

$$\sigma(y) = y^{(1)} + y^{(2)} + \cdots + y^{(h)}$$

with respect to K_0, the determinant of the matrix

$$LL' = (\sigma(x_0^{2h-k-l})) = D$$

is seen to be the square of the difference product of $\xi_1, \ldots, \xi_h$; hence, it is the discriminant of equation (1) divided by the factor P_0^{2h-2}. The determinant of D is therefore a nonvanishing element of K_0. The inverse matrix D^{-1} then exists, and equation (3) yields

$$(S_1 S_2 \cdots S_h) = (\sigma(y x_0^{h-1}) \sigma(y x_0^{h-2}) \cdots \sigma(y)) D^{-1}.$$

The traces appearing on the right-hand side are rational functions of $x_1, \ldots, x_n$. Hence, in order to prove the assertion of the preceding paragraph, we need only show that, for $y = f(z + w)$, the degrees of the numerators and the denominators of these traces do not exceed a bound independent of w.

Recalling the proof of Theorem 5 in Section 11, we replace the Abelian function $f_0(z)$ occurring there by $f(z)$ and retain the other notation. Then the functions $g_0(z + w)$ and $h_0(z + w)$ become equivariant Jacobian functions whose principal matrix A_0 is independent of w. More generally, we consider the Abelian function $f(z + w)(f_0(z))^{h-k}$, $k = 1, \ldots, h$, where $f_0(z)$ denotes the function introduced in the present theorem. It then follows that the corresponding principal matrix A_0 depends in an easily specified manner upon k, but A_0 is still independent of w. Hence, in condition (22) of Section 11 the principal matrices A and A_0 are independent of the choice of w once $f(z), f_0(z), f_1(z), \ldots, f_n(z)$ and k are fixed. Therefore, determining q by (23), condition (22) can be satisfied by a number r independent of w. Equation (24) then yields an algebraic relation between yx_0^{h-k} and $x_1, \ldots, x_n$ whose total degree does not exceed a bound independent of w. Then the numerators as well as the denominators of all reduced coefficients of the irreducible equation of yx_0^{h-k} over K_0 have total degrees bounded uniformly with respect to w. Thus the required statement about the traces is proved.

Now the proof of Theorem 1 is easily completed. We multiply equation (2) by the common denominator of $S_1, \ldots, S_h$. For a sufficiently large natural number r, chosen independently of w, we denote the products $yx_1^{k_1} \cdots x_n^{k_n}$, $(k_1, \ldots, k_n = 0, \ldots, r)$ and $x_0^{k_0}x_1^{k_1} \cdots x_n^{k_n}$, $(k_0 = 0, \ldots, h-1;$ $k_1, \ldots, k_n = 0, \ldots, r)$ in a fixed order by $H_1(z), \ldots, H_m(z)$. We obtain an equation

$$\chi_1(w)H_1(z) + \cdots + \chi_m(w)H_m(z) = 0, \tag{4}$$

where m is independent of w, and the coefficients $\chi_1(w), \ldots, \chi_m(w)$ do not vanish simultaneously for any choice of w. Those products $H_1(z), \ldots, H_m(z)$ which contain y may depend on w. In place of z in (4) we successively insert m independently varying columns $z_{(1)}, \ldots, z_{(m)}$ and also consider w as a variable column. The entries of the $m \times m$ square matrix $(H_k(z_{(l)}))$ are then meromorphic functions of $z_{(1)}, \ldots, z_{(m)}$ and w. The rank ν of this matrix obviously satisfies $0 < \nu < m$. We may assume that the subdeterminant formed by the first ν rows and columns is not identically zero. We then expand the subdeterminant formed by the first $\nu + 1$ rows and columns with respect to its last column. Again, replacing $z_{(\nu+1)}$ by z and selecting suitable constant columns for $z_{(1)}, \ldots, z_{(\nu)}$, we obtain a new equation of the form (4). Now the coefficients $\chi_1(w), \ldots, \chi_m(w)$ are polynomials in $f(w + z_{(1)}), \ldots, f(w + z_{(\nu)})$ not all of which vanish identically in w. Moreover, referring to (2), each of the functions $f(w + z_{(l)})$, $l = 1, \ldots, \nu$ can again be expressed rationally in terms of $f_0(w), f_1(w), \ldots, f_n(w)$. Summing up all terms in (4) containing y, the coefficient of y cannot vanish identically since otherwise x_0 would have degree less than h over K_0. Hence (4) is solvable with respect to y, and we obtain $f(z + w)$ as a rational function of $f_0(z), \ldots, f_n(z)$ and $f_0(w), \ldots, f_n(w)$, where the denominator does not vanish identically in z and w. We obtain the statement of Theorem 1 by putting $f(z) = f_l(z)$, $l = 0, \ldots, n$.

According to Theorem 1 of Section 12, the generators of K can be selected as quotients of theta functions. In this case there arises the problem of determining a more explicit form of the rational functions occurring in the addition theorem. Furthermore, we are interested in how their coefficients depend upon the period matrix $(T\ W)$. In case $n = 1$, these questions are answered completely by the explicit form of the addition theorem for elliptic functions formulated in Chapter 1 by means of the $\wp$-function. For arbitrary n, a completely satisfactory explicit form of the addition theorem is not yet known. These questions are related to the so-called Riemann theta-formula which will not be discussed here.

If $f(z)$ is an element of K, the same is true of $f(-z)$. Therefore, $f(-z)$ can be expressed rationally in terms of the generators $f_0(z), \ldots, f_n(z)$. Applying this result to the $n + 1$ functions $f_0(-z), \ldots, f_n(-z)$, we obtain a supplement to the addition theorem which may be called the *inversion theorem*. Combining both theorems yields the *subtraction theorem* which expresses $f_k(z - w)$ rationally in terms of $f_0(z), \ldots, f_n(z)$ and $f_0(w), \ldots, f_n(w)$.

Our next goal is to formulate the addition theorem and the supplementary inversion theorem as properties which characterize the Picard varieties $\mathfrak{N}$. Let $x_0, \ldots, x_m$ be homogeneous coordinates on $\mathfrak{N}$, and let the equations $x_k = g_k(z)$, $k = 0, \ldots, m$ describe the biregular mapping of $\mathfrak{N}$ onto the period torus $\mathfrak{T}$. Since the ratios of the $m + 1$ Jacobian functions $g_k(z)$ may be expressed birationally in terms of the generators $f_0(z), \ldots, f_n(z)$ of K, Theorem 1 yields a corresponding addition theorem involving the ratios of the functions $g_k(z + w)$ for variable columns z and w. Put $z + w = u$ and let z, w vary in sufficiently small neighborhoods of any two points z_0, w_0 of $\mathfrak{T}$. Let $\mathfrak{p}$, $\mathfrak{q}$ be the points on $\mathfrak{N}$ corresponding to z_0, w_0. Furthermore, let the point $\mathfrak{r}$ of $\mathfrak{N}$ be the image of $u_0 = z_0 + w_0$. If we consider sufficiently small neighborhoods on $\mathfrak{N}$, the system of local coordinates on $\mathfrak{N}$ depends biregularly on z. Therefore, local coordinates in a neighborhood of $\mathfrak{r}$ are regular functions of u, hence regular functions of z and w in suitable neighborhoods of z_0 and w_0, and hence also regular functions of any local coordinates on $\mathfrak{N}$ in neighborhoods of $\mathfrak{p}$ and $\mathfrak{q}$. For every fixed w the addition theorem therefore yields a birational regular mapping of $\mathfrak{N}$ onto itself to which there corresponds the translation carrying z to $z + w$ on the period torus. Since passing from z to $-z$ is also a birational regular transformation according to the inversion theorem, we thus obtain a group of birational regular mappings of the Picard variety onto itself to which there corresponds, on $\mathfrak{T}$, exactly the group of all translations. The group of translations is the factor group $\mathfrak{Z}/\Omega$ of the space $\mathfrak{Z}$, considered as additive vector group with respect to the discrete subgroup Ω. Passing from the period torus $\mathfrak{T}$ to the Picard variety $\mathfrak{N}$, we may consider $\mathfrak{N}$ as the underlying set of this group, where the group operations are given by the rational formulas of the addition theorem and the inversion theorem. Conversely, this group property determines a class of irreducible nonsingular algebraic varieties. They are called *Picard varieties* after their discoverer, who supplied a major portion of the proofs.

Theorem 2: Let there be given an irreducible nonsingular algebraic variety $\mathfrak{N}$ in projective space carrying a group structure and let the group operations be given by regular functions when expressed in terms of local inhomogeneous coordinates. Then $\mathfrak{N}$ is a Picard variety.

Proof: If $\mathfrak{x}$ and $\mathfrak{y}$ are points of $\mathfrak{N}$ the group product and the inverse will be denoted by $\mathfrak{x}\mathfrak{y}$ and $\mathfrak{x}^{-1}$. We begin by showing that $\mathfrak{N}$ is a commutative group. Let n be the dimension of $\mathfrak{N}$.

The commutator $[\mathfrak{x}, \mathfrak{y}] = \mathfrak{x}\mathfrak{y}\mathfrak{x}^{-1}\mathfrak{y}^{-1}$ is continuous as a function of $\mathfrak{x}$ and $\mathfrak{y}$. It is even uniformly continuous since $\mathfrak{N}$ is compact. If $\mathfrak{e}$ is the unit element of the group $\mathfrak{N}$, we have $[\mathfrak{x}, \mathfrak{e}] = \mathfrak{e}$. Consequently, for every neighborhood $\mathfrak{U}$ of $\mathfrak{e}$ in $\mathfrak{N}$ there exists a sufficiently small neighborhood $\mathfrak{B}$ of $\mathfrak{e}$ such that the commutator $[\mathfrak{x}, \mathfrak{y}]$ is contained in $\mathfrak{U}$ whenever $\mathfrak{y}$ is an element of $\mathfrak{B}$ and $\mathfrak{x}$ is any element of $\mathfrak{N}$. We select $\mathfrak{U}$ so small that it can be covered by a single system of local inhomogeneous coordinates $t_1, \ldots, t_n$. For every fixed $\mathfrak{y}$ in $\mathfrak{B}$, the local coordinates representing the commutator $[\mathfrak{x}, \mathfrak{y}]$ are regular functions of any system of local coordinates at $\mathfrak{x}$, where $\mathfrak{x}$ ranges over all of $\mathfrak{N}$. Since $\mathfrak{N}$ is compact, the maximum principle for regular functions shows that the local coordinates of $[\mathfrak{x}, \mathfrak{y}]$, hence the commutator $[\mathfrak{x}, \mathfrak{y}]$ itself, is constant for every fixed $\mathfrak{y}$ in $\mathfrak{B}$ and $\mathfrak{x}$ ranging over $\mathfrak{N}$. In particular, setting $\mathfrak{x} = \mathfrak{e}$ we obtain $[\mathfrak{e}, \mathfrak{y}] = \mathfrak{e}$ and hence $[\mathfrak{x}, \mathfrak{y}] = \mathfrak{e}$. This means that the entire neighborhood $\mathfrak{B}$ of $\mathfrak{e}$ belongs to the center $\mathfrak{C}$ of the group $\mathfrak{N}$. In view of the continuity of $[\mathfrak{x}, \mathfrak{y}]$, the center $\mathfrak{C}$ is a closed subgroup of $\mathfrak{N}$. On the other hand, if $\mathfrak{y}_0$ is any point of $\mathfrak{C}$ and $\mathfrak{z}$ ranges over $\mathfrak{B}$, the point $\mathfrak{y} = \mathfrak{z}\mathfrak{y}_0$ satisfies

$$[\mathfrak{x}, \mathfrak{y}] = \mathfrak{x}\mathfrak{z}\mathfrak{y}_0\mathfrak{x}^{-1}\mathfrak{y}_0^{-1}\mathfrak{z}^{-1} = \mathfrak{x}\mathfrak{z}\mathfrak{y}_0\mathfrak{z}^{-1}\mathfrak{x}^{-1}\mathfrak{y}_0^{-1} = [\mathfrak{x}\mathfrak{z}, \mathfrak{y}_0] = \mathfrak{e}$$

for every $\mathfrak{x}$ in $\mathfrak{N}$. Hence $\mathfrak{C}$ is an open subset of $\mathfrak{N}$. The irreducible algebraic variety $\mathfrak{N}$ is connected. The nonvoid open and closed subset $\mathfrak{C}$ of $\mathfrak{N}$ therefore coincides with $\mathfrak{N}$. Hence $\mathfrak{N}$ is in fact commutative. For any three elements $\mathfrak{x}, \mathfrak{y}, \mathfrak{z}$ of $\mathfrak{N}$ we have thus proved that

$$(\mathfrak{x}\mathfrak{y})\mathfrak{z} = (\mathfrak{x}\mathfrak{z})\mathfrak{y}. \tag{5}$$

Now we must apply basic facts from the theory of Lie groups. We form columns ξ, η, ζ, and ρ from the local coordinates of the respective points $\mathfrak{x}, \mathfrak{y}, \mathfrak{z}$, and $\mathfrak{r} = \mathfrak{x}\mathfrak{y}$. We let $\mathfrak{y}$ and $\mathfrak{z}$ range over a sufficiently small neighborhood of $\mathfrak{e}$ and let $\mathfrak{x}$ range over a sufficiently small neighborhood of an arbitrarily fixed point $\mathfrak{x}_0$ of $\mathfrak{N}$. Furthermore, let the point $\mathfrak{z} = \mathfrak{e}$ correspond to the column $\zeta = 0$. Then we obtain an equation

$$\rho = f(\xi, \eta), \tag{6}$$

where f denotes a column consisting of n regular functions $f_1, \ldots, f_n$ of the local coordinates ξ and η. In the indicated neighborhoods, equation (5) is then given by the functional equation

$$f(f(\xi, \eta), \zeta) = f(f(\xi, \zeta), \eta).$$

Forming the partial derivatives with respect to the components $\zeta_1, \ldots, \zeta_n$ of ζ we obtain

$$f_{\eta_l}(f(\xi, \eta), \zeta) = \sum_{r=1}^{n} f_{\xi_r}(f(\xi, \zeta), \eta) f_{r\eta_l}(\xi, \zeta), \qquad (l = 1, \ldots, n) \tag{7}$$

where the subscripts ξ_r and η_l denote the derivatives of f with respect to the components of the first and second argument of f respectively.

Introducing the matrices

$$f_\xi(\xi, \eta) = \left(\frac{\partial f_k(\xi, \eta)}{\partial \xi_l}\right), \qquad f_\eta(\xi, \eta) = \left(\frac{\partial f_k(\xi, \eta)}{\partial \eta_l}\right),$$

equation (7) is transformed into

$$f_\eta(f(\xi, \eta), \zeta) = f_\xi(f(\xi, \zeta), \eta) f_\eta(\xi, \zeta). \tag{8}$$

Inserting $\zeta = 0$ means $\mathfrak{z} = \mathfrak{e}$ and $\mathfrak{x}\mathfrak{z} = \mathfrak{x}$. Hence, in view of (6), equation (8) yields the fundamental formula

$$f_\eta(\rho, 0) = f_\xi(\xi, \eta) f_\eta(\xi, 0). \tag{9}$$

Since $\mathfrak{x}\mathfrak{y} = \mathfrak{r}$ can be solved for $\mathfrak{y} = \mathfrak{x}^{-1}\mathfrak{r}$, the Jacobian determinant $|f_\eta(\xi, \eta)|$ does not vanish at any point in the neighborhoods under consideration. From (9) we therefore conclude

$$(f_\eta(\xi, 0))^{-1} = (f_\eta(\rho, 0))^{-1} f_\xi(\xi, \eta). \tag{10}$$

We now use the entries of the matrix

$$(f_\eta(\xi, 0))^{-1} = P = (p_{kl})$$

as coefficients of the n linear differential forms

$$\delta_k = \sum_{l=1}^{n} p_{kl}\, d\xi_l, \qquad (k = 1, \ldots, n).$$

We will show that each of the integrability conditions

$$\frac{\partial p_{kl}}{\partial \xi_m} = \frac{\partial p_{km}}{\partial \xi_l}, \qquad (k, l, m = 1, \ldots, n) \tag{11}$$

is satisfied. Differentiating (9) with respect to η_j, $(j = 1, \ldots, n)$, and then putting $\eta = 0$, we obtain

$$\sum_{m=1}^{n} f_{\eta\xi_m}(\xi, 0) f_{m\eta_j}(\xi, 0) = f_{\xi\eta_j}(\xi, 0) f_\eta(\xi, 0), \qquad (j = 1, \ldots, n).$$

Denoting the inverse of P by

$$P^{-1} = Q = (q_{kl}) = f_\eta(\xi, 0),$$

we derive the explicit form

$$\sum_{m=1}^{n} q_{kl\xi_m} q_{mj} = \sum_{r=1}^{n} q_{kj\xi_r} q_{rl}, \qquad (k, l, j = 1, \ldots, n).$$

Using the fact that P and Q are mutually inverse we conclude that

$$q_{rs\xi_m} = \sum_{j,t=1}^{n} q_{rj\xi_t} q_{ts} p_{jm}, \qquad (r, s, m = 1, \ldots, n).$$

$$\sum_{r,s=1}^{n} p_{kr} q_{rs\xi_m} p_{sl} = \sum_{r,s=1}^{n} p_{kr} q_{rs\xi_l} p_{sm}, \qquad (k, m, l = 1, \ldots, n). \tag{12}$$

On the other hand, the relation $PQ = E$ also yields

$$(dP)Q + P(dQ) = 0, \qquad dP = -P(dQ)P.$$

Hence

$$p_{kl\xi_m} = -\sum_{r,s=1}^{n} p_{kr} q_{rs\xi_m} p_{sl}.$$

Equation (12) therefore implies (11). Hence, the differential forms δ_k are, indeed, integrable.

Integration of

$$dz = P d\xi = (f_\eta(\xi, 0))^{-1}\, d\xi \tag{13}$$

in a neighborhood of $\mathfrak{x}_0$ on $\mathfrak{R}$ now yields n functions $z_1, \ldots, z_n$ regular with respect to the local coordinates ξ. Let $\mathfrak{x}_1$ be another point of $\mathfrak{R}$ and let the column τ represent local coordinates in a neighborhood of $\mathfrak{x}_1$. If the neighborhoods of $\mathfrak{x}_0$ and $\mathfrak{x}_1$ have a nonempty intersection, $\mathfrak{D}$, then the respective local coordinates ξ and τ are related by a biregular equation $\tau = \phi(\xi)$. Putting $\phi(\rho) = \lambda$, the column

$$\lambda = \phi(f(\xi, \eta)) = g(\tau, \eta)$$

is seen to be regular with respect to τ and η, and we derive the equations

$$d\tau = \phi_\xi(\xi)\, d\xi, \qquad \lambda_\eta = g_\eta(\tau, \eta) = \phi_\xi(\rho) f_\eta(\xi, \eta).$$

In particular, for $\eta = 0$, we obtain

$$(g_\eta(\tau, 0))^{-1}\, d\tau = (f_\eta(\xi, 0))^{-1}\, d\xi.$$

Thus it is shown that the differential dz is invariant under the coordinate transformation from ξ to τ in $\mathfrak{D}$. Hence this differential is well-defined throughout $\mathfrak{R}$. Now if C denotes an arbitrary path of integration on $\mathfrak{R}$, then the n integrals defined by

$$I(C) = \int_C dz$$

are uniquely determined and remain unchanged when C is replaced by a homotopic path.

Now let C_1 be a path from $\mathfrak{e}$ to a point $\mathfrak{x}_1$ and let $\mathfrak{y} = \mathfrak{x}_2$ be located sufficiently close to $\mathfrak{e}$. The mapping $\mathfrak{r} = \mathfrak{x}\mathfrak{x}_2$ then transforms C_1 into the image curve $C_1^* = C_1\mathfrak{x}_2$ which joins $\mathfrak{x}_2$ to the point $\mathfrak{x}_1\mathfrak{x}_2 = \mathfrak{x}_3$. In view of (6) and (10), we obtain

$$(f_\eta(\rho, 0))^{-1}\, d\rho = (f_\eta(\xi, 0))^{-1}\, d\xi,$$

and hence

$$I(C_1) = I(C_1^*). \tag{14}$$

Since $\mathfrak{R}$ is compact this formula remains valid for arbitrary $\mathfrak{x}_2$. If C_2 is any curve joining $\mathfrak{e}$ to $\mathfrak{x}_2$, we compose C_2 and C_1^* and obtain a curve C_3 joining $\mathfrak{e}$ to $\mathfrak{x}_3$. The curve C_3 may be deformed without changing the integral. From (14) we derive

$$I(C_1) + I(C_2) = I(C_3). \tag{15}$$

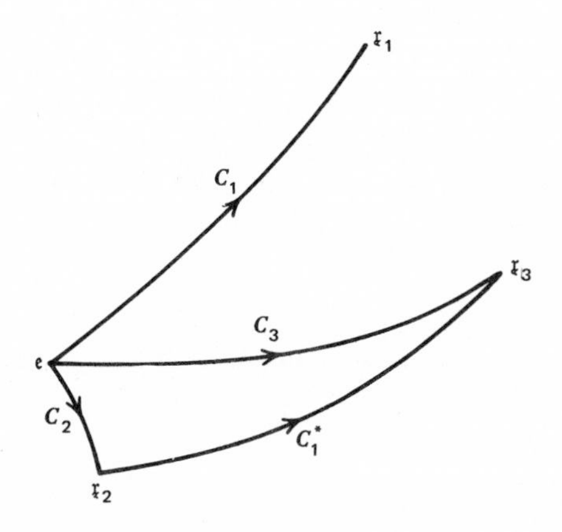

Figure 4

Here the local coordinates of $\mathfrak{x}_3$ are regular functions of those of $\mathfrak{x}_1$ and $\mathfrak{x}_2$. We will show in a moment that for variable $\mathfrak{x}$, $\mathfrak{y}$, and $\mathfrak{x}\mathfrak{y} = \mathfrak{r}$ the ratios of all projective coordinates of $\mathfrak{r}$ can be expressed rationally in terms of the projective coordinates of $\mathfrak{x}$ and $\mathfrak{y}$. Furthermore, equation (15) will turn out to be a generalization of the addition theorem for elliptic integrals of the first kind.

First, we extend the arguments which led to the $\wp$-function in Chapter 1 (Volume I) to the present case. For that purpose we introduce the universal covering space $\mathfrak{R}^*$ of $\mathfrak{R}$ by considering all curves on $\mathfrak{R}$ starting at $\mathfrak{e}$ and leading to an arbitrary point $\mathfrak{x}$ of $\mathfrak{R}$. To every class of homotopic curves with endpoint $\mathfrak{x}$ we assign a point $\mathfrak{x}^*$ of $\mathfrak{R}^*$ lying above $\mathfrak{x}$. The neighborhoods on $\mathfrak{R}^*$ are introduced in an obvious manner. The integral

$$z(\mathfrak{x}^*) = \int_{\mathfrak{e}}^{\mathfrak{x}^*} dz \tag{16}$$

evidently defines n single-valued functions $z_1, \ldots, z_n$ which are independent of the path of integration joining $\mathfrak{e}$ and $\mathfrak{x}^*$. Thus a mapping of $\mathfrak{R}^*$ into $\mathfrak{Z}$ is defined, and we will now study it in more detail. Again let ξ denote local coordinates in a neighborhood of the projection $\mathfrak{x}$ of $\mathfrak{x}^*$. In view of (13), the Jacobian determinant of the mapping in question is not equal to zero throughout this neighborhood. Hence the inverse mapping is also locally regular. Because of the compactness of $\mathfrak{R}$ we conclude, as in Section 7 of Chapter 1 (Volume I), that $\mathfrak{R}^*$ is mapped onto the whole of $\mathfrak{Z}$. Furthermore, if we have $z(\mathfrak{x}^*) = z(\mathfrak{y}^*)$ for any two points $\mathfrak{x}^*$ and $\mathfrak{y}^*$ of $\mathfrak{R}^*$, we select curves C^* and D^* on $\mathfrak{R}^*$ joining $\mathfrak{e}$ with $\mathfrak{x}^*$ and $\mathfrak{y}^*$. Their images C and D in $\mathfrak{Z}$ have endpoints 0 and z in common. Hence C and D are homotopic on $\mathfrak{Z}$ because $\mathfrak{Z}$ is simply connected. Since the closed curve $C^{-1}D$ can be contracted to z while the point z is kept fixed and since the mapping of $\mathfrak{R}^*$ onto $\mathfrak{Z}$ is locally invertible, it follows that the endpoints $\mathfrak{x}^*$ and $\mathfrak{y}^*$ of C^* and D^* must also coincide. Hence the mapping of $\mathfrak{R}^*$ onto $\mathfrak{Z}$ defined by (16) is biregular.

Now let $\mathfrak{x}^*$ and $\mathfrak{y}^*$ be any two points of $\mathfrak{R}^*$ located above the same projection $\mathfrak{x}$. We obtain

$$\omega = z(\mathfrak{y}^*) - z(\mathfrak{x}^*) = \int_{\mathfrak{e}}^{\mathfrak{e}^*} dz \tag{17}$$

for a point $\mathfrak{e}^*$ of $\mathfrak{R}^*$ lying above $\mathfrak{e}$. Conversely, if $\mathfrak{e}^*$ is any point of $\mathfrak{R}^*$ with projection $\mathfrak{e}$, we compose the projections of two curves on $\mathfrak{R}^*$ leading from $\mathfrak{e}$ to $\mathfrak{e}^*$ and from $\mathfrak{e}$ to $\mathfrak{x}^*$ and obtain the projection of a curve on $\mathfrak{R}^*$ joining $\mathfrak{e}$ with a point $\mathfrak{y}^*$ over $\mathfrak{x}$. This point $\mathfrak{y}^*$ satisfies (17). The columns ω obtained for arbitrary points $\mathfrak{e}^*$ are called *periods*. They form an additive group Ω which is isomorphic to the homotopy group of $\mathfrak{R}$. Obviously the group Ω is commutative, hence the same is true of the homotopy group.

Since the points $\mathfrak{e}^*$ have no accumulation point on $\mathfrak{R}^*$, the continuity of the mapping of $\mathfrak{Z}$ onto $\mathfrak{R}^*$ implies that the periods, ω, also have no accumulation point on $\mathfrak{Z}$. The vector group Ω is therefore discrete and hence a lattice. Denoting $\mathfrak{Z}/\Omega$ by $\mathfrak{T}$, we obtain a biregular mapping of $\mathfrak{R}$ onto $\mathfrak{T}$. This implies the compactness of $\mathfrak{T}$. But the quotient $\mathfrak{T} = \mathfrak{Z}/\Omega$ is compact only if Ω has a basis of $2n$ linearly independent periods over the reals. Hence $\mathfrak{T}$ is a period torus. Two points $\mathfrak{x}^*$ and $\mathfrak{y}^*$ of $\mathfrak{R}^*$ have the same projection if and only if the congruence relation $z(\mathfrak{x}^*) \equiv z(\mathfrak{y}^*)$ holds with respect to the period group.

Finally, let $x_0, \ldots, x_m$ be homogeneous coordinates in the projective space $\mathfrak{Q}_m$ where $\mathfrak{R}$ is defined and let x_0 be different from zero at the point $\mathfrak{e}$. In view of (13) and (16), the m ratios $x_k/x_0 = t_k$, $k = 1, \ldots, m$ are regular functions of z, at least in a neighborhood of $z = 0$. Given any curve on $\mathfrak{Z}$ starting from 0, the local coordinates at any point of the image of that curve on $\mathfrak{R}^*$ are regular functions of z. Since $\mathfrak{R}$ has no singularities, the ratios t_k are meromorphic functions of any local coordinates at any point of $\mathfrak{R}$. Hence these ratios are also meromorphic functions of z. Furthermore, since for any period ω the two points z and $z + \omega$ yield the same point on $\mathfrak{R}$, each ratio t_k belongs to the Abelian function field K corresponding to the lattice Ω. The field K cannot be degenerate since, otherwise, Theorems 5 and 6 of Section 11 would imply that the dimension of the algebraic variety $\mathfrak{R}$ is less than n. Finally, if we express $x_0, \ldots, x_m$ as equivariant Jacobian functions we conclude that $\mathfrak{R}$ is a Picard variety corresponding to the Abelian function field K. Thus the proof of Theorem 2 is complete.

Using (16) and (17), formula (15) can be expressed as

$$z(\mathfrak{x}) + z(\mathfrak{y}) \equiv z(\mathfrak{x}\mathfrak{y}),$$

where $\mathfrak{x}$ and $\mathfrak{y}$ are any two points of $\mathfrak{R}$. According to Theorem 1, the inhomogeneous coordinates t_k, $(k = 1, \ldots, m)$, at the point $\mathfrak{x}\mathfrak{y}$ are rational functions of the inhomogeneous coordinates at $\mathfrak{x}$ and $\mathfrak{y}$. Thus the group operations are birational transformations on $\mathfrak{R}$, which was not assumed in Theorem 2. This result elicits the following remark. Let $\mathfrak{R}$ be an irreducible nonsingular algebraic variety which need not be a Picard variety. A function on $\mathfrak{R}$ is called *meromorphic* if it is meromorphic with respect to any system of local coordinates. Evidently, every element of the function field K corresponding to the given algebraic variety is indeed meromorphic on $\mathfrak{R}$. Conversely, every meromorphic function on $\mathfrak{R}$ turns out to be an element of K. This converse requires a more detailed investigation which we will not pursue here. In this connection we should refer to Theorem 4 in Section 1 of Chapter 2 (Volume I) and Theorem 7 in Section 14 of Chapter 1 (Volume I).

Theorem 2 is closely connected to a result noted without proof by Weierstrass in his lectures on Abelian functions. This result is concerned with meromorphic functions of n variables which satisfy an algebraic addition theorem of the form indicated in Theorem 1. The assertion is that such meromorphic functions are either Abelian functions or singular Abelian functions. Here it should be remarked that the notion of singular Abelian functions is distinct from the notion of degenerate Abelian functions previously introduced in the present chapter. In the case $n = 1$, these singular functions are precisely the singular elliptic functions treated in Section 16 of Chapter 1 (Volume I). For arbitrary n one is led to the so-called *quasi-Abelian functions* which were investigated in more detail by Severi. We will not pursue this subject further.

Finally we will discuss a property which characterizes the n differentials $dz_1, \ldots, dz_n$ introduced in (13). Again denote local coordinates in a neighborhood of a given point of the Picard variety $\mathfrak{R}$ by $\xi_1, \ldots, \xi_n$ and form the Jacobian matrix

$$\left(\frac{\partial z_k}{\partial \xi_l}\right) = \left(\frac{\partial \xi_k}{\partial z_l}\right)^{-1} = P.$$

Since each ξ_k, $(k = 1, \ldots, n)$, is a ratio of the projective coordinates $x_0, \ldots, x_m$, and since the partial derivatives of an Abelian function in K also belong to K, each entry of the matrix P is an element of K. The coefficient of each $d\xi_l$, $(l = 1, \ldots, n)$ in the n differentials

$$dz_k = \sum_{l=1}^{n} \frac{\partial z_k}{\partial \xi_l} d\xi_l, \qquad (k = 1, \ldots, n)$$

is therefore an element of the algebraic function field corresponding to $\mathfrak{R}$ and is regular with respect to any system of local coordinates. Hence these differentials satisfy exactly the condition that was used in Chapter 4 (Volume II) to define differentials of the first kind for algebraic function fields of one variable. Therefore the present differentials will also be called *differentials of the first kind*. This definition only requires the assumption that $\mathfrak{R}$ is an irreducible nonsingular algebraic variety in projective space. If $\mathfrak{R}$ is a Picard variety of dimension $n = 1$ we already know that, up to an arbitrary constant factor, there exists only one differential of the first kind. This was investigated in detail in Chapter 1 (Volume I). Now let n be arbitrary and let dz_0 be a differential of the first kind for the Picard variety $\mathfrak{R}$. Then each partial derivative

$$\frac{\partial z_0}{\partial z_l} = \sum_{k=1}^{n} \frac{\partial z_0}{\partial \xi_k} \frac{\partial \xi_k}{\partial z_l}, \qquad (l = 1, \ldots, n)$$

is a regular function of z everywhere, and moreover it is an Abelian function contained in K. Hence these partial derivatives are constant, and we obtain

$$dz_0 = c_1\, dz_1 + \cdots + c_n\, dz_n$$

with constant coefficients $c_1, \ldots, c_n$. On the other hand, since the determinant $|P|$ is different from zero, the n differentials $dz_1, \ldots, dz_n$ are linearly independent over the complex numbers. Therefore, they form a basis for the complex vector space of differentials of the first kind. The n integrals of the first kind $z_1, \ldots, z_n$ then map the covering space $\mathfrak{R}^*$ of $\mathfrak{R}$ biregularly onto the entire complex space $\mathfrak{Z}$, thus attaining the same result as in the special case $n = 1$. At the same time we obtain a uniformization of $\mathfrak{R}$ in the large. Uniformization by automorphic functions as well as the investigation of differentials of the first kind has yielded complete and satisfactory results concerning algebraic functions of one variable and arbitrary genus. These were outlined in Chapters 3 and 4. A complete extension of these theories to several variables is not possible since, in particular, we lack the methods of conformal mappings. But to a lesser extent some of the arguments of Chapter 3 can be generalized. This will be the subject of Chapter 6.

6

Modular functions of several variables

1. Automorphic functions in several variables

In searching for a generalization of automorphic functions to the case of several variables, three types of difficulties appear. These difficulties are of an algebraic, analytic, and topological nature.

The algebraic difficulties are connected with the fact that a birational transformation of an algebraic variety of dimension $n > 1$ need not result in a mapping which is one to one everywhere. After preliminary investigations by Zariski, Hironaka recently proved that every irreducible algebraic variety is birationally equivalent to a nonsingular algebraic variety. We could carry out our future considerations with such a nonsingular variety provided these considerations permit restriction to birational regular transformations.

In Chapter 2 (Volume I), for $n = 1$, we started with an algebraic function w that was obtained as the solution of an irreducible algebraic equation $P(w, z) = 0$. Next we formed the universal covering surface, $\mathfrak{S}$, of the corresponding Riemann surface $\mathfrak{R}$. By the uniformization theorem, which had been proved at that time, $\mathfrak{S}$ could be mapped conformally onto the whole s-sphere, the whole s-plane or the unit disk $|s| < 1$, depending on whether the genus, p, of $\mathfrak{R}$ is $p = 0$, $p = 1$, or $p > 1$. This mapping of $\mathfrak{S}$ onto a schlicht normal domain resulted in an isomorphism of the fundamental group of $\mathfrak{R}$ and a certain discontinuous group, Γ, of fractional linear transformations of the uniformizing variable s. Then the field of all automorphic functions with respect to Γ was precisely the algebraic function field of one variable generated by z and w. The proof of the uniformization theorem essentially required potential theory and the Dirichlet principle. In searching for a generalization to several variables, we find unsurmountable difficulties in this direction. Namely, for $n = 1$ potential theory is merely another formulation of a branch of analytic function theory of one variable, whereas for $n > 1$ this is no longer the case because of the nature of the relevant differential equations.

In order to bypass these difficulties, we only consider irreducible nonsingular algebraic varieties $\mathfrak{M}$ whose universal covering space, $\mathfrak{M}^*$, can be mapped biregularly onto a schlicht domain $\mathfrak{G}$ in complex n-space $\mathfrak{Z}$. Now even the topological classification of the variety $\mathfrak{M}$ and the domain $\mathfrak{G}$ give rise to very difficult questions. In the case $n = 1$, the value of the genus p

alone determines which normal domain is under consideration. In addition, two compact Riemann surfaces are homeomorphic if and only if they have the same genus. However, even in the case $n = 2$ the problem of determining a complete system of topological invariants is unsolved.

For the Picard varieties considered in the last two sections of Chapter 5 we had $\mathfrak{G} = \mathfrak{Z}$. We now study the case where $\mathfrak{G}$ is a proper subdomain of $\mathfrak{Z}$. In addition, if $n = 1$, then from the simple connectivity of $\mathfrak{M}^*$ and $\mathfrak{G}$ it follows by the Riemann mapping theorem that $\mathfrak{G}$ can be mapped conformally onto the unit disk $\mathfrak{E}$. Moreover, the group of all biregular mappings of $\mathfrak{E}$ onto itself consists of all fractional linear transformations

$$w = \frac{\bar{p}z + \bar{q}}{qz + p} \qquad (p\bar{p} - q\bar{q} = 1). \tag{1}$$

These form a Lie group depending on three real parameters. Thus every simply connected proper subdomain of the complex plane also admits a real three parameter group of biregular self-mappings. However, for $n > 1$ there are bounded simply connected domains in $\mathfrak{Z}$, the so-called *rigid domains*, which permit no biregular self-mapping besides the identity. On the other hand, simple examples can be given of domains which possess such mappings. The domains, $\mathfrak{G}$, which we consider and which are biregularly equivalent to the universal covering space $\mathfrak{M}^*$, are not rigid if $\mathfrak{M}$ itself is not simply connected. This is due to the fact that the mapping of $\mathfrak{M}^*$ onto $\mathfrak{G}$ provides a faithful representation of the fundamental group of $\mathfrak{M}$ by means of biregular self-mappings of $\mathfrak{G}$.

We cannot commence the foundation of the theory of automorphic functions in several variables by the methods of algebraic geometry because of the largely unsolved difficulties indicated above. Instead, we start with a domain $\mathfrak{G}$ in complex space $\mathfrak{Z}$ and consider a group, Γ, of biregular self-mappings

$$w = \gamma(z) \tag{2}$$

of $\mathfrak{G}$. Here w stands for a column of n complex functions $w_1, \ldots, w_n$. A meromorphic function $f(z)$ defined on $\mathfrak{G}$ is said to be an *automorphic function with respect to* Γ provided the functional equation $f(w) = f(\gamma(z)) = f(z)$ holds for all γ in Γ. Obviously, these automorphic functions form a field. Let Γ be discontinuous in $\mathfrak{G}$ so that, by the definition given in Section 3 of Chapter 3 (Volume II), for each point ζ of $\mathfrak{G}$ the totality of image points $\gamma(\zeta)$ has no limit point in $\mathfrak{G}$. Since every domain $\mathfrak{G}$ in $\mathfrak{Z}$ can be covered by countably many closed n-disks contained within $\mathfrak{G}$, it follows immediately that Γ has at most countably many elements γ. In the sequel we also make the essential assumption that $\mathfrak{G}$ is bounded. Then the n mapping functions $w_1(z), \ldots, w_n(z)$ are uniformly bounded with respect to γ on $\mathfrak{G}$, and by the generalization of the Cauchy integral formula given by (6) in Section 3 of Chapter 5 these functions are also uniformly continuous in γ at every point z of $\mathfrak{G}$. We define the Jacobian determinant

$$j_\gamma = j_\gamma(z) = \left| \frac{\partial w_k}{\partial z_l} \right|$$

which is a regular, nowhere vanishing function of z on $\mathfrak{G}$ for every group element γ. Now, in analogy with Section 5 of Chapter 3 (Volume II), we can easily generalize the definition and the proof of convergence of the Poincaré series.

Theorem 1: The series

$$S = \sum_\gamma j_\gamma \bar{j}_\gamma$$

is uniformly convergent on every compact subset of $\mathfrak{G}$.

Proof: Let ζ be a fixed point in $\mathfrak{G}$. The elements of Γ with fixed point ζ form a finite subgroup, Δ, whose order will be denoted by h. Let r be a positive number such that the n-disk $\mathfrak{K}$ given by $|z - \zeta| < r$ lies in $\mathfrak{G}$. Let $\mathfrak{K}_\gamma$ be its image under (2). First, we show that for a sufficiently small r two images $\mathfrak{K}_\alpha$ and $\mathfrak{K}_\beta$ have a point in common only if the group element $\beta^{-1}\alpha = \delta$ belongs to Δ. Namely, if there is a sequence of $r \to 0$ and a corresponding sequence $\delta = \delta_r$ of elements of Γ such that $\mathfrak{K}_\delta$ and $\mathfrak{K}$ have a point ζ_r in common for each δ_r, then it follows from the uniform continuity of $w(z)$ with respect to γ at ζ that $\delta_r(\zeta)$ tends to ζ. However, since Γ is discontinuous in $\mathfrak{G}$, we must have $\delta_r(\zeta) = \zeta$ for sufficiently small r; therefore, δ belongs to Δ.

If we introduce real Euclidean coordinates by taking the real and imaginary parts of z, then the ratio of the volume elements corresponding to mapping (2) is given by

$$\frac{(dv)_\gamma}{dv} = j_\gamma \bar{j}_\gamma = I_\gamma(z).$$

If V is the inner content of $\mathfrak{G}$ and r is sufficiently small we have

$$hV \geqslant \sum_\gamma \int_{\mathfrak{K}_\gamma} dv = \sum_\gamma \int_{\mathfrak{K}} (dv)_\gamma = \sum_\gamma \int_{\mathfrak{K}} I_\gamma(z)\, dv \geqslant (\pi r^2)^n \sum_\gamma I_\gamma(\zeta).$$

Here we used Theorem 1 of Section 7 of Chapter 2 (Volume I) to estimate the last integral. Thus the series S is convergent throughout $\mathfrak{G}$. Furthermore, if $|\eta - \zeta| < r/2$ then the uniform convergence of S in the domain $|z - \eta| < r/2$ follows from the inequality

$$\int_{\mathfrak{K}} I_\gamma(z)\, dv > \int_{|z-\eta|<r/2} I_\gamma(z)\, dv \geqslant \left(\frac{\pi r^2}{4}\right)^n I_\gamma(\eta).$$

Finally, by a routine argument we obtain the complete statement of the theorem.

The definition of the special Poincaré series which was given in Section 5 of Chapter 3 (Volume II) may now be carried over without any difficulty. Namely, we set

$$f_k(z) = \sum_{\gamma} (j_\gamma(z))^k \qquad (k = 2, 3, \ldots). \tag{3}$$

As a consequence of Theorem 1, $f_k(z)$ is regular throughout $\mathfrak{G}$. Furthermore, let μ, ν be two elements of Γ and $\mu\nu = \lambda$. Then by use of the multiplication theorem for Jacobian determinants we obtain the formula

$$j_{\mu\nu}(z) = j_\lambda(z) = j_\mu(\nu(z)) j_\nu(z) \tag{4}$$

from $\lambda(z) = \mu(\nu(z))$. Since, for a fixed ν, $\gamma\nu$ runs over all elements of Γ when γ runs over all of Γ, the transformation formula

$$f_k(\nu(z)) = (j_\nu(z))^{-k} f_k(z) \qquad (k = 2, 3, \ldots)$$

holds. Accordingly, by an *automorphic form of weight k corresponding to* Γ we mean a regular function $f(z)$ on $\mathfrak{G}$ which satisfies the functional equation

$$f(\gamma(z)) = (j_\gamma(z))^{-k} f(z)$$

for each γ in Γ. Here k is assumed to be integral.

In particular, if $n = 1$ and Γ is a hyperbolic polygon group, then the deeper investigation in Section 6 of Chapter 3 (Volume II) shows that every automorphic function corresponding to Γ can be expressed as a rational combination of finitely many fixed $f_k(z)$. Here z must be subjected to a preliminary fractional linear transformation. The corresponding result cannot be proved for arbitrary n and Γ under the present, very general, hypotheses. Therefore, we will at first abandon the restriction to the special Poincaré series, and in analogy with the functions treated earlier in the case $n = 1$, introduce the general Poincaré series by the Ansatz

$$\Phi(z) = \sum_{\gamma} (j_\gamma(z))^k \phi(\gamma(z)). \tag{5}$$

Here k is an integer greater than 1 and $\phi(z)$ is any regular bounded function on $\mathfrak{G}$. Application of (4) and Theorem 1 shows that $\Phi(z)$ is also an automorphic form of weight k corresponding to Γ.

Theorem 2: There are $n + 1$ polynomials $\phi_0(z), \ldots, \phi_n(z)$ in the variables $z_1, \ldots, z_n$ and a number k such that the corresponding Poincaré series $\Phi_0(z), \ldots, \Phi_n(z)$ formed by means of formula (5) satisfy the following conditions: the function $\Phi_0(z)$ is not identically zero and the n quotients

$$g_l(z) = \frac{\Phi_l(z)}{\Phi_0(z)} \qquad (l = 1, \ldots, n)$$

are analytically independent.

Proof: We consider the totality of points, $\gamma(\zeta)$, that are equivalent with respect to Γ to an arbitrarily selected point ζ of $\mathfrak{G}$. As a result of Theorem 1, the positive values $I_\gamma(z) = j_\gamma \bar{j}_\gamma$ with $z = \zeta$ take on a maximum for at least one γ, say for $\gamma = \nu$. Thus the inequality

$$I_{\gamma\nu}(\zeta) \leqslant I_\nu(\zeta)$$

is satisfied for all γ; by (4), this is equivalent to the statement

$$I_\gamma(\eta) \leqslant 1, \qquad \eta = \nu(\zeta). \tag{6}$$

We select all those elements $\gamma = \mu$ of Γ for which the condition $I_\mu(z) = 1$ is satisfied identically in z. This means that the mapping $w = \mu(z)$ preserves content. By Theorem 1 these elements form a subgroup, M, of finite order, m, in Γ. Then the function $j_\mu(z)$ is a constant of absolute value 1; therefore, by (4) and the group property of M, an mth root of unity. In particular, $j_\varepsilon(z) = 1$ for the identity, ε, of Γ.

Clearly, conditions (6) hold with equality for all $\gamma = \mu$ in M. However, in addition, they may still hold for finitely many γ which do not belong to M. We enumerate the elements of Γ by decreasing values of $I_\gamma(\eta)$. Here $\gamma_1, \ldots, \gamma_m$ are taken to be the elements of M, and we assume that $\gamma = \gamma_k$, $I_\gamma(\eta) = 1$ $(k = 1, \ldots, s)$, $I_\gamma(\eta) < 1$ $(k = s + 1, \ldots, t)$, $\sum_{k>t} I_\gamma(\eta) < 1$ for certain natural integers s, t and $m \leqslant s \leqslant t$.

If $s > m$, then the regular function $j_\gamma(z)$ with $\gamma = \gamma_s$ is not constant in z. As a result of the maximum principle, the continuity of $I_\gamma(z)$, and Theorem 1, we can find a point ζ^* in $\mathfrak{G}$ sufficiently close to η so that the maximum of the values $I_\gamma(\zeta^*)$ is greater than 1, but for $\gamma = \gamma_k$ and $k > s$ the inequality $I_\gamma(\zeta^*) < 1$ again holds. If this maximum is attained for $\gamma = \rho$ then from (4) we obtain the following conditions which are analogous to (6)

$$I_\gamma(\eta^*) \leqslant 1, \qquad \eta^* = \rho(\zeta^*).$$

But now, equality holds at most $s - 1$ times. In this way we can clearly reduce s to the value m.

We can consequently find a point ξ in $\mathfrak{G}$ at which the inequality $I_\gamma(\xi) < 1$ holds for all γ not belonging to M. Because of the continuity of $I_\gamma(z)$ and Theorem 1, there is a positive number $c < 1$ and a sufficiently small n-disk $\mathfrak{K}$ in $\mathfrak{G}$ defined by $|z - \xi| \leqslant r$ such that the inequalities

$$I_\gamma(z) < c \qquad (\gamma = \gamma_k;\, k = m + 1, m + 2, \ldots) \tag{7}$$

all hold throughout this n-disk. Since the function $\gamma(z) - z$ does not vanish identically for $\gamma \neq \varepsilon$ we can also assume ξ chosen so that ξ is not a fixed point for the $m - 1$ content preserving mappings $w = \gamma_k(z)$ $(k = 2, \ldots, m)$.

In (5) we take $k = ml$ with $l = 2, 3, \ldots$, and define

$$\Phi(z, l) = \sum_{\gamma} (j_\gamma(z))^{ml} \phi(\gamma(z)).$$

Then, since

$$(j_\mu(z))^m = 1$$

for the m elements μ of M, it follows from (7) that the formula

$$\lim_{l \to \infty} \Phi(z,l) = \sum_{\mu} \phi(\mu(z)) \tag{8}$$

holds uniformly on $\mathfrak{K}$.

We select any polynomial $P(z)$ which satisfies the m conditions

$$P(\xi) = 1, \qquad P(\gamma_k(\xi)) = 0; \qquad (k = 2, \ldots, m)$$

and put

$$\begin{aligned} &\phi_0(z) = m^{-1}, \qquad \phi_k(z) = (z_k - \xi_k)(P(z))^2; \qquad && (k = 1, \ldots, n) \\ &\Phi_k(z, l) = \sum_{\gamma} (j_\gamma(z))^{ml} \phi_k(\gamma(z)); && (k = 0, \ldots, n). \end{aligned} \tag{9}$$

By (8), $\Phi_0(z, l)$ tends to 1 uniformly on $\mathfrak{K}$ as $l \to \infty$, and is therefore unequal to 0 on $\mathfrak{K}$ for all sufficiently large l. We define

$$g_k(z, l) = \frac{\Phi_k(z, l)}{\Phi_0(z, l)} \qquad (k = 1, \ldots, n).$$

From (8) we obtain the formula

$$\lim_{l \to \infty} g_k(z, l) = \sum_{\mu} \phi_k(\mu(z)) = \psi_k(z) \qquad (k = 1, \ldots, n)$$

uniformly on $\mathfrak{K}$. By Weierstrass's theorem the partial derivatives of the $g_k(z, l)$ now tend uniformly to those of the $\psi_k(z)$ on the interior of $\mathfrak{K}$. From (9), however, the Jacobian matrix of $\psi_1, \ldots, \psi_n$ with respect to $z_1, \ldots, z_n$ is seen to be just the unit matrix for $z = \xi$. Therefore, the Jacobian determinant of $g_1(z, l), \ldots, g_n(z, l)$ is unequal to 0 at $z = \xi$ for sufficiently large l. For such a value of l, the $n + 1$ functions $\Phi_k(z) = \Phi_k(z, l)$ $(k = 0, \ldots, n)$ fulfill the required conditions, and Theorem 2 is thereby proved.

In the sequel let $g_1, \ldots, g_n$ be any n automorphic functions with respect to Γ which are algebraically independent. The existence of such a system of functions is guaranteed by Theorem 2. The next question about the field, K, of all functions automorphic with respect to Γ is whether it has transcendence degree n and is a simple algebraic extension of the field of rational functions of $g_1, \ldots, g_n$. In Section 6 of Chapter 3 (Volume II) this question is answered in the affirmative for the case of hyperbolic polygon groups. We cannot expect a corresponding answer without some further restrictions on Γ since, for example, up to now Γ might only consist of the identity. In the investigation of automorphic functions of one variable, it was particularly important that we restricted ourselves mainly to the case of a compact fundamental region. We will now make a similar restriction.

In case $n = 1$ we can draw upon the device of noneuclidean geometry in order to construct normal fundamental regions of a particularly simple type. Namely, we do the following. If ζ is not a fixed point of Γ, then we select a point, $\gamma(z)$, from each equivalence class whose noneuclidean distance from ζ is as small as possible. Thus we obtain the normal fundamental region $\mathfrak{F}$ with center ζ. In particular, $\zeta = 0$ is not a fixed point if and only if the elements $\gamma \neq \varepsilon$ of Γ given by (1) all have coefficients $q \neq 0$. In the case at hand, this clearly means that the subgroup M consists of just the identity. If we now select $\zeta = 0$ as the center, it follows from (14) in Section 6 of Chapter 3 (Volume II) that $\mathfrak{F}$ is determined by the conditions

$$I_\gamma(z) \leqslant 1, \tag{10}$$

which also played a role just now in the proof of Theorem 2. This suggests that, for arbitrary n, we define a fundamental region $\mathfrak{F}$ for Γ by means of (10) if Γ contains no volume preserving mapping besides ε. Then we obtain $\mathfrak{F}$ if we select a point with maximal value for $I_\gamma(z)$ from each equivalence class. If M does not consist of the identity alone, then the region defined by (10) is not yet a fundamental region since it is taken into itself by all mappings in M. It is now easy, by another simple rule, to choose one of the m equivalent points, $\mu(z)$, so that a fundamental region is actually formed. We will not, however, pursue the matter since for our purposes a precise fundamental region will not be needed. Instead we only need a closed region $\mathfrak{B}$ in $\mathfrak{G}$ whose images $\mathfrak{B}_\gamma$ cover $\mathfrak{G}$; moreover, this covering need not necessarily be simple. We now make the assumption, which is essential for future considerations, that $\mathfrak{B}$ is compact in $\mathfrak{G}$.

With this assumption we can determine a point from each equivalence class so that these representatives have no limit point on the boundary of $\mathfrak{G}$. Namely, we select each point from $\mathfrak{B}$. Conversely, if the group, Γ, has this property, then there is a positive lower bound for the distances from the selected points to the boundary of $\mathfrak{G}$. Thus we can immediately find a compact region $\mathfrak{B}$ of the required type. We then say that $\mathfrak{G}/\Gamma$ is compact. Incidentally, it is easy to see by means of Theorem 1 that $\mathfrak{G}/\Gamma$ is compact if and only if the region defined by (10) is compact; this can then be taken as the region $\mathfrak{B}$.

Now let $\mathfrak{A}$ be any compact subset of $\mathfrak{G}$. We will show that only a finite number of image regions $\mathfrak{B}_\gamma$ can have a point in common with $\mathfrak{A}$. If this is not the case, then there is a sequence $\gamma_1, \gamma_2, \ldots$ of elements γ in Γ and a sequence of points $\zeta_1, \zeta_2, \ldots$ in $\mathfrak{B}$ such that the image points $\gamma_k(\zeta_k) = \eta_k$ $(k = 1, 2, \ldots)$ all lie in $\mathfrak{A}$. Because of the compactness of $\mathfrak{A}$ and $\mathfrak{B}$ we can establish that the η_k and ζ_k tend to limits η and ζ in $\mathfrak{A}$ and $\mathfrak{B}$. However, by means of the uniform continuity of the functions $\gamma_k(z)$, the points $\gamma_k(\zeta)$ then also tend to η, whereas Γ is discontinuous.

By the way, since $\mathfrak{G}$ is an open region in $\mathfrak{Z}$, it follows from the compactness of $\mathfrak{B}$ that Γ possesses infinitely many elements.

2. *Algebraic relations between automorphic functions*

First we present a simple generalization of Schwarz's lemma which will play a basic role in subsequent considerations.

Theorem 1: Let $\phi(z)$ be a power series which is convergent and has absolute value $\leqslant M$ in the n-disk $|z| \leqslant r$. If no terms of order $0, 1, \ldots, h-1$ appear in this power series, then we have

$$|\phi(z)| \leqslant M\left(\frac{|z|}{r}\right)^h \qquad (|z| \leqslant r).$$

Proof: The statement is correct for $z = 0$. For a given z satisfying $0 < |z| \leqslant r$ we introduce the following function of the single complex variable, λ,

$$\phi(\lambda z) = \psi(\lambda).$$

This function clearly has a representation as a convergent power series in the disk $|\lambda| \leqslant r/|z|$ and vanishes with order at least h at $\lambda = 0$. Thus the function

$$\chi(\lambda) = \lambda^{-h}\psi(\lambda)$$

is also regular in the interior of the same disk, and

$$|\chi(\lambda)| \leqslant M\,|\lambda|^{-h} \qquad \left(0 < |\lambda| \leqslant \frac{r}{|z|}\right).$$

Therefore by the maximum principle, we even have

$$|\chi(\lambda)| \leqslant M\left(\frac{|z|}{r}\right)^h \qquad \left(|\lambda| \leqslant \frac{r}{|z|}\right).$$

Since $\phi(z) = \psi(1) = \chi(1)$, the assertion follows. Incidentally, this statement was also derived in the same way in the proof of Theorem 2 in Section 7 of the previous chapter.

Now let $\mathfrak{B}$ be a compact subset of $\mathfrak{G}$ whose images $\mathfrak{B}_\gamma$ cover all of $\mathfrak{G}$. If $f(z)$ is a meromorphic function on $\mathfrak{G}$ and a an arbitrary point of $\mathfrak{G}$, then

$$f(z) = \frac{p_a(z)}{q_a(z)}, \tag{11}$$

where p_a and q_a are two power series in the variables $z_a = z - a$ which are regular and relatively prime throughout the closed n-disk $\mathfrak{K}_a$ given by $|z - a| \leqslant r_a$. We put $r_a/e = \rho_a$ and take $\mathfrak{L}_a$ to be the smaller concentric closed n-disk $|z - a| \leqslant \rho_a$. Now we can find finitely many points $a_1, \ldots, a_m$ such that the $\mathfrak{L}_a$ $(a = a_1, \ldots, a_m)$ cover $\mathfrak{B}$, and we let $\mathfrak{A}$ be the compact subset of $\mathfrak{G}$ formed from the union of all the $\mathfrak{K}_a$ $(a = a_1, \ldots, a_m)$. Clearly, $\mathfrak{A}$ contains $\mathfrak{B}$. Furthermore, we determine finitely many elements $\gamma = \gamma_1, \ldots, \gamma_p$ of Γ such that the corresponding p images $\mathfrak{B}_\gamma$, in turn, cover $\mathfrak{A}$. That this is possible follows from the considerations at the end of the previous section. We also let b stand for any one of the points $a_1, \ldots, a_m$ and form the intersection, $\mathfrak{D}(\gamma, a, b)$, of the regions $\mathfrak{K}_b$ and $(\mathfrak{K}_a)_\gamma$. Now let $f(z)$ be automorphic and consider those intersections which are nonvoid. If we put $z = \gamma(v)$, then

$$\frac{p_b(z)}{q_b(z)} = \frac{p_a(v)}{q_a(v)}$$

on the corresponding intersection; therefore, we also have

$$q_b(z) = j_{\gamma ab}(z) q_a(v), \tag{12}$$

where the function $j_{\gamma ab}(z)$ is a unit on every connected component of $\mathfrak{D}(\gamma, a, b)$. If instead of a single automorphic function we have finitely many such functions, then for sufficiently small radii r_a we can arrange that the $\mathfrak{K}_a$ work for all these functions.

Theorem 2: If $\mathfrak{G}/\Gamma$ is compact, then every set of $n + 1$ automorphic functions is algebraically dependent.

Proof: Let the $n + 1$ functions $f, f_1, \ldots, f_n$ all be automorphic. Corresponding to (11) and (12), we now set

$$f_k = \frac{p_{ka}}{q_{ka}}, \qquad q_{kb}(z) = j_{k\gamma ab}(z) q_{ka}(v).$$

We select two natural numbers μ and τ so that the following inequalities

$$|j_{\gamma ab}| < e^\mu, \qquad \left| \prod_{k=1}^{n} j_{k\gamma ab} \right| < e^\tau \tag{13}$$

are satisfied on all nonempty intersections $\mathfrak{D}(\gamma, a, b)$ with $\gamma = \gamma_1, \ldots, \gamma_p$

and $a, b = a_1, \ldots, a_m$. Furthermore, let

$$s = m\tau^n + 1 \tag{14}$$

and let $t = t_h$ $(h = 1, 2, \ldots)$ be the largest nonnegative integer such that

$$st^n < mh^n. \tag{15}$$

Thus

$$mh^n \leqslant s(t+1)^n < (s+1)(t+1)^n. \tag{16}$$

Since t tends to ∞ when h tends to ∞, we can select h so large that the inequality

$$s > \left(\frac{\mu s}{t} + \tau\right)^n m$$

follows from (14). By (15) we then have

$$\mu s + \tau t < h. \tag{17}$$

We set

$$Q_a = q_a^s \prod_{k=1}^n q_{ka}^t, \qquad I_{\gamma ab} = j_{\gamma ab}^s \prod_{k=1}^n j_{k\gamma ab}^t. \tag{18}$$

We form a polynomial $g(x, x_1, \ldots, x_n)$ in the $n + 1$ variables $x, x_1, \ldots, x_n$ which has undetermined coefficients and in which x appears with degree s and each of the other variables appears with degree t. Then the function

$$P_a = Q_a g(f, f_1, \ldots, f_n) \tag{19}$$

is always regular on $\mathfrak{R}_a$.

We now require that this function as well as all of its partial derivatives of orders $1, \ldots, h - 1$ vanish at the point $z = a$, and for each fixed a we obtain C homogeneous linear equations for the coefficients of the polynomial g where

$$C = \sum_{l=0}^{h-1} \binom{n + l - 1}{l} = \binom{n + h - 1}{n} = \prod_{k=1}^{n} \frac{h + k - 1}{k} < h^n.$$

If we take $a = a_1, \ldots, a_m$, then we obtain B such equations where

$$B = mC \leqslant mh^n,$$

whereas the number of coefficients is

$$A = (s + 1)(t + 1)^n.$$

As a result of (16), we have $A > B$. We can therefore construct a nonzero polynomial $g(x, x_1, \ldots, x_n)$ which satisfies all the required conditions and does not vanish identically.

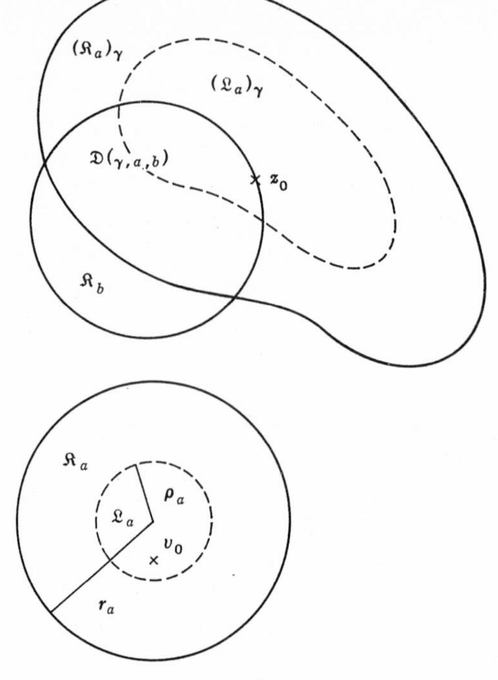

Figure 5

Let the maximum M of $|P_a(z)|$ for $a = a_1, \ldots, a_m$ and z on $\mathfrak{K}_a$ be attained at $z = z_0$ and $a = b$. Consequently, we have

$$|P_b(z_0)| = M, \qquad |P_a(z)| \leqslant M.$$

The point z_0 lies in $\mathfrak{A}$ and, therefore, also in a domain $(\mathfrak{L}_a)_\gamma$. We put $z_0 = \gamma(v_0)$ (Figure 5). Then Theorem 1 yields the estimate

$$|P_a(v_0)| \leqslant M\left(\frac{\rho_a}{r_a}\right)^h = Me^{-h}.$$

From (12), (13), (18), and (19) we obtain

$$P_b(z_0) = I_{\gamma ab}(z_0)P_a(v_0),$$
$$M \leqslant Me^{\mu s + \tau t - h}$$

from which, by (17), the equations $M = 0$ and consequently,

$$g(f, f_1, \ldots, f_n) = 0 \tag{20}$$

follow. Thus the theorem is proved.

Theorem 3: If $\mathfrak{G}/\Gamma$ is compact, then the field of automorphic functions, K, is an algebraic function field of transcendence degree n.

Proof: By Theorem 2 of Section 1, we can select n algebraically independent automorphic functions $f_1, \ldots, f_n$. Theorem 2 of the present section assures us that any other function f in K satisfies an algebraic equation

$$g_0 f^\nu + g_1 f^{\nu-1} + \cdots + g_\nu = 0 \tag{21}$$

where $g_0, g_1, \ldots, g_\nu$ are polynomials in $f_1, \ldots, f_n$ and g_0 is not identically 0. From now on we can assume that $g_0 = 1$ if we write f in place of $g_0 f$. For fixed f we let λ be an upper bound for the degrees of $g_1, \ldots, g_\nu$ in each of the variables $f_1, \ldots, f_n$. Now we think of the n-disks $\mathfrak{R}_a$ defined above as formed simultaneously for the n functions $f_1(z), \ldots, f_n(z)$, without taking $f(z)$ into consideration. Utilizing the old notation, we now put

$$\prod_{k=1}^{n} q_{ka}^{\lambda} = r_a, \quad r_a^l g_l = h_{la} \quad (l = 1, \ldots, \nu), \quad r_a f = s_a,$$

whereby (21) becomes

$$s_a^\nu + \sum_{l=1}^{\nu} h_{la} s_a^{\nu-l} = 0. \tag{22}$$

Since the functions $r_a g_l$ and, consequently, h_{la} $(l = 1, \ldots, \nu)$ are regular on $\mathfrak{R}_a$, the function s_a, as a root of the algebraic equation (22), is bounded there. On the other hand, s_a is meromorphic on $\mathfrak{R}_a$ and, therefore, even regular.

If we now define $p_a = s_a$, $q_a = r_a$, then (11) is satisfied; therefore the same domain $\mathfrak{R}_a$ is admissible for the function $f(z)$. Of course, p_a and q_a need not now be relatively prime. However, this is also unnecessary in the present case for the proof of Theorem 2. Namely, (12) is satisfied for

$$j_{\gamma ab} = \prod_{k=1}^{n} j_{k\gamma ab}^{\lambda}.$$

Then whereas μ still depends on f, the numbers m and τ are independent of f, and as a result of (14) the same is also true of s. Since the degree of the polynomial $g(x, x_1, \ldots, x_n)$ is at most s in the first variable x, it follows that f is of bounded degree with respect to the field, Λ, of rational functions of $f_1, \ldots, f_n$. Hence it follows that K is a simple algebraic extension of Λ, and Theorem 3 is thereby proved.

The abelian function fields treated in the previous chapter are still not covered by Theorems 2 and 3 because we repeatedly appealed to the boundedness of $\mathfrak{G}$ in Section 1. However, in the case of abelian functions, it is easy to see that the consequences of this boundedness are obtained automatically if we assume the existence of a nondegenerate abelian function. Thus in this case, we can also utilize the above method of proof and obtain a new, simple proof of Theorem 5 in Section 11 of the previous chapter. However, this time the Jacobian functions do not appear. The theorem about meromorphic functions on algebraic varieties, which was mentioned in Section 13 of the previous chapter, can also be derived in this fashion.

Theorem 4: Let $\mathfrak{G}/\Gamma$ be compact. Then there are $n + 2$ automorphic forms $\psi_0, \ldots, \psi_{n+1}$ of equal weight such that every automorphic function can be represented as a quotient of two homogeneous polynomials of the same degree in $\psi_0, \ldots, \psi_{n+1}$ whose denominator does not vanish identically in z.

Proof: For the n functions $f_1, \ldots, f_n$ utilized in Theorem 3 we take, in particular,

$$f_k = g_k = \Phi_k/\Phi_0 \qquad (k = 1, \ldots, n),$$

where g_k, Φ_k, Φ_0 were defined in Theorem 2 of Section 1. Furthermore, let f be selected so that the field of all automorphic functions is generated by $f, f_1, \ldots, f_n$. Then the first part of the proof of Theorem 3 may also be carried out with the Ansatz

$$q_{ka} = \Phi_0 \quad (k = 1, \ldots, n), \qquad r_a = \Phi_0^{n\lambda}, \qquad s_a = r_a f,$$

where r_a and s_a turn out to be automorphic forms of the same weight. Furthermore, if we define

$$\psi_0 = \Phi_0^{n\lambda}, \qquad \psi_k = \psi_0 f_k \quad (k = 1, \ldots, n), \qquad \psi_{n+1} = \psi_0 f,$$

then

$$f_k = \psi_k/\psi_0 \quad (k = 1, \ldots, n), \qquad f = \psi_{n+1}/\psi_0, \qquad \psi_0 \neq 0,$$

and $\psi_0, \ldots, \psi_{n+1}$ have the desired properties.

In a certain sense, Theorem 4 is a generalization of Theorem 5 of Section 6 of Chapter 3 (Volume II). However, it should be remarked that there we only used special Poincaré series, whereas now the automorphic form ψ_{n+1} does not appear as a Poincaré series. By a more precise investigation we can show that the use of Poincaré series alone suffices if instead of $n + 2$ of them we allow a sufficiently large number, $m + 1$, of them. In this connection a question is suggested concerning the extent to which the earlier considerations about Picard varieties may be carried over. It can be answered without additional difficulties if, in addition, we assume that the group Γ possesses no fixed points on $\mathfrak{G}$. Theorem 1 of Section 12 of the previous chapter may be carried over to the present case by a refinement of the method used in proving Theorem 2 of the previous section. In doing this we must replace the period torus $\mathfrak{Z}/\Omega$ by $\mathfrak{G}/\Gamma$. Then the Poincaré series appear in place of the theta functions. We will not pursue this matter but merely remark that the irreducible nonsingular algebraic varieties appearing here do not have the group property investigated in Section 13 of the previous chapter. There is also no relation between dz and the differentials of first kind. Because of this these varieties do not command as much interest as Picard varieties. On the other hand, we do not know how they can be defined by purely algebraic properties. In particular, we have no method of deciding whether a given algebraic variety of higher dimension can be uniformized by automorphic functions.

In order to continue the theory, the most important problem is to characterize those bounded domains $\mathfrak{G}$ which possess a discontinuous group of regular self-mappings with compact fundamental region. In this general form the problem is unsolved for $n > 1$. In case $n = 1$, the Riemann mapping theorem shows that we can take the unit disk $\mathfrak{E}$ in place of any simply connected domain $\mathfrak{G}$. Thus we know that there are groups with the desired properties. Namely, all polygon groups without parabolic elements. On the other hand, we know three distinct ways to construct polygon groups. The first method arises from uniformization of algebraic curves of genus $p > 1$, but does not supply a constructive method for actually representing the concrete group. The second method is constructive and begins with non-euclidean polygons whose sides correspond to one another pairwise by noneuclidean motions. This was carried out in an important special case in Section 9 of Chapter 3 (Volume II). The third method is also constructive and of an arithmetic nature in that we begin with the units of an indefinite quaternion-algebra over the rational number field. Only the third of these methods, which was founded by Poincaré, has been carried over successfully to the case of several variables.

Should $\mathfrak{G}/\Gamma$ be compact, then in no case may Γ be of finite order. Now, as was mentioned earlier, for every $n > 1$ there exist rigid domains which are bounded and simply connected. On the other hand, without additional hypothesis we know of no general method for producing bounded simply connected domains which admit infinitely many invertible regular self-mappings. Thus we are forced to make further restrictions in order to penetrate more deeply. It has proved expedient to require that the group Ω of all invertible regular self-mappings of $\mathfrak{G}$ acts transitively. This means that for every two points ξ and η of $\mathfrak{G}$ there is at least one mapping in Ω which carries ξ to η. The domain is then called *homogeneous.* On the other hand, two bounded domains in complex space will be called *equivalent* if they can be mapped onto one another in a regular invertible fashion.

In the case $n = 1$ the Riemann mapping theorem imples that every bounded simply connected domain $\mathfrak{G}$ is homogeneous and equivalent to the unit disk $\mathfrak{E}$. Thus the group Ω for $\mathfrak{E}$ is given by (1) in Section 1. As E. Cartan has shown, in case $n = 2$, every bounded homogeneous domain is equivalent to precisely one of the domains given by $|z_1| < 1$, $|z_2| < 1$ and $|z_1|^2 + |z_2|^2 < 1$. For the first domain it is easy to show that Ω consists of all mappings which carry both the unit disks into themselves according to (1) of Section 1, possibly interchanging them. For the second domain, we obtain Ω from a theorem of Poincaré if we replace z_1, z_2 by z_1/z_0, z_2/z_0 and determine all homogeneous linear transformations with complex coefficients which leave the Hermitian form $z_1\bar{z}_1 + z_2\bar{z}_2 - z_0\bar{z}_0$ invariant. In the case $n = 3$, Cartan gave without proof four homogeneous bounded normal domains to which all others are equivalent. In each case the determination of $\mathfrak{G}$ and Ω may be carried out by application of the methods of the theory of Lie groups. The substantial difficulties which appear here for arbitrary n were recently overcome by Pjatetzki–Schapiro. In order to simplify the investigation, Cartan had previously placed a stronger restriction on $\mathfrak{G}$. Namely, for each point ζ of $\mathfrak{G}$ there should be exactly one mapping with the single fixed point ζ which yields the identity upon being applied twice. The domain $\mathfrak{G}$ is then called *symmetric*. It can be shown for all n that homogeneity follows from symmetry. For $n = 1, 2, 3$ the converse also holds, although this is no longer true for $n = 5$.

The results of Cartan yield a lucid and simple classification of bounded symmetric domains. For this purpose it is sufficient to represent only the irreducible bounded symmetric domains, that is, those which are not equivalent to the Cartesian product of two such domains of smaller dimension. It can be shown that there are six types of equivalence classes of irreducible bounded symmetric domains which, themselves, correspond to simple Lie groups. Two of these types appear just once for $n = 16$ and $n = 27$, whereas the remaining types appear for infinitely many n. Among the latter four types of Cartan domains, we will consider later only that one which is especially important because of its relationship to Abelian functions. Let $n = [p(p + 1)/2]$ $(p = 1, 2, \ldots)$, and let $Z = (z_{kl})$ be a complex $p \times p$ matrix subject to the two conditions

$$E - Z'\overline{Z} > 0, \qquad Z = Z'. \tag{23}$$

The n independent entries z_{kl} $(1 \leqslant k \leqslant l \leqslant p)$ are the variables on the bounded domain $\mathfrak{G}$ given by (23). Obviously for $p = 1$, we have $n = 1$ and $\mathfrak{G} = \mathfrak{E}$.

Now it is of great importance for function theory that we can, by number-theoretic means, find discontinuous groups Γ for which $\mathfrak{G}/\Gamma$ is compact. Here $\mathfrak{G}$ may be any of the four general types, as well as the reducible domains formed from them. Namely, it follows from Theorem 3 that the automorphic functions with respect to Γ form an algebraic function field of transcendence degree n. We see in this way that the theory developed earlier in the chapter is not void.

Finally, without supplying the details, we now give a useful example for arbitrary n. In particular, among the Cartan types is to be found the domain $\mathfrak{G}$ given by the inequality

$$|z_1|^2 + \cdots + |z_n|^2 < 1; \tag{24}$$

that is, the domain furnished by the interior of the real $2n$ dimensional unit ball. Now let P be a totally real algebraic number field whose degree h is greater than one. Denote the conjugates of a generator ρ of P by $\rho^{(1)} = \rho$, $\rho^{(2)}, \ldots, \rho^{(h)}$. We now select $n + 1$ additional numbers $\alpha_0, \alpha_1, \ldots, \alpha_n$ in P of which $\alpha_1, \ldots, \alpha_n$ are totally positive whereas

$$\alpha_0 > 0, \qquad \alpha_0^{(2)} < 0, \ldots, \alpha_0^{(h)} < 0.$$

Finally, by adjoining $\sqrt{-\beta}$, with β a totally positive number in P, we form the totally imaginary extension Σ of P with relative degree two. We now consider the Hermitian form

$$Q = \alpha_1 w_1 \bar{w}_1 + \alpha_2 w_2 \bar{w}_2 + \cdots + \alpha_n w_n \bar{w}_n - \alpha_0 w_0 \bar{w}_0$$

and the totality of homogeneous linear transformations with integral coefficients from Σ which leave Q invariant. If we set

$$z_k = (w_k\sqrt{\alpha_k})/(w_0\sqrt{\alpha_0}) \qquad (k = 1, \ldots, n),$$

we obtain a group Γ of projective transformations which carry the domain $\mathfrak{G}$ given by (24) into itself. With the assistance of the reduction theory of quadratic forms, which will not be explained here, we can show that Γ acts discontinuously on $\mathfrak{G}$ and $\mathfrak{G}/\Gamma$ is compact.

The assumptions about α_0 may, at first, appear odd, but a detailed discussion reveals that they are necessary. Namely, if all h conjugates of α_0 were negative, then Γ would be a finite group and therefore $\mathfrak{G}/\Gamma$ would be noncompact. On the other hand, if more than one conjugate were positive, then, under more refined investigation, it would turn out that Γ is not discontinuous in $\mathfrak{G}$. A special case arises for $n = 2$ if P is the real quadratic field generated by $\sqrt{2}$, Σ the biquadratic number field resulting from adjunction of $\sqrt{-1}$, $\alpha_1 = \alpha_2 = 1$, and $\alpha_0 = \sqrt{2}$. Then the problem is to determine the units with integers from Σ of the ternary Hermitian form

$$Q = w_1\bar{w}_1 + w_2\bar{w}_2 - \sqrt{2}\, w_0\bar{w}_0.$$

We cannot, however, supply a proof of the fact that Γ is discontinuous and $\mathfrak{G}/\Gamma$ is compact because this would require deeper auxiliary material from geometric and algebraic number theory.

3. *Symplectic geometry*

We will now investigate the bounded symmetric domains of the type determined by (23) in the previous section. In the case $n = 1$ this becomes the unit disk $\mathfrak{E}$. The fractional linear transformation

$$z = i\frac{1 + w}{1 - w}$$

maps the unit disk $1 - w\bar{w} > 0$ in the w-plane one-to-one and onto the upper half plane defined by $(1/2i)(z - \bar{z}) > 0$. We will now present a generalization of this.

Let $W = (w_{kl})$ and $Z = (z_{kl})$ be complex $p \times p$ matrices. We decompose $Z = X + iY$ into its real part X and its imaginary part Y and consider the conditions

$$\frac{1}{2i}(Z' - \bar{Z}) > 0, \qquad Z = Z'. \tag{1}$$

These conditions, which can also be expressed as $X = X'$, $Y = Y' > 0$, define a domain $\mathfrak{H}$ in the n-dimensional complex space of the independent variables $z_{kl} = x_{kl} + iy_{kl}$ $(1 \leqslant k \leqslant l \leqslant p;\ n = p(p + 1)/2)$. In Section 5 of Chapter 4 (Volume II) we called $\mathfrak{H}$ the *generalized upper half plane* or the *upper half plane of degree p*. The definition of the generalized unit disk will be modeled after (23) of the previous section as follows:

$$E - W'\bar{W} > 0, \qquad W = W', \tag{2}$$

where the n independent complex variables are now given by the w_{kl} $(1 \leqslant k \leqslant l \leqslant p)$. We also denote the generalized unit disk by $\mathfrak{E}$ when p is fixed.

Theorem 1: The generalized unit disk is mapped biregularly onto the generalized upper half plane by the formula

$$Z = i(E + W)(E - W)^{-1}. \tag{3}$$

Proof: Let W be any point in $\mathfrak{E}$ and let the complex column $\mathfrak{x}$ be a solution of $(E - W)\mathfrak{x} = 0$. Then the following equations hold:

$$\mathfrak{x} = W\mathfrak{x}, \qquad \bar{\mathfrak{x}} = \bar{W}\bar{\mathfrak{x}}, \qquad \mathfrak{x}' = \mathfrak{x}'W',$$

$$\mathfrak{x}'(E - W'\bar{W})\bar{\mathfrak{x}} = \mathfrak{x}'\bar{\mathfrak{x}} - (\mathfrak{x}'W')(\bar{W}\bar{\mathfrak{x}}) = 0.$$

Since by (2) the Hermitian matrix $E - W'\bar{W}$ is positive definite, it follows that $\mathfrak{x} = 0$. Thus $|E - W| \neq 0$, and the inverse matrix of $E - W$ which appears in (3) actually exists. Furthermore, $E + W$ and $E - W$ commute with one another, and W is symmetric. Thus the complex matrix Z defined by (3) is also symmetric; consequently, the second condition in (1) is fulfilled. The same is also true of the first condition as is obvious from the formula

$$\frac{1}{2i}(Z-\overline{Z}) = \tfrac{1}{2}(E-W')^{-1}\{(E+W')(E-\overline{W})+(E-W')(E+\overline{W})\}(E-\overline{W})^{-1}$$
$$= (E-W')^{-1}(E-W'\overline{W})(E-\overline{W})^{-1}.$$

Thus (3) carries each point W of $\mathfrak{E}$ to a unique point Z of $\mathfrak{H}$.

Conversely, let Z be an arbitrary point of $\mathfrak{H}$ and x a solution of

$$(Z+iE)x = 0.$$

It follows that

$$ix = -Zx, \qquad i\bar{x} = \overline{Z}\ , \qquad -ix' = x'Z',$$

$$\frac{1}{2i}\,x'(Z'-\overline{Z})\bar{x} = \frac{1}{2i}(x'Z')\bar{x} - \frac{1}{2i}\,x'(\overline{Z}\bar{x}) = -x'\bar{x},$$

and since the Hermitian matrix $(1/2i)(Z'-\overline{Z})$ is positive definite by (1), it again follows that $x=0$. Thus $|Z+iE| \neq 0$, and the complex matrix

$$(Z-iE)(Z+iE)^{-1} = W \tag{4}$$

exists and is symmetric because of the symmetry of Z and the mutual commutativity of $Z+iE$ and $Z-iE$. Finally, since

$$E - W'\overline{W} = (Z'+iE)^{-1}\{(Z'+iE)(\overline{Z}-iE)-(Z'-iE)(\overline{Z}+iE)\}(\overline{Z}-iE)^{-1}$$
$$= 4(Z'+iE)^{-1}\frac{1}{2i}(Z\ -\overline{Z})(\pi - iE)^{-1},$$

it follows from (1) that the point W lies in $\mathfrak{E}$. Expression (4) also yields

$$Z - iE = WZ + iW = ZW + iW, \qquad Z(E-W) = i(E+W)$$

and, therefore, expression (2). The converse also holds.

Thus it is shown that (3) defines a one-to-one mapping of $\mathfrak{E}$ onto $\mathfrak{H}$ and that (4) furnishes the inverse mapping. The mapping is obviously birational and since the determinants $|E-W|$ and $|Z+iE|$ are nonzero throughout $\mathfrak{E}$ and $\mathfrak{H}$, it is biregular everywhere. This completes the proof of Theorem 1.

The relation of the symplectic group to the generalized upper half plane was pointed out in Section 5 of Chapter 4 (Volume II). We will carry out the appropriate small calculation once again in an easily surveyed form by beginning with two $p \times p$ complex matrices F, G and putting

$$R = \begin{pmatrix} F \\ G \end{pmatrix}, \qquad J = \begin{pmatrix} 0 & E \\ -E & 0 \end{pmatrix}. \tag{5}$$

Since J is skew-symmetric and $(1/2i)J$ Hermitian, the same holds for $R'JR$ and $(1/2i)R'J\bar{R}$. We now demand that both conditions

$$R'JR = 0, \qquad \frac{1}{2i}R'J\bar{R} > 0 \tag{6}$$

be satisfied. As a consequence of (5), these conditions can be expressed as

$$F'G = G'F, \qquad \frac{1}{2i}(F'\bar{G} - G'\bar{F}) > 0. \tag{7}$$

From considerations utilized in the proof of Theorem 1, as well as in Section 10 of Chapter 5, it follows from (7) that $|G| \neq 0$. Hence we may introduce $Z = FG^{-1}$. By (7) the matrix Z lies in $\mathfrak{H}$. Conversely, for each point Z in $\mathfrak{H}$ and for an arbitrary invertible matrix G we can define the matrix $F = ZG$, whereupon (7) is satisfied. By definition, the elements

$$L = \begin{pmatrix} A & B \\ C & D \end{pmatrix} \tag{8}$$

of the symplectic group Δ are the real solutions L of the matrix equation $L'JL = J$. Clearly, the matrix LR also satisfies both conditions (6) when R satisfies these conditions and L is any symplectic matrix. This means precisely that the symplectic transformation,

$$Z^* = (AZ + B)(CZ + D)^{-1}, \tag{9}$$

maps the generalized upper half plane onto itself biregularly; thereupon, the inverse transformation is given by

$$Z = (D'Z^* - B')(-C'Z^* + A')^{-1}.$$

In view of (3) we introduce the $(2n) \times (2n)$ matrix

$$V = \begin{pmatrix} iE & iE \\ -E & E \end{pmatrix},$$

which has the inverse

$$V^{-1} = \frac{1}{2i}\begin{pmatrix} E & -iE \\ E & iE \end{pmatrix},$$

and the group $\Omega = V^{-1}\Delta V$. Then we obtain

$$V'JV = \begin{pmatrix} iE & -E \\ iE & E \end{pmatrix}\begin{pmatrix} -E & E \\ -iE & -iE \end{pmatrix} = \begin{pmatrix} 0 & 2iE \\ -2iE & 0 \end{pmatrix} = 2iJ, \tag{10}$$

$$\frac{1}{2i}V'J\bar{V} = \frac{1}{2i}\begin{pmatrix} iE & -E \\ iE & E \end{pmatrix}\begin{pmatrix} -E & E \\ iE & iE \end{pmatrix} = \begin{pmatrix} -E & 0 \\ 0 & E \end{pmatrix}. \tag{11}$$

Furthermore, we put

$$V^{-1}R = \begin{pmatrix} F_0 \\ G_0 \end{pmatrix}, \qquad V^{-1}LV = L_0 = \begin{pmatrix} A_0 & B_0 \\ C_0 & D_0 \end{pmatrix}.$$

Then it follows from (10) and (11) that (7) becomes

$$F_0'G_0 = G_0'F_0, \qquad G_0'\bar{G}_0 - F_0'\bar{F}_0 > 0.$$

In particular, we then have $|G_0| \neq 0$; whereupon, we can introduce $W_0 = F_0G_0^{-1}$ and the relations become

$$W_0 = W_0', \qquad E - W_0'\bar{W}_0 > 0.$$

This discussion contains in homogeneous form the transformation carried out in the proof of Theorem 1. Beyond this, we now discern that the transformations

$$W^* = (A_0W + B_0)(C_0W + D_0)^{-1} \tag{12}$$

formed from the elements L_0 of Ω, map the generalized unit disk onto itself biregularly.

Furthermore, if we introduce the matrices

$$N = \begin{pmatrix} -E & 0 \\ 0 & E \end{pmatrix}, \qquad K = -N^{-1}J = \begin{pmatrix} 0 & E \\ E & 0 \end{pmatrix},$$

then from $L = VL_0V^{-1} = \bar{V}\bar{L}_0\bar{V}^{-1}$ and $L'JL = J$, we obtain the following equations by using (10) and (11):

$$L_0'JL_0 = J, \qquad L_0'N\bar{L}_0 = N.$$

Thus we also have

$$L_0'JL_0 = J, \qquad KL_0 = \bar{L}_0K. \tag{13}$$

Conversely, for each complex solution L_0 of (13) the matrix $L = VL_0V^{-1}$ is real symplectic. The first condition in (13) implies that the matrix L_0 is an element of the complex symplectic group. In addition L_0 must also satisfy the second condition. Since the first condition is equivalent to $L_0JL_0' = J$, we obtain the following relations between the four blocks of L_0:

$$A_0B_0' = B_0A_0', \qquad C_0D_0' = D_0C_0', \qquad A_0D_0' - B_0C_0' = E.$$

The second condition in (13) clearly yields

$$\begin{pmatrix} C_0 & D_0 \\ A_0 & B_0 \end{pmatrix} = \begin{pmatrix} \bar{B}_0 & \bar{A}_0 \\ \bar{D}_0 & \bar{C}_0 \end{pmatrix}, \qquad A_0 = \bar{D}_0, \qquad B_0 = \bar{C}_0.$$

If we put $D_0 = P$, $C_0 = Q$, then (12) carries over to

$$W^* = (\bar{P}W + \bar{Q})(QW + P)^{-1} \tag{14}$$

where the matrices P and Q must fulfill both conditions

$$PQ' = QP', \qquad \bar{P}P' - \bar{Q}Q' = E. \tag{15}$$

We introduce a parametric representation of the solutions P, Q of (15) in the following way. First, it follows from the second equation in (15) that $|P| \neq 0$. Therefore we can set $P^{-1}Q = W_0$ and obtain $W_0 = W_0'$ from the first equation. Substituting $Q = PW_0$, the second equation (15) is transformed into

$$\bar{P}(E - \bar{W}_0 W_0')P' = E,$$

from which it follows that

$$E - W_0'\bar{W}_0 = (\bar{P}'P)^{-1} > 0. \tag{16}$$

Accordingly, we can select any point of $\mathfrak{E}$ for W_0 and then any solution of (16) for P, whereupon we must set $Q = PW_0$. Furthermore, if we set $H = W_0'\bar{W}_0$, then all the eigenvalues of the Hermitian matrix H belong to the interval $0 \leqslant \lambda < 1$ and consequently the series

$$(E - H)^{-1/2} = E + \frac{1}{2}H + \frac{1 \cdot 3}{2 \cdot 4}H^2 + \frac{1 \cdot 3 \cdot 5}{2 \cdot 4 \cdot 6}H^3 + \cdots \tag{17}$$

must converge. The Hermitian matrix defined by (17) is then a particular solution P of (16). Thus we obtain as a general solution of (15)

$$P = U(E - W_0'\bar{W}_0)^{-1/2}, \qquad Q = PW_0 \tag{18}$$

with any unitary matrix U and arbitrary W_0 in $\mathfrak{E}$. Thus U and W_0 enter as parameters. If we replace W_0 with $-\bar{W}_0$, then the mapping (14) carries the given point W_0 in $\mathfrak{E}$ to 0 and the inverse mapping

$$W = (P'W^* - \bar{Q}')(-Q'W^* + \bar{P}')^{-1}$$

takes 0 to W_0. Accordingly, it has also been shown that Ω acts transitively on $\mathfrak{E}$.

In the case $\rho = 1$ it follows from (14) and (15) that Ω is the group of all biregular self mappings of the unit disk given by (1) of Section 1. Accordingly, we can conjecture that for arbitrary ρ the generalized unit disk also admits no self mapping except these. That this is in fact the case can be seen as follows. Since Ω is transitive on $\mathfrak{E}$ we need only determine those biregular self-mappings of the generalized unit disk which keep the point 0 fixed. In view of (14) and (18) we must show that these are all given by the linear transformations

$$W^* = U'WU \tag{19}$$

with U a unitary matrix. This proof is carried out in two steps. First, using auxiliary material from function theory, we show by considering the ratio of volumes that the desired mappings must be homogeneous and linear in the n independent variables w_{kl}. It then follows from purely algebraic considerations that these mappings must be of the form (19). We will not carry out these steps in detail and will merely remark that upon transition to $\mathfrak{H}$ the symplectic transformations (9) yield all the biregular self-mappings of the generalized upper half-plane.

The mapping (9) is the identity if and only if the equation

$$ZCZ + ZD = AZ + B$$

holds for all Z in $\mathfrak{H}$. In fact, we can even take any complex symmetric matrix for Z. From this it follows as in Section 5 of Chapter 4 (Volume II) that we must have $B = C = 0$ and $A = D = \pm E$. Just as in the case of a single variable, two distinct symplectic matrices yield the same transformation if and only if they differ by the scalar factor ± 1.

In particular, if we select $U = \pm iE$ in (19), then we obtain the mapping $W^* = -W$ which is an involution whose only fixed point is $W = 0$. We can show that no other element of Ω has the same property. In view of the transitivity of Ω, the domain $\mathfrak{E}$ thus turns out to be a symmetric space in the sense of Cartan.

Now we are interested in those subgroups Γ of Ω which are discontinuous in $\mathfrak{E}$ and which are designated as circle groups in the case $p = 1$. We generalize Theorem 2 of Section 3 of Chapter 3 (Volume II) by proving the following statement.

Theorem 2: A subgroup of Ω acts discontinuously of $\mathfrak{E}$ if and only if it is discrete.

Proof: If a subgroup Γ of Ω is not discrete, then it contains a sequence of distinct matrices which tend towards the unit matrix. Consequently, in the corresponding mappings given by (14) we find that $P \to E$, $Q \to 0$; therefore, we also have $\bar{Q}P^{-1} \to 0$. On the other hand, since $W^* = \bar{Q}P^{-1}$ is the image of $W = 0$, we see that Γ is not discontinuous on $\mathfrak{E}$.

In order to derive the converse, it is sufficient in view of Theorem 1 to establish the corresponding result for $\mathfrak{H}$ and the symplectic group Δ. Thus let Γ be a subgroup of Δ which is not discontinuous on $\mathfrak{H}$. Then there exists a point Z in $\mathfrak{H}$ and a sequence of distinct elements, L, of Γ such that the image points $Z^* = Z_L$ of Z, which arise in accordance with (9), converge to a point Z_0 in $\mathfrak{H}$. If we replace (9) with the homogeneous substitution

$$\begin{pmatrix} F_L \\ G_L \end{pmatrix} = \begin{pmatrix} A & B \\ C & D \end{pmatrix} \begin{pmatrix} F \\ G \end{pmatrix}$$

with $|G| \neq 0$ and $F = ZG$, then we also have $F_L = Z_L G_L$ as well as

$$G_L = (CZ + D)G.$$

Letting Y and Y_L stand for the imaginary parts of the matrices $Z = X +$the and Z_L, it follows from (5), (6), (7), and the invariance of $R'J\bar{R}$ that $Y i$ relation

$$G'Y\bar{G} = G'_L Y_L \bar{G}_L$$

holds; therefore, we have

$$(20) \qquad \begin{aligned} Y &= (CZ + D)'Y_L(C\bar{Z} + D) \\ Y_L^{-1} &= (C\bar{Z} + D)Y^{-1}(CZ + D)'. \end{aligned}$$

Introducing the notation $CX + D = K$, we obtain

$$CZ + D = K + iCY,$$

$$\begin{aligned} (C\bar{Z} + D)Y^{-1}(CZ + D)' &= (K - iCY)Y^{-1}(K' + iYC') \\ &= KY^{-1}K' + CYC', \end{aligned}$$

since

$$KC' - CK' = (CX + D)C' - C(XC' + D') = DC' - CD' = 0.$$

Thus we obtain the formula

$$(21) \qquad Y_L^{-1} = KY^{-1}K' + CYC', \qquad K = CX + D,$$

which will also be important in the sequel.

Since the sequence Z_L tends to the point Z_0 in $\mathfrak{H}$, the diagonal entries of Y_L^{-1} are bounded when we pass to the limit. On the other hand, the real symmetric matrices Y and Y^{-1} are both positive definite, and (21) shows that the entries of all rows of K, C, and D also remain bounded. Finally, the boundedness of all entries of the matrix $AZ + B$ and its imaginary part AY follows from the equation

$$AZ + B = Z_L(CZ + D).$$

Thus the entries of A and B and consequently L are all bounded. The matrices L then have a limit point in the closed group Δ, and Γ is not discrete. This completes the proof of Theorem 2.

In the investigations of the circle groups in Chapter 3 (Volume II), we made full use of the invariant noneuclidean metric. There exists a corresponding Riemannian metric for homogeneous domains of any dimension and, in particular, for symmetric domains; but in general this metric differs from the noneuclidean metric. For the case which is of interest to us, that of the generalized upper half plane, this metric is given in the following manner. In these considerations we let $dZ = (dz_{kl})$ denote the matrix formed from the differentials dz_{kl} and $\sigma(M)$ denote the trace of the square matrix M.

Theorem 3: The quadratic differential form

$$(22)\quad (ds)^2 = \sigma(Y^{-1}(dZ)Y^{-1}(d\bar{Z})) = \sigma(Y^{-1}(dX)Y^{-1}(dX)) + \sigma(Y^{-1}(dY)Y^{-1}(dY))$$

is positive definite throughout $\mathfrak{H}$ and invariant under Δ.

Proof: From the symplectic transformation

$$Z_L = (AZ + B)(CZ + D)^{-1}$$

we obtain

$$Z_L(CZ + D) = AZ + B, \qquad (dZ_L)(CZ + D) = (A - Z_LC)(dZ),$$
$$(CZ + D)'Z_L = (AZ + B)',$$

$$(23)\quad (CZ + D)'(dZ_L)(CZ + D) = ((ZC' + D')A - (ZA' + B')C)(dZ) = dZ.$$

In conjunction with (20) this yields

$$(C\bar{Z} + D)Y^{-1}\,dZ = Y_L^{-1}(dZ_L)(CZ + D),$$
$$(CZ + D)Y^{-1}(d\bar{Z}) = Y_L^{-1}(d\bar{Z}_L)(\bar{Z}C + D),$$
$$(CZ + D)Y^{-1}(d\bar{Z})Y^{-1}\,dZ = Y_L^{-1}(d\bar{Z}_L)Y_L^{-1}(dZ_L)(CZ + D),$$
$$\sigma(Y^{-1}(d\bar{Z})Y^{-1}(dZ)) = \sigma(Y_L^{-1}(d\bar{Z}_L)Y_L^{-1}(dZ_L)).$$

Thus the invariance under Δ of the line-element defined by (22) is demonstrated.

In particular, setting $Z = iE$, we have

$$(24)\quad (ds)^2 = \sigma((dZ)(d\bar{Z})) = \sum_{k=1}^{p}\{(dx_{kk})^2 + (dy_{kk})^2\} + 2\sum_{1\leqslant k<l\leqslant p}\{(dx_{kl})^2 + (dy_{kl})^2\},$$

which is clearly positive definite. Because of the transitivity of Δ on $\mathfrak{H}$, the differential form $(ds)^2$ is positive definite throughout $\mathfrak{H}$. This completes the proof.

The line element ds of the so-called *symplectic geometry* is defined by Theorem 3. For $p = 1$ this is just the noneuclidean line element introduced in Section 2 of Chapter 3 (Volume II) and, in a manner analogous to the proof given there, we can also show that for aribtrary p this line element is determined up to a constant factor by the required invariance under Δ. A few more of the previous considerations can also be generalized meaningfully. In particular, with the help of the symplectic metric given by (22), we can carry over the construction of a normal fundamental region of a circle group given in Section 3 of Chapter 3 (Volume II) to the case of a discontinuous subgroup of the symplectic group. However, we will not do this since this method is not sufficiently adaptable to the particular arithmetic properties of the groups to be considered in the sequel.

The euclidean volume element in $2n = p(p+1)$ dimensional real space with coordinates x_{kl}, y_{kl} $(1 \leqslant k \leqslant l \leqslant p)$ is given by

$$dv = \prod_{1 \leqslant k \leqslant l \leqslant p} (dx_{kl}\, dy_{kl}).$$

We will now determine the corresponding symplectic volume element.

Theorem 4: The symplectic volume element is given by

$$dw = 2^{n-p}\, |Y|^{-p-1}\, dv.$$

Proof: In Riemannian differential geometry, if the line element is given by

$$(ds)^2 = \sum_{k,l=1}^{\nu} g_{kl}\, dx_k\, dx_l,$$

then it is known that the corresponding volume element is

$$dw = |g_{kl}|^{1/2}\, dx_1 \cdots dx_\nu.$$

Therefore, it follows from (24) that the result to be proved holds at the point $Z = iE$ of $\mathfrak{H}$. Because of the transitivity of Δ and Theorem 3, we need only demonstrate the invariance of the right-hand side of (25) under symplectic transformations. If we abbreviate

$$CZ + D = Q,$$

then (20) and (23) yield the relations

$$Y_L^{-1} = \bar{Q} Y^{-1} Q', \qquad Q'(dZ_L)Q = dZ; \tag{26}$$

and in particular,

$$|Y_L|^{-p-1} = |\bar{Q}Q|^{p+1}\, |Y|^{-p-1}.$$

On the other hand, since $X = \frac{1}{2}(Z + \bar{Z})$, $Y = (1/2i)(Z - \bar{Z})$, the ratio $(dv_L)/(dv)$ of the euclidean volume elements is equal to the square of the absolute value of the Jacobian determinant of Z_L with respect to Z. Therefore, it is sufficient to show that this Jacobian determinant has the value $|Q|^{-p-1}$. This follows from (26) in the following manner.

If S and Q are $p \times p$ matrices with indeterminant entries and, moreover, if S is symmetric, then $T = Q'SQ$ defines a homogeneous linear transformation of the n independent entries of S. The entries in T are homogeneous polynomials of the second degree in the entries of Q. The determinant of this transformation is then a homogeneous polynomial of degree $2n$, which we denote by $f(Q)$. If we select $Q = Q_1 Q_2$, then it follows that the multiplicative law

$$f(Q_1 Q_2) = f(Q_1) f(Q_2)$$

holds identically for the entries of Q_1 and Q_2. Setting $Q_2 = |Q_1|\, Q_1^{-1}$ and $Q_1 = Q$, we obtain

$$|Q|^{2n} f(E) = f(|Q|\, E) = f(Q) f(|Q|\, Q^{-1}).$$

Since $f(E) = 1$, and it is known that the determinant $|Q|$ is irreducible as a polynomial in the entries of Q, it follows that $f(Q)$ must be a power of $|Q|$. Comparing degrees, we find that

$$f(Q) = |Q|^{2n/p} = |Q|^{p+1},$$

and the proof of Theorem 4 is complete.

We will not pursue the problem of representing discontinuous subgroups Γ of Ω or Δ which have a compact fundamental region. This is a question which is of interest for its own sake, and like the example introduced at the end of Section 2, it can be treated by utilization of the arithmetic theory of quadratic forms. From now on we will consider in detail only the case which appears to be especially important because of its connection with the theory of Abelian functions. This means we are dealing with the elliptic modular group and its generalizations. Now for the groups under consideration, $\mathfrak{H}/\Gamma$ is no longer compact. Therefore, the proofs given in Section 2 cannot be carried over without change; thus a new chain of reasoning arises.

4. Abelian functions and modular functions

Let $\mathfrak{R}$ be a Picard variety of dimension p. According to our considerations at the end of the preceding chapter, there exist p linearly independent differentials of the first kind $dz_1, \ldots, dz_p$ corresponding to the algebraic function field K of $\mathfrak{R}$. These differentials are uniquely determined up to an arbitrary invertible homogeneous linear transformation $dz \rightarrow Q\,dz$ with constant complex coefficients. The elements of K are then Abelian functions of z. Once we have selected a basis of the period lattice, we obtain a fixed period matrix C. Any subsequent choice of a nondegenerate function in K yields a corresponding principal matrix A.

It is an interesting algebraic problem to determine for a given Riemann matrix C all corresponding principal matrices, that is, the totality of integral skew-symmetric invertible matrices A satisfying

$$CA^{-1}C' = 0, \qquad \frac{1}{2i}\,\bar{C}A^{-1}C' > 0.$$

We will not pursue this problem. But we will show that, for given integral skew-symmetric invertible A, a Riemann matrix C can be determined so that every principal matrix corresponding to C has the form λA with positive scalar λ. In view of Theorem 3 in Section 10 of the preceding chapter, we may start from the normal forms

$$(1)\quad A = \begin{pmatrix} 0 & T \\ -T & 0 \end{pmatrix}, \qquad C = (TW), \qquad W = W', \qquad \frac{1}{2i}(W' - \overline{W}) > 0$$

where T is a diagonal matrix. Let

$$A_0 = \begin{pmatrix} A_1 & B' \\ -B & A_2 \end{pmatrix}$$

by any fixed skew-symmetric matrix which satisfies the equation $CA_0C' = 0$ identically when W ranges over $\mathfrak{H}$. We obtain the identity

$$TA_1T + TB'W - WBT + WA_2W = 0$$

which, in particular, implies $A_1 = 0$ and $A_2 = 0$. Putting $BT = (b_{kl})$ and choosing W as a diagonal matrix with variable diagonal entries, we obtain $b_{kl} = 0$ for $k \neq l$. Hence BT is a diagonal matrix. Finally, since the equations $b_{kk}w_{kl} = w_{kl}b_{ll}$ must be satisfied, we obtain $B = \mu T^{-1}$ with scalar μ and hence $A_0 = \mu A^{-1}$. Therefore, if A_0 is not of the form μA^{-1}, the relation $CA_0C' = 0$ cannot hold identically in W. In this case it represents an algebraic hypersurface in $\mathfrak{H}$. Since there exist only countably many integral skew-symmetric matrices, an element W of $\mathfrak{H}$ can be determined such that the Riemann matrix (TW) admits as principal matrices precisely the integral matrices λA with

$$A = \begin{pmatrix} 0 & T \\ -T & 0 \end{pmatrix}, \qquad \lambda > 0.$$

In particular, choosing λ^{-1} to be the greatest common divisor of the elements of A, we may assume that the entries of A are relatively prime. According to our previous normalization, this means that the first diagonal element of T has the value $t_1 = 1$. However, it should be noted that in the investigation of Picard varieties outlined in Section 12 of the preceding chapter T was replaced by $3T$.

By fixing the normal forms (1), the matrix W is uniquely determined only up to an arbitrary modular substitution of level T. According to equation (5) in Section 10 of the preceding chapter, such a substitution is defined by

$$(2)\qquad W^* = (M_4'W + M_2'T)(T^{-1}M_3'W + T^{-1}M_1'T)^{-2}$$

where the unimodular matrix

$$\begin{pmatrix} M_1 & M_2 \\ M_3 & M_4 \end{pmatrix} = U$$

satisfies $U'AU = A$. Here the matrix

$$M = \begin{pmatrix} M_4' & M_2'T \\ T^{-1}M_3' & T^{-1}M_1'T \end{pmatrix} = \begin{pmatrix} 0 & T \\ E & 0 \end{pmatrix}^{-1} U' \begin{pmatrix} 0 & T \\ E & 0 \end{pmatrix} \tag{3}$$

is symplectic and the matrices U form the homogeneous modular group of level T defined by the equation $U'AU = A$. In view of (2) and (3), the modular substitutions of level T form a group $\Gamma(T)$ which will be called the *inhomogeneous modular group of level* T. Here it should be noted that U and $-U$ yield the same mapping (2). In particular, for $T = E$ we obtain the *modular group of degree* p. In view of (3), the group $\Gamma(T)$ is also discrete and hence discontinuous in $\mathfrak{H}$ according to Theorem 2 of the preceding section. It is expedient to carry out our future investigations in $\mathfrak{H}$ rather than in $\mathfrak{E}$ since transferring them to $\mathfrak{E}$ would conceal the simple arithmetical structure of modular substitutions. We will now substantiate the importance of automorphic functions corresponding to $\Gamma(T)$ for the theory of Abelian functions. The classical case $p = 1$ will serve as the starting point.

Using inhomogeneous coordinates u, v, the Picard variety in that particular case may be defined by

$$v^2 = 4u^3 - g_2 u - g_3, \tag{4}$$

where the complex coefficients g_2, g_3 must satisfy the condition

$$g_2^3 - 27g_3^2 \neq 0.$$

If ω ranges over all periods of the integral of the first kind

$$z = \int_\infty^u \frac{du}{v},$$

then, according to Section 12 of Chapter 1 (Volume I), the following relations hold:

$$g_2 = 60 \sum_{\omega \neq 0} \omega^{-4}, \qquad g_3 = 140 \sum_{\omega \neq 0} \omega^{-6}. \tag{5}$$

Passing from dz to $q\,dz$ with arbitrary constant $q \neq 0$, the quantities ω, g_2, g_3 must be replaced by $q\omega$, $q^{-4}g_2$, $q^{-6}g_3$, whereas the invariant

$$j = g_2^3(g_2^3 - 27g_3^2)^{-1}$$

remains unchanged.

We select fundamental periods ω_1, ω_2 such that the imaginary part of the ratio $\omega_2/\omega_1 = w = x + iy$ satisfies $y > 0$. Then j turns out to be a function $j(w)$ of w which is invariant with respect to all modular substitutions

$$w^* = \frac{aw + b}{cw + d}. \tag{6}$$

Putting $\omega = \nu\omega_1 + \mu\omega_2$ in (5), the pair ν, μ must range over all pairs of integers except 0, 0, and we obtain

$$\frac{\omega_1^4}{60} g_2 = \sum_{\mu,\nu}' (\mu w + \nu)^{-4}, \qquad \frac{\omega_1^6}{140} g_3 = \sum_{\mu,\nu}' (\mu w + \nu)^{-6}. \tag{7}$$

Introducing new fundamental periods by setting $\omega_2^* = a\omega_2 + b\omega_1$, $\omega_1^* = c\omega_2 + d\omega_1$ yields the equations

$$\frac{\omega_2^*}{\omega_1^*} = w^*, \qquad \frac{\omega_1^*}{\omega_1} = cw + d.$$

Since g_2 and g_3 do not depend on the period basis, upon application of modular substitutions the right-hand sides of (7) satisfy the transformation formula

$$(w^*) = (cw + d)^k(w) \tag{8}$$

with $k = 4$ and $k = 6$. Moreover, as we will presently prove, the functions of w defined by (7) are regular in the upper half plane.

A function $\phi(w)$, which is regular in $\mathfrak{H}$, is called a *modular form of weight* k if relation (8) holds for every modular substitution (6) and if $\phi(w)$ is bounded on the fundamental region $\mathfrak{B}$ of the elliptic modular group introduced in Section 9 of Chapter 1 (Volume I). Here the integer k is assumed to be even. Generalizing (7), we obtain the so-called *Eisenstein series*

$$\gamma_k(w) = \sum_{\mu,\nu}' (\mu w + \nu)^{-k} \qquad (k = 4, 6, 8, \ldots) \tag{9}$$

which are then modular forms of weight k. This is proved in the following way. First each series $\gamma_k(w)$ is uniformly convergent in $\mathfrak{B}$ and, more generally, in every half strip

$$-c_1 \leqslant x \leqslant c_1, \qquad 0 < c_2 \leqslant y.$$

This is seen from the estimate

$$\left|\frac{\mu i + \nu}{\mu w + \nu}\right| = \left|1 + \frac{i - w}{w + \nu\mu^{-1}}\right| \leqslant 1 + \frac{c_1 + 1}{y} + 1 \leqslant 2 + \frac{c_1 + 1}{c_2}$$

and from Theorem 1 in Section 12 of Chapter 1 (Volume I). Since each term of the series $\gamma_k(w)$ is regular on $\mathfrak{H}$, Weierstrass' theorem implies that $\gamma_k(w)$, itself, is a regular function on $\mathfrak{H}$. Also, each $\gamma_k(w)$ is bounded on $\mathfrak{B}$. We finally obtain the transformation formula (8) from the relation

$$\omega_1^{-k}\gamma_k(w) = \sum_{\omega \neq 0} \omega^{-k}.$$

In view of a result established at the end of Section 12 of Chapter 1 (Volume I), each $\gamma_k(w)$, $k = 4, 6, 8, \ldots$ can be expressed as an isobaric polynomial in $\gamma_4(w)$ and $\gamma_6(w)$ with positive rational coefficients. We will return to this algebraic result later in the discussion.

Since $w^* = w + 1$ is a modular substitution, relation (8) implies that every modular form has period 1. Introducing the variable

$$e^{2\pi i w} = q,$$

the modular form $\phi(w)$ becomes a regular function of q within the punctured unit disk $0 < |q| < 1$. Moreover, this function stays bounded as $|q| \to 0$. In view of Riemann's theorem it can be analytically continued over $q = 0$ and is regular at this point. Hence we obtain a Fourier expansion

$$\phi(w) = \sum_{n=0}^{\infty} a_n e^{2\pi i n w} = \sum_{n=0}^{\infty} a_n q^n, \tag{10}$$

which contains no negative powers of q and which is absolutely convergent in $\mathfrak{H}$. Conversely, relation (10) implies the boundedness of $\phi(w)$ within the fundamental region $\mathfrak{B}$ of the modular group.

In order to determine the Fourier coefficients a_n of the Eisenstein series, we start from the well known partial fractions expansion

$$\pi \cot(\pi w) = w^{-1} + \sum_{\nu=1}^{\infty}\{(w + \nu)^{-1} + (w - \nu)^{-1}\}.$$

We also recall the relation

$$i \cot(\pi w) = \frac{1 + q}{1 - q} = 1 + 2\sum_{n=1}^{\infty} q^n,$$

which holds for w in the upper half plane. Differentiating $(k - 1)$ times with respect to w we obtain

$$\sum_{\nu=-\infty}^{\infty} (w + \nu)^{-k} = \frac{(-2\pi i)^k}{(k - 1)!} \sum_{n=1}^{\infty} n^{k-1} q^n, \qquad (k = 2, 3, \ldots).$$

Replacing w by μw and summing over $\mu = 1, 2, \ldots$ yields

$$\sum_{\mu=1}^{\infty} \sum_{\nu=-\infty}^{\infty} (\mu w + \nu)^{-k} = \frac{(-2\pi i)^k}{(k - 1)!} \sum_{n=1}^{\infty} \sigma_{k-1}(n) q^n, \qquad (k = 2, 3, \ldots),$$

where

$$\sigma_{k-1}(n) = \sum_{t \mid n} t^{k-1}, \qquad (n = 1, 2, \ldots)$$

is the sum of the powers t^{k-1} taken over all positive divisors t of n. We observe that the summation in (9) also includes the terms with $\mu \leqslant 0$. Then we obtain the desired Fourier expansion

$$\gamma_k(w) = 2\zeta(k) + \frac{2(2\pi i)^k}{(k-1)!} \sum_{n=1}^{\infty} \sigma_{k-1}(n) q^n, \qquad (k = 4, 6, \ldots).$$

Here the well known relations

$$2\zeta(k) = 2\sum_{n=1}^{\infty} n^{-k} = -\frac{(2\pi i)^k}{k!} B_k, \qquad (k = 2, 4, \ldots),$$

involving the Bernoulli numbers B_k are used.

For later purposes it is expedient to rewrite the Eisenstein series by combining the pairs μ, ν and $-\mu$, $-\nu$ and by extracting the greatest common divisor (μ, ν). We define the sum

$$E_k(w) = \sum_{\mu,\nu}'' (\mu w + \nu)^{-k}, \qquad (k = 4, 6, \ldots), \tag{11}$$

by restricting μ, ν to all relatively prime pairs of integers satisfying the additional conditions $\mu > 0$, ν arbitrary or $\mu = 0$, $\nu = 1$. Then we obtain

$$\gamma_k(w) = 2\zeta(k) E_k(w), \qquad E_k(w) = 1 - \frac{2k}{B_k} \sum_{n=1}^{\infty} \sigma_{k-1}(n) q^n.$$

Hence the Fourier coefficients of $E_k(w)$ are rational numbers which have the numerator of B_k as common denominator. In view of (7) we obtain

$$g_2 = \frac{4}{3}\left(\frac{\pi}{\omega_1}\right)^4 E_4, \qquad g_3 = \frac{8}{27}\left(\frac{\pi}{\omega_1}\right)^6 E_6,$$

$$E_4(w) = 1 + 240q + \cdots, \qquad E_6 = 1 - 504q + \cdots, \tag{12}$$

$$g_2^3 - 27g_3^2 = \frac{64}{27}\left(\frac{\pi}{\omega_1}\right)^{12} (E_4^3 - E_6^2) = \frac{64}{27}\left(\frac{\pi}{\omega_1}\right)^{12} (1728q + \cdots);$$

hence,

$$1728\, j(w) = q^{-1} + b_0 + b_1 q + \cdots, \tag{13}$$

with rational coefficients $b_0, b_1, \ldots$. Furthermore, a more detailed investigation shows that these coefficients are also natural numbers. However, we will not pursue this matter here.

According to the definition of $j(w)$, this function is regular in the entire upper half plane and invariant under modular substitutions. In Section 9 of Chapter 3 (Volume II) it was called the *elliptic modular function*. In view of Theorem 1 in Section 8 of the same chapter, this function has the important property of separating every two points of $\mathfrak{H}$ by its values if these points are not equivalent with respect to the modular group. More generally, we will consider functions $\psi(w)$ which are meromorphic in $\mathfrak{H}$ and invariant under the modular group. In particular, it is evident that every arbitrary meromorphic function of the variable j assumes these properties upon substituting $j = j(w)$. This degree of generality is not appropriate in the present situation. Therefore, we require that the poles of $\psi(w)$ form only finitely many equivalence classes; that is, the function $\psi(w)$ has only finitely many poles in $\mathfrak{B}$. In this case, since $\psi(w)$ is a regular function of q within the punctured disk $0 < |q| < r$, for sufficiently small r, it can be expanded in a convergent Laurent series in that domain. In addition, we require that this Laurent series contain only finitely many negative powers of q; in view of (13), this is true for the particular function $j(w)$. Hence $\psi(w)$, considered as a function of q, can be analytically continued over the point $q = 0$; and it either has a pole or is regular there. Every function satisfying all these conditions is called an *elliptic modular function.* The elliptic modular functions obviously form a field. We will show later that it consists precisely of the rational functions of $j(w)$.

We now turn to the generalization to the case of an arbitrary Picard variety $\mathfrak{N}$ of p dimensions in projective space $\mathfrak{Q}_m$. In analogy with (4), we must set up a system of finitely many algebraic equations $P = 0$ in the $m + 1$ homogeneous coordinates which are satisfied exactly at the points of $\mathfrak{N}$ and we must study the coefficients appearing in these equations. In this connection Theorem 1 of Section 12 of the preceding chapter is decisive. It states that the theta functions $\Theta_r(T, W, z)$ are suitable coordinates which are not only entire functions of z but also regular functions of W in the generalized upper half plane $\mathfrak{H}$. It is of essential importance that, for variable W, certain fixed normal forms of the required algebraic equations $P = 0$ can be determined which define the corresponding Picard variety for every W in $\mathfrak{H}$ without exception. Thus we can also study how the coefficients which arise depend on W and how they are transformed upon application of arbitrary modular substitutions of level T. In fact, this procedure represents a possible approach to the desired generalization of the elliptic modular functions. Primarily, this goal requires considerations leading to the transformation theory of theta functions. However, because of its voluminous nature, this approach will not be presented here.

For this reason we give an independent introduction to the required modular functions of level T. First, we define modular forms as we did in the case $p = 1$. For this purpose we notice that the Jacobian determinant of the mapping (6) is given by

$$\frac{dw^*}{dw} = (cw + d)^{-2}.$$

For the sake of brevity, we write the modular substitution of level T given by (2) in the form

$$W^* = (AW + B)(CW + D)^{-1}.$$

Since both period matrix and principal matrix do not occur in this connection, the notation is unobjectionable. As was shown in Section 3 for arbitrary symplectic transformations, the Jacobian determinant of W^* with respect to W is given by $|CW + D|^{-p-1}$. Therefore, it seems reasonable to define a *modular form* $\phi(W)$ *of level* T *and weight* k by the following two conditions. It must be regular on $\mathfrak{H}$ and satisfy the transformation formula

$$\phi(W^*) = |CW + D|^k \phi(W) \tag{14}$$

upon application of any modular substitution of level T. In order that the power of the determinant be well defined, we only admit integral values of k. Furthermore, we note that the modular substitution remains unchanged if A, B, C, D are simultaneously multiplied by -1. In this case the right-hand side of (14) is multiplied by $(-1)^{pk}$. Hence, we assume k to be even if p is odd.

The analogy with the case $p = 1$ suggests that we require, in addition, the boundedness of $\phi(W)$ in a suitable fundamental region of $\Gamma(T)$. However, it will be proven that this additional condition is unnecessary in the theory of modular forms for $p > 1$. As a matter of fact, this boundedness will follow from the definition of modular forms given in the preceding paragraph. In comparison with the elliptic case, the case $p > 1$ also admits a simplification in defining modular functions of level T. In case $p > 1$, the modular functions level T are simply the meromorphic functions on $\mathfrak{H}$ which are invariant with respect to $\Gamma(T)$. The main theorem then states that the modular functions of level T form an algebraic function field of transcendence degree $n = p(p + 1)/2$. The proof requires some preparatory work and will be completed in the last section.

It is convenient to first establish the theory in detail in the case $T = E$ and then to reduce the general case to that particular case by an algebraic argument. We will study the theory of the special case in the next section. The remainder of the present section will be devoted to explaining how the theory of Abelian functions yields a more extensive generalization of modular functions. We start from Section 13 of the preceding chapter. There the Picard varieties $\mathfrak{N}$ are characterized as those irreducible nonsingular algebraic varieties in projective space which are furnished with a group structure such that the group operations can be expressed everywhere on $\mathfrak{N}$ by regular functions of local coordinates. Then every group element provides a birational and regular self-mapping of $\mathfrak{N}$.

We are now interested in regular mappings of $\mathfrak{N}$ onto or into itself, where we do not assume that these mappings are one-to-one. Again, we introduce the universal covering space $\mathfrak{R}^*$ of $\mathfrak{R}$ and select a fixed basis of the p-dimensional vector space of differentials of the first kind. Then the formula

$$z(\mathfrak{x}^*) = \int_{\mathfrak{e}}^{\mathfrak{x}^*} dz \tag{15}$$

defines a biregular mapping of $\mathfrak{R}^*$ onto $\mathfrak{Z}$. Clearly the given mapping $\mathfrak{A}$ of $\mathfrak{R}$ can be lifted to a mapping $\mathfrak{A}^*$ of $\mathfrak{R}^*$ onto or into itself by first defining this mapping locally and then using analytic continuation and the monodromy theorem. Applying (15), we pass to $\mathfrak{Z}$ and also obtain a mapping

$$\underline{z} = \phi(z) \tag{16}$$

of $\mathfrak{Z}$ onto or into itself. Here the p elements of the column $\phi(z)$ are entire functions of $z_1, \ldots, z_p$. If the point $\mathfrak{x}^*$ on $\mathfrak{R}^*$ describes a path whose projection on $\mathfrak{R}$ is closed, then the integral $z(\mathfrak{x}^*)$ increases by a period ω lying in the period group which was previously denoted by Ω. The image of that path under the mapping $\mathfrak{A}^*$ again has a closed projection on $\mathfrak{R}$; hence, $\underline{z}$ also increases by a period $\underline{\omega}$ in Ω. Therefore, we obtain

$$\underline{z} + \underline{\omega} = \phi(z + \omega), \qquad \phi(z + \omega) - \phi(z) = \underline{\omega}. \tag{17}$$

These relations are valid identically in z once the period ω is arbitrarily fixed. In view of the continuity of $\phi(z)$ and the discreteness of Ω, the period $\underline{\omega}$ depends on ω, but not on z. Each of the partial derivatives of $\phi(z)$ then yields an entire Abelian function which, consequently, must be constant. Hence, the mapping $\phi(z)$ itself is linear and we may write

$$\underline{z} = Mz + a \tag{18}$$

with a constant column a and a constant matrix M of degree p.

The particular case of (18) given by $M = E$ is disposed of by the addition theorem treated in Section 13 of the preceding chapter. From now on we may therefore assume $a = 0$; whence, $\mathfrak{A}$ leaves the point $\mathfrak{e}$ of $\mathfrak{R}$ fixed. We will not exclude the case where the determinant $|M|$ vanishes. In this case $\mathfrak{R}$ is mapped by $\mathfrak{A}$ onto a subset of a smaller dimension. By (16), (17), and (18) we obtain

$$\underline{\omega} = M\omega.$$

Let $\omega_1, \ldots, \omega_{2p}$ denote the columns of a fixed arbitrarily selected period matrix C, and let $\underline{C}$ be the matrix formed by the corresponding columns $\underline{\omega}_k$, $k = 1, \ldots, 2p$. We conclude that

$$\underline{C} = MC.$$

On the other hand, since each $\underline{\omega}_k$ also belongs to Ω we have a relation

$$\underline{C} = CL,$$

where L is a matrix of degree $2p$ with integral entries. Hence, we obtain the important equality

$$MC = CL. \tag{19}$$

Applying complex conjugation we derive $\bar{M}\bar{C} = \bar{C}L$, whence

$$\begin{pmatrix} M & 0 \\ 0 & \bar{M} \end{pmatrix} P = PL, \qquad P = \begin{pmatrix} C \\ \bar{C} \end{pmatrix}. \tag{20}$$

Because $|P| \neq 0$, the relation $|L| \neq 0$ holds if and only if $|M| \neq 0$, and $L = 0$ if and only if $M = 0$.

Conversely, if a complex matrix M of degree p satisfies equation (19) with an integral matrix L, our previous considerations show that putting $\underline{z} = Mz$ yields a mapping $\mathfrak{A}$ of the desired type. If $f(z)$ is an Abelian function corresponding to Ω, then so is $f(Mz)$. Hence, if $f_0(z), \ldots, f_p(z)$ are generators of the Abelian function field, each of the functions $f_0(Mz), \ldots, f_p(Mz)$ can be expressed rationally in terms of these generators, and $\mathfrak{A}$ proves to be a rational mapping.

Every matrix M possessing the indicated property is called a *multiplier* of the Riemann matrix C or of the period lattice Ω. Hence the problem of determining all mappings $\mathfrak{A}$ is reduced to the determination of all multipliers. In view of (19), these multipliers form a ring which contains E as unit element and hence all matrices kE with arbitrary integral scalar k. Therefore, the ring of multipliers contains a subring isomorphic to the ring of integers. A more detailed investigation shows that if $C = (T \quad W)$ is the normal form of the period matrix and if we fix T and select a sufficiently general W in $\mathfrak{H}$, then all multipliers have the trivial form kE. However, it is precisely those particular lattices which admit nontrivial multipliers that are of special interest.

To obtain further information we consider the case $p = 1$. Here we may write

$$M = (\mu), \qquad C = (1 \quad w), \qquad L = \begin{pmatrix} d & b \\ c & a \end{pmatrix}$$

with integers a, b, c, d. Then (19) yields the equations

$$\mu(1 \quad w) = (1 \quad w)\begin{pmatrix} d & b \\ c & a \end{pmatrix}, \qquad aw + b = \mu w, \qquad cw + d = \mu,$$

$$(a - \mu)(d - \mu) = bc, \qquad \mu^2 - (a + d)\mu + ad - bc = 0.$$

If μ is not a rational integer we obtain $c \neq 0$, $ad - bc \neq 0$. Since the period ratio w is not real and μ satisfies a quadratic equation with rational integral coefficients, it follows that w is an imaginary quadratic irrational number. In addition μ is an algebraic integer in the field generated by w, and μ is not a real number. Conversely, one can show that, for any given imaginary quadratic irrational number w, the ring of all multipliers μ does not consist only of the rational integers k, but is a nontrivial order in the field generated by w. Therefore, in the case where nonreal multipliers exist we say that *the lattice admits complex multiplication.* Here we should recall Fagnano's discovery which was dealt with in Section 1 of Chapter 1 (Volume I). At that point we had already mentioned that Fagnano's formulas express, in complex notation, the multiplication of the lemniscate integral by $1 + i$ and $1 - i$. Now we notice that in the case of the lemniscate integral the period ratio $w = i$ is, in fact, an imaginary-quadratic irrational number, and we obtain

$$(1 \pm i)(1 \quad i) = (1 \quad i)\begin{pmatrix} 1 & \mp 1 \\ \pm 1 & 1 \end{pmatrix}.$$

The theory of complex multiplication, developed in particular by Kronecker, leads to most remarkable results whose main applications are to number theory and, therefore, will not be treated here. As a final comment on this subject, we observe that the explicit formulation of the mapping $\mathfrak{A}$ in case $p = 1$ requires development of the transformation theory of elliptic functions and also leads to the investigation of corresponding properties of the elliptic modular functions.

Again let p be arbitrary. Then the most important problem is to determine all possible rings of multipliers occurring for the various suitable choices of Riemann matrices C. This problem proves to be more accessible if we admit, more generally, matrices with rational entries as solutions L of (19). Clearly, we now obtain all the corresponding new solutions M from the previous solutions by division by natural numbers. The ring of multipliers thus obtained is an algebra of finite dimension over the rationals. In view of (20), this algebra has a faithful representation by rational matrices L, which will be called the *multiplier algebra* of C. We will now see that this algebra admits an easily described involution. Here an involution of an algebra $\mathfrak{a}$ consisting of elements $\alpha, \beta, \ldots$ is understood to be a mapping $\alpha \rightarrow \alpha^*$ of $\mathfrak{a}$ onto itself satisfying the rules $(\alpha^*)^* = \alpha$ and $(\alpha + \beta)^* = \alpha^* + \beta^*$, $(\alpha\beta)^* = \beta^*\alpha^*$.

Theorem 1: Let C be a Riemannian matrix and A a corresponding principal matrix. Then the equation

$$L^* = A^{-1}L'A$$

defines an involution of the multiplier algebra of C, and we have

$$\sigma(LL^*) > 0$$

for every multiplier $L \neq 0$. Here $\sigma(LL^*)$ denotes the trace of LL^*.

Proof: Putting

$$i\bar{C}A^{-1}C' = H,$$

the period relations imply $H > 0$; hence, in particular, $|H| \neq 0$, and

$$i\bar{P}A^{-1}P' = \begin{pmatrix} H & 0 \\ 0 & -\bar{H} \end{pmatrix}. \tag{21}$$

On the other hand, by (20) we have

$$\bar{P}'\begin{pmatrix} \bar{M}' & 0 \\ 0 & M' \end{pmatrix} = L'\bar{P}',$$

and in view of (21) we obtain

$$\begin{pmatrix} \bar{H} & 0 \\ 0 & -H \end{pmatrix}\begin{pmatrix} \bar{M}' & 0 \\ 0 & M' \end{pmatrix}\begin{pmatrix} \bar{H}^{-1} & 0 \\ 0 & -H^{-1} \end{pmatrix}P = -iPA^{-1}L'\bar{P}'\begin{pmatrix} \bar{H}^{-1} & 0 \\ 0 & -H^{-1} \end{pmatrix}P$$
$$= PA^{-1}L'A.$$

Introducing the notation

$$M^* = \bar{H}\bar{M}'\bar{H}^{-1}, \qquad L^* = A^{-1}L'A,$$

it follows that

$$M^*C = CL^*.$$

Hence, if L belongs to the multiplier algebra of C, then so does L^*. It is immediately verified that $(L^*)^* = L$ and that the other properties required of an involution are also satisfied. Thus we have proved that the mapping $L \to L^*$ is an involution. Furthermore, according to (20) and (21) we obtain

$$\bar{P}LL^*\bar{P}^{-1} = \begin{pmatrix} \bar{M} & 0 \\ 0 & M \end{pmatrix}\bar{P}A^{-1}P'\begin{pmatrix} M' & 0 \\ 0 & \bar{M}' \end{pmatrix}(\bar{P}A^{-1}P')^{-1}$$
$$= \begin{pmatrix} \bar{M}HM'H^{-1} & 0 \\ 0 & M\bar{H}\bar{M}'\bar{H}^{-1} \end{pmatrix}$$

and hence,

$$\sigma(LL^*) = 2\sigma(\bar{M}HM'H^{-1}).$$

Now we observe that both matrices H^{-1} and $K = \bar{M}HM'$ are Hermitian matrices satisfying $H^{-1} > 0$ and $K \geqslant 0$, where $K \neq 0$ for $M \neq 0$. Transforming both matrices H and K simultaneously into diagonal form, we conclude that $\sigma(KH^{-1}) > 0$ for $M \neq 0$. Finally, relation (20) implies that $M = 0$ only if $L = 0$. Thus the proof of Theorem 1 is complete.

In view of Theorem 1, in order to set up a complete system of multiplier algebras we need only investigate algebras of finite rank over the rationals admitting an involution with the indicated positivity property. The detailed investigation was carried out by Albert and applies sophisticated theorems of number theory.

To every given multiplier algebra $\mathfrak{a}$, we now assign a subspace of $\mathfrak{H}$ and a subgroup of the modular group of level T in the following way. We consider all Riemann matrices $C = (T \quad W)$ whose multiplier algebra either coincides with $\mathfrak{a}$ or contains $\mathfrak{a}$, where the corresponding principal matrix is fixed as

$$A = \begin{pmatrix} 0 & T \\ -T & 0 \end{pmatrix}.$$

Then it can be shown that the admissible points W of $\mathfrak{H}$ again constitute a symmetric domain $\mathfrak{H}(\mathfrak{a})$ whose dimension is less than n if $\mathfrak{a}$ is not isomorphic to the field of rational numbers. Decomposing this symmetric domain into irreducible factors, we see that three of the Cartan types may actually occur. Furthermore, if U is an element of the modular group of level T, whence U is unimodular and $U'AU = A$, then we put $\underline{C} = (T \quad \underline{W})$, where $\underline{W}$ is obtained from W by applying the modular substitution of level T defined by (2) and (3). Since, in that case, the equation

$$CU = (TM_1 + WM_3)T^{-1}\underline{C}$$

holds, we see by applying (19) that the matrix $U^{-1}LU$ is an element of the multiplier algebra of $\underline{C}$. Moreover, if U commutes with all elements L of $\mathfrak{a}$, and hence belongs to the so-called "commutator algebra" of $\mathfrak{a}$, it follows that $\underline{W}$ also belongs to $\mathfrak{H}(\mathfrak{a})$. Let $\Gamma(T, \mathfrak{a})$ denote the intersection of $\Gamma(T)$ with the commutator algebra of $\mathfrak{a}$. The elements of this generalized modular group $\Gamma(T, \mathfrak{a})$ represent modular substitutions of level T which map $\mathfrak{H}(\mathfrak{a})$ onto itself. A more detailed investigation shows that $\Gamma(T, \mathfrak{a})$ acts discontinuously on $\mathfrak{H}(\mathfrak{a})$ and possesses a fundamental region in this domain which can be constructed by use of the reduction theory of quadratic forms.

Of special interest is the case when $\mathfrak{a}$ is isomorphic to an algebraic number field. In this case Theorem 1 implies that $\mathfrak{a}$ is either a totally imaginary extension of a totally real number field or is totally real itself. Assuming in the second case that p is also equal to the degree of $\mathfrak{a}$ over the rationals, the group $\Gamma(T, \mathfrak{a})$ coincides with the so called *Hilbert modular group* for a suitable choice of T. The corresponding automorphic functions are called *Hilbert's modular functions* although it was Hilbert's students Blumenthal and Hecke who first penetrated deeply into the theory of these functions. More recently, further progress has been achieved, but there is much left to be done. Finally, we ought to mention that Pjatetzki–Schapiro has considered the modular functions corresponding to an arbitrary multiplier algebra.

Our subsequent investigations will at first be restricted to the proper modular functions of degree p corresponding to the modular group $\Gamma(E)$. Only at the end will we return to an arbitrary principal matrix instead of $A = J$. We will not consider the modular groups $\Gamma(T, \mathfrak{a})$ and their corresponding functions in the sequel.

5. *The fundamental region of the modular group*

Let Γ denote the modular group of degree p and $\mathfrak{H}$ the corresponding generalized upper half plane. Our immediate goal is to construct a particular fundamental region in $\mathfrak{H}$ for the group Γ of mappings

$$W^* = (AW + B)(CW + D)^{-1}$$

which will be expedient for our purposes.

We start with $p = 1$ and refer to Section 9 of Chapter 1 (Volume I). Let ω_1, ω_2 be fundamental periods of an elliptic function with $\omega_2/\omega_1 = w$ located in the upper half plane $\mathfrak{H}$. By applying all homogeneous modular substitutions

$$\omega_1^* = a\omega_1 + b\omega_2, \qquad \omega_2^* = c\omega_1 + d\omega_2, \tag{1}$$

with integral a, b, c, d and $ad - bc = 1$, we obtain all pairs of fundamental periods whose ratios $\omega_2^*/\omega_1^* = w^*$ again lie in $\mathfrak{H}$. For a given lattice a

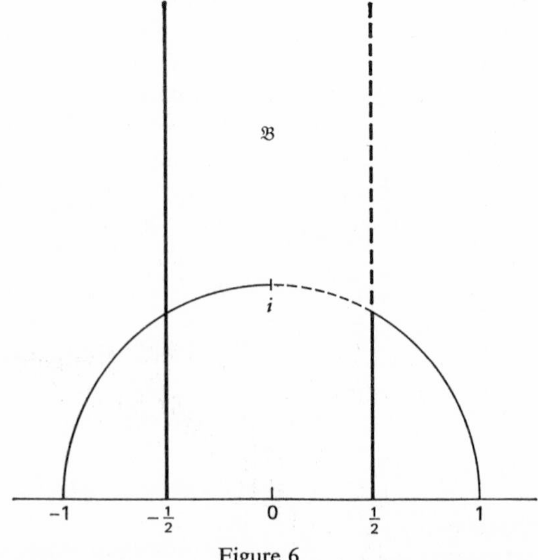

Figure 6

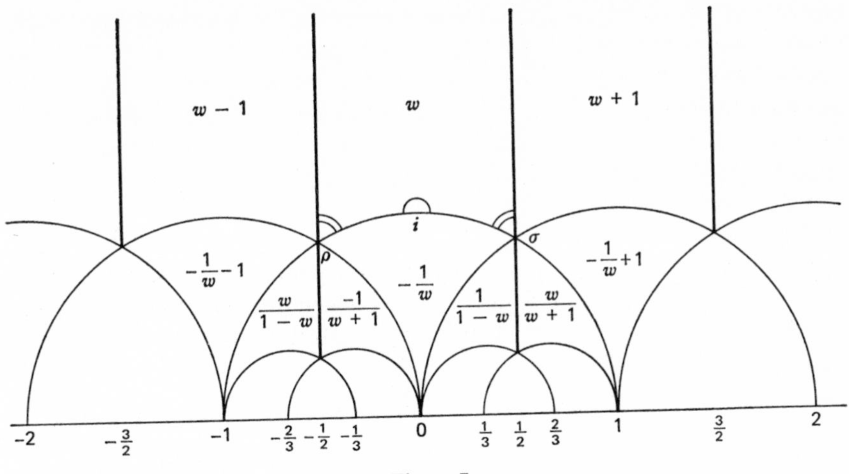

Figure 7

reduced pair ω_1^*, ω_2^* is determined by two conditions. First, the absolute value of ω_1^* is chosen as small as possible, and second, among those pairs with this smallest value $|\omega_1^*|$, the absolute value of ω_2^* is also selected as small as possible.

Now, assume the pair ω_1, ω_2 is already reduced. The inequality

$$|a + bw| \geqslant 1$$

must then be satisfied for all relatively prime pairs a, b, and the inequality $|c + w| \geqslant |w|$ must also hold for all integral c. Putting $w = x + iy$, the conditions $a = 0$, $b = 1$ and $c = \pm 1$ yield the inequalities

$$x^2 + y^2 \geqslant 1, \qquad -\tfrac{1}{2} \leqslant x \leqslant \tfrac{1}{2}$$

which, as was shown in Section 9 of Chapter 1 (Volume I), determine the region (Figure 6) also denoted by $\mathfrak{B}$ in that discussion. The previous considerations imply that $\mathfrak{B}$ is a fundamental region of the elliptic modular group. More precisely, it was shown that every point of $\mathfrak{H}$ is equivalent to exactly one point of the region $\mathfrak{F}$ which is obtained from $\mathfrak{B}$ by omitting that part of the boundary which belongs to the right half plane. This part of the boundary is indicated by a broken line in the figure. Now we will identify those images of $\mathfrak{B}$ which are adjacent to $\mathfrak{B}$, that is, which have at least one boundary point in common with $\mathfrak{B}$. The two vertices of $\mathfrak{B}$ are given by

$$\sigma = e^{\pi i/3} = \frac{1 + \sqrt{-3}}{2}, \qquad \rho = \sigma^2 = e^{2\pi i/3} = \frac{-1 + \sqrt{-3}}{2}$$

The substitutions indicated in Figure 7 yield altogether nine neighboring regions not including $\mathfrak{B}$ itself. One realizes that i is a fixed point with respect to $w^* = -w^{-1}$, whereas ρ is a fixed point with respect to either of the substitutions $w^* = -w^{-1} - 1$ and $w^* = -(w+1)^{-1}$. By calling special attention to the fixed points, these three substitutions can be put in the form

$$\frac{w^* - i}{w^* + i} = -\frac{w - i}{w + i}, \qquad \frac{w^* - \rho}{w^* - \bar{\rho}} = \rho\frac{w - \rho}{w - \bar{\rho}}, \qquad \frac{w^* - \rho}{w^* - \bar{\rho}} = \bar{\rho}\frac{w - \rho}{w - \bar{\rho}}.$$

Furthermore, identifying equivalent boundary points of $\mathfrak{B}$, we obtain complete neighborhoods of any boundary point which is different from ρ, σ, and i. At the point i only, an application of the substitution

$$\left(\frac{w - i}{w + i}\right)^2 = t$$

yields a complete neighborhood of $t = 0$, and at the point ρ we must correspondingly apply

$$\left(\frac{w - \rho}{w - \bar{\rho}}\right)^3 = t.$$

As Gauss recognized, there is a close relation between the definition of $\mathfrak{B}$ and the reduction theory of positive definite binary quadratic forms founded by Lagrange. Setting

$$\omega_1\bar{\omega}_1 = \lambda, \qquad \omega_1\bar{\omega}_2 + \omega_2\bar{\omega}_1 = -2\mu, \qquad \omega_2\bar{\omega}_2 = \nu$$

and denoting two real variables by u and v, the expression

$$(2) \qquad \begin{aligned} |\omega_1 u - \omega_2 v|^2 &= (\omega_1 u - \omega_2 v)(\bar{\omega}_1 u - \bar{\omega}_2 v) \\ &= \lambda u^2 + 2\mu uv + \nu v^2 \end{aligned}$$

is seen to be a positive definite quadratic form with determinant

$$\lambda\nu - \mu^2 = \left(\frac{\omega_1\bar{\omega}_2 - \omega_2\bar{\omega}_1}{2i}\right)^2.$$

Conversely, given any three real numbers λ, μ, ν satisfying $\lambda > 0$ and $\lambda\nu - \mu^2 > 0$, these equations determine ω_1, ω_2 uniquely up to a common factor of absolute value 1; that is, we have

$$|\omega_1| = +\sqrt{\lambda}, \qquad \omega_2/\omega_1 = \lambda^{-1}(-\mu + \sqrt{\mu^2 - \lambda\nu}).$$

Using the coefficients in (1), we introduce the unimodular substitution of determinant 1

$$(3) \qquad u = au^* - cv^*, \qquad v = -bu^* + dv^*.$$

We obtain

$$\omega_1 u - \omega_2 v = \omega_1^* u^* - \omega_2^* v^*.$$

Moreover, putting

$$\begin{pmatrix} a & -c \\ -b & d \end{pmatrix} = U, \qquad \begin{pmatrix} \lambda & \mu \\ \mu & \nu \end{pmatrix} = S,$$

the homogeneous linear substitution with matrix U transforms the quadratic form (2) with matrix S into a properly equivalent quadratic form whose matrix is given by $U'SU$. Part of the task of reduction theory is to distinguish by suitable conditions representatives of these equivalence classes which constitute a region with nice boundary in the space of all real symmetric positive definite matrices S. According to Lagrange this is done by first selecting among all properly equivalent matrices

$$U'SU = \begin{pmatrix} \lambda^* & \mu^* \\ \mu^* & \nu^* \end{pmatrix}$$

those representatives which assume the minimal value of λ^* and then selecting among those with minimal λ^* a representative with minimal value of ν^*. Because $\lambda^* = \omega_1^*\overline{\omega_1^*}$ and $\nu^* = \omega_2^*\overline{\omega_2^*}$, this procedure amounts to the same thing as the determination of a reduced pair ω_1^*, ω_2^* in the previous sense. If S is already reduced according to Lagrange's definition, the inequalities

$$|\omega_2 \pm \omega_1| \geqslant |\omega_2| \geqslant |\omega_1|$$

are satisfied. In view of the proof of Theorem 3 in Section 9 of Chapter 1 (Volume 1), These inequalities characterize the pair ω_1, ω_2 as a reduced pair. The reduction conditions may be written as

$$-\lambda \leqslant 2\mu \leqslant \lambda \leqslant \nu, \tag{4}$$

and they were stated by Lagrange in this form. Since these inequalities are homogeneous, we can restrict ourselves to the case where the determinant satisfies

$$|S| = \lambda\nu - \mu^2 = 1.$$

Again, introducing $\omega_2\omega_1^{-1} = w = x + iy$ we obtain

$$\nu = (x^2 + y^2)\lambda, \qquad \mu = -x\lambda,$$

and hence $\lambda = y^{-1}$. This yields the parametric representation

$$S = \begin{pmatrix} \dfrac{1}{y} & \dfrac{-x}{y} \\[2ex] \dfrac{-x}{y} & y + \dfrac{x^2}{y} \end{pmatrix}$$

whose generalization will now be treated.

Theorem 1: A symplectic matrix S is positive symmetric if and only if there exists an element $W = X + iY$ of $\mathfrak{H}$ satisfying

$$S = \begin{pmatrix} Y^{-1} & -Y^{-1}X \\ -XY^{-1} & Y + XY^{-1}X \end{pmatrix}. \tag{5}$$

Subjecting W to an arbitrary symplectic transformation

$$W^* = (AW + B)(CW + D)^{-1} \tag{6}$$

with matrix

$$M = \begin{pmatrix} A & B \\ C & D \end{pmatrix}$$

corresponds to the transformation

$$S = M'S^*M. \tag{7}$$

Proof: Let S be symmetric and symplectic. For scalar λ, we obtain

$$(S + \lambda iJ)J(S - \lambda iJ) = (1 - \lambda^2)J, \qquad |S + \lambda iJ|^2 = (1 - \lambda^2)^{2p}. \tag{8}$$

Now, assume, in addition, $S > 0$. Since iJ is Hermitian and S positive Hermitian, there exist an invertible complex matrix Q and a real diagonal matrix R satisfying $S = Q'\bar{Q}$ and $iJ = Q'R\bar{Q}$. Denoting the diagonal elements of R by $r_1, \ldots, r_{2p}$, we derive from (8) and

$$S + \lambda iJ = Q'(E + \lambda R)\bar{Q}$$

that p of the numbers r_k $(k = 1, \ldots, 2p)$ are equal to 1 and the other p numbers are equal to -1. Hence, the Hermitian matrix $S + iJ$ has rank p and is nonnegative. This result suggests the Ansatz

$$S + iJ = (E \quad L)'Y^{-1}(E \quad \bar{L}), \qquad Y > 0, \tag{9}$$

where Y^{-1} is the symmetric matrix formed by the first p rows and columns of S. By use of (8) with $\lambda = -1$, we obtain

$$(E \quad L)J(E \quad L)' = 0.$$

Hence $L' = L$. Taking the imaginary parts in (9) yields

$$Y^{-1}(\bar{L} - L) = 2iE.$$

Therefore, we may put

$$L = -(X + iY) = -W$$

with a real matrix X and an element W of $\mathfrak{H}$. Since, in this case, we have

$$L'Y^{-1}\bar{L} = W'Y^{-1}\bar{W} = Y + XY^{-1}X,$$

equation (9) implies assertion (5).

Putting $L = L_1^{-1}L_2$ and applying the substitution

$$L^* = L_1D' - L_2C', \qquad L_2^* = -L_1B' + L_2A',$$

equation (6) becomes

$$(L_1^*L_2^*) = (L_1L_2)M^{-1}, \qquad L_1^*W^* = -L_2^*$$

by use of the symmetry of W^*. Introducing $(L_1 \quad L_2) = P$ and $(L_1^* \quad L_2^*) = P^*$ we conclude that $P = P^*M$ and

$$\frac{i}{2}\,\bar{P}JP' = \frac{i}{2}\,\bar{L}_1(E - \bar{W})J(E - W)'L_1' = \bar{L}_1YL_1'.$$

Then (9) can be written as

$$S + iJ = P'\left(\frac{i}{2}\,\bar{P}JP'\right)^{-1}\bar{P}.$$

By use of P^*, we write $S^* + iJ$ in an analogous fashion and observe the formulas $M = \bar{M}$, $MJM' = M'JM = J$. Then it follows that

$$M'(S^* + iJ)M = S + iJ, \qquad M'S^*M = S;$$

whence, (7) is also proved.

The connection between the construction of the fundamental region of the elliptic modular group and the reduction theory of binary quadratic forms suggests looking for a generalization to arbitrary values of p. Therefore, our next step will be to explain the reduction theory of positive definite quadratic forms of an arbitrary number of variables. This theory was created in full generality by Minkowski after Hermite had done some preliminary work. In recapitulating this theory we must omit the proofs of its important theorems. We will only give the results and some comments.

Let $x_1, \ldots, x_m$ be real variables forming the column x and let S be the matrix of a positive definite quadratic form in these variables. The coefficients, that is, the entries of S are supposed to be real numbers. For the sake of brevity the form $x'Sx$ will be denoted by $S[x]$. By considering the n elements s_{kl} ($1 \leqslant k \leqslant l \leqslant m$; $n = m(m+1)/2$) in fixed order as rectangular Cartesian coordinates, the positive symmetric matrices S form a convex open domain $\mathfrak{S}$ in n-dimensional Euclidean space. Every unimodular transformation $x \to Ux$ yields a corresponding linear mapping

$$S \to U'SU = S[U] = T \tag{10}$$

of $\mathfrak{S}$ onto itself where, obviously, U and $-U$ yield the same mapping. It also can be easily verified that any two such mappings $S \to S[U_1]$ and $S \to S[U_2]$ satisfy $S[U_1] = S[U_2]$ identically on $\mathfrak{S}$ if and only if $U_2 = \pm U_1$. Therefore, let Λ denote the factor group of the unimodular group obtained by identification of U and $-U$. Then a faithful representation of Λ is given by (10). A fundamental region $\mathfrak{R}$ for this group can be determined by the following conditions. Let $\breve{\mathfrak{u}}_1, \ldots, \breve{\mathfrak{u}}_m$ be the columns of U and assume the columns $\breve{\mathfrak{u}}_1, \ldots, \breve{\mathfrak{u}}_{r-1}$ are already selected for an integer r satisfying $1 \leqslant r \leqslant m$. Then consider all unimodular matrices U possessing these $r - 1$ fixed first columns and select among the corresponding columns $\breve{\mathfrak{u}}_r$ a representative so that the value $S[\breve{\mathfrak{u}}_r] = s_r$ is as small as possible. By multiplying $\breve{\mathfrak{u}}_2, \ldots, \breve{\mathfrak{u}}_m$ by suitable scalar factors ± 1 we obviously can satisfy the $m - 1$ additional conditions $\breve{\mathfrak{u}}_1' S \breve{\mathfrak{u}}_r \geqslant 0$ $(r = 2, \ldots, m)$. Then it appears that the indicated conditions for $\pm U$, in general, determine a unique representative of the class of all quadratic forms $T[x]$ equivalent to $S[x]$. This representative is called *reduced*. Again denoting the representative T by S, the region $\mathfrak{R}$ of all reduced S is described by the following homogeneous linear inequalities. Let $\mathfrak{e}_1, \ldots, \mathfrak{e}_m$ stand for the columns of the unit matrix. Then we have

$$S[\mathfrak{g}_r] \geqslant S[\mathfrak{e}_r] \qquad (r = 1, \ldots, m) \tag{11}$$

for all integral columns $\mathfrak{g}_r \neq \pm \mathfrak{e}_r$ whose last $m - r + 1$ entries are relatively prime. Furthermore, we have

$$\mathfrak{e}_1' S \mathfrak{e}_r \geqslant 0 \qquad (r = 2, \ldots, m). \tag{12}$$

Now, we may prove the important theorem that, among the infinitely many inequalities (11) and (12), only finitely many are essential and these imply all the others. This means that $\mathfrak{R}$ is a convex pyramid in n-dimensional space whose vertex is located at the origin.

Let $s_k = s_{kk}$ $(k = 1, \ldots, m)$ be the diagonal elements of S. As a simple consequence of (11) we obtain the particular inequalities

$$s_k \leqslant s_{k+1}, \qquad -s_k \leqslant 2s_{kl} \leqslant s_k \qquad (1 \leqslant k \leqslant l \leqslant m). \tag{13}$$

Moreover, Minkowski drew a deeper conclusion from (11) by showing that the inequality

$$s_1 s_2 \cdots s_m \leqslant \gamma_m \, |S| \tag{14}$$

holds with a suitable positive number γ_m depending only on m. In case $m = 1$, equality obviously holds in (14) with $\gamma_1 = 1$; whereas, in case $m = 2$, the inequalities (13) imply the estimate

$$\frac{|S|}{s_1 s_2} = 1 - \frac{s_{12}^2}{s_1 s_2} \geqslant 1 - \frac{s_1}{4 s_2} \geqslant \frac{3}{4}.$$

This shows that (14) is satisfied with $\gamma_2 = 4/3$. By the way, one easily verifies that this is the optimal value of γ_2. For arbitrary m the corresponding optimal value of γ_m is not explicitly known. It should be noted in this connection that the inequality

$$|S| \leqslant s_1 s_2 \cdots s_m$$

holds for arbitrary $S > 0$ and is derived from Hadamard's determinant theorem without any difficulties.

By the so-called Jacobian transformation, that is the method of completing the square, we obtain

$$S[x] = t_1 y_1^2 + t_2 y_2^2 + \cdots + t_m y_m^2, \qquad y_k = x_k + q_{k,k+1} x_{k+1} + \cdots + q_{km} x_m \quad (k = 1, \ldots, m)$$

where $t_1, \ldots, t_m$ are positive and q_{kl} $(1 \leqslant k < l \leqslant m)$ are real. Let T denote the diagonal matrix with diagonal elements $t_1, \ldots, t_m$ and Q the triangular matrix with entries q_{kl} where, in addition, we define $q_{kk} = 1$ $(k = 1, \ldots, m)$ and $q_{kl} = 0$ $(1 \leqslant l < k \leqslant m)$. Then it follows that

$$S = T[Q]. \tag{15}$$

Here both the positive diagonal matrix T and the real triangular matrix Q are uniquely determined by the positive symmetric matrix S; conversely, any such matrices T and Q yield a unique point S in $\mathfrak{S}$. In addition, as a consequence of (13) and (14) it follows that, for reduced S, the inequalities

$$\frac{t_k}{t_{k+1}} \leqslant c, \qquad -c \leqslant q_{kl} \leqslant c \qquad (1 \leqslant k < l \leqslant m) \tag{16}$$

are always satisfied with a positive number c depending only on m.

For arbitrarily fixed $c > 0$ let $\mathfrak{R}(c)$ be the closed region in $\mathfrak{S}$ defined by (16). Then there exists a sufficiently large value of c only depending on m such that $\mathfrak{R}$ is entirely contained in $\mathfrak{R}(c)$. The advantage of this result is that, on one hand, $\mathfrak{R}(c)$ has the very simple explicit definition (16) and, on the other hand, we can find some properties of $\mathfrak{R}$ by proving the analogous properties of $\mathfrak{R}(c)$. This is illustrated by the following consideration. Since $\mathfrak{R}$ is a fundamental region of Λ in $\mathfrak{S}$, the images $\mathfrak{R}_U$ of $\mathfrak{R}$ under the various mappings (10) yield a covering without gaps and overlaps of $\mathfrak{S}$ by pyramids. All these pyramids have their vertices at the origin of n-dimensional space. It appears that only finitely many images $\mathfrak{R}_U$ are neighbors of $\mathfrak{R}$, that is, have any point in common with $\mathfrak{R}$. Now, this important result is an immediate consequence of the more general result that for any given positive c only finitely many images of $\mathfrak{R}(c)$ under all mappings (10) have a point in common with $\mathfrak{R}(c)$. In particular, it follows that every point of $\mathfrak{R}$ is covered by the images of $\mathfrak{R}(c)$ only finitely-many times. Because of this property it is possible to replace the fundamental region $\mathfrak{R}$ by the more convenient region $\mathfrak{R}(c)$ in some of our investigations.

Obviously, now we should construct a fundamental region in $\mathfrak{H}$ for the modular group of degree p by use of Theorem 1 and the previously sketched reduction theory of Minkowski for $m = 2p$. In case $p = 1$, this was already done at the beginning of the present section. However, the following should be noted. The homogeneous elliptic modular group consists of all unimodular matrices

$$U = \begin{pmatrix} a & b \\ c & d \end{pmatrix}$$

with determinant 1, which we shall call *proper unimodular matrices.* Minkowski's reduction theory, however, is based on the full unimodular group. Now, if p is odd, the determinants of U and $-U$ are equal except for sign, and we may restrict ourselves to proper unimodular transformations when reducing a quadratic form $S[x]$. If p is even, we obtain a fundamental region for the proper unimodular group from $\mathfrak{R}$ by omitting the last condition in (12). In case $p = 1$, this leads precisely to Lagrange's conditions (4) and the fundamental region $\mathfrak{B}$ of the elliptic modular group. In case $p > 1$, another difficulty arises from the fact that a unimodular matrix with $2p$ rows, in general, is not a modular matrix of degree p if it has determinant 1. This difficulty is overcome in the following way.

Let $\mathfrak{S}^*$ denote the set of all matrices S which are simultaneously positive symmetric and symplectic. In view of Theorem 1, the domain $\mathfrak{S}^*$ is birationally mapped onto $\mathfrak{H}$, where the real and imaginary parts of the elements $w_{kl} = x_{kl} + iy_{kl}$ $(1 \leqslant k \leqslant l \leqslant p)$ of $W = X + iY$ are taken as coordinates in $\mathfrak{H}$. If S is any point of $\mathfrak{S}^*$, we conclude that

$$S = U'RU$$

with suitable unimodular U and reduced R, which need not be symplectic. Introducing the unimodular matrix

$$UJU' = L, \tag{17}$$

the relation $SJS = J$ implies that

$$L'RL = R^{-1}. \tag{18}$$

Now, let $V = (v_{kl})$ be the matrix of the linear substitution $x_k \rightarrow x_{2p-k+1}$ $(k = 1, \ldots, 2p)$; that is, $v_{kl} = 1$ for $k + l = 2p + 1$, and $v_{kl} = 0$ otherwise. Then, putting

$$R = T[Q]$$

with a diagonal matrix T and a triangular matrix Q, we obviously obtain

$$R^{-1}[V] = (T^{-1}[V])[V(Q^{-1})'V],$$

where the diagonal matrix $T^{-1}[V]$ contains the same diagonal elements as T^{-1} in opposite order, and $V(Q^{-1})'V$ is a triangular matrix of the same form as Q.

Now by use of (16), it follows that for suitable c independent of S both R and $R^{-1}[V]$ are located in $\mathfrak{R}(c)$. Then, in view of the previously discussed property of $\mathfrak{R}(c)$, using (18) we derive the fundamental result that the matrices LV, hence the matrices L themselves, belong to a finite set independent of S. For each admissible L select a fixed unimodular U_0 satisfying

$$U_0 J U_0' = L.$$

Then, if (17) is satisfied by the same L we obviously obtain

$$U_0^{-1} U = M$$

with a modular matrix M of degree p. Furthermore

$$U = U_0 M, \qquad S = (R[U_0])[M].$$

Hence, $S[M^{-1}]$ lies in one of the finitely many pyramids $\mathfrak{R}_{U_0}$ obtained from Minkowski's fundamental region $\mathfrak{R}$ by applying the mappings $R \to R[U_0]$ which correspond to the finite set of matrices L. The intersection of $\mathfrak{S}^*$ with the union of these pyramids $\mathfrak{R}_{U_0}$ then yields a fundamental region of Γ if we again pass to $\mathfrak{H}$ by use of Theorem 1. But, for function theoretical applications it is not expedient to base further investigations on this fundamental region because the determination of the matrices U_0 causes some complications. Therefore, it is more useful to proceed in the way which we now describe.

Putting $w = x + iy$, $w^* = x^* + iy^*$ and assuming that w and w^* are related by the elliptic modular substitution $w^* = (aw + b)(cw + d)^{-1}$, we obtain

$$y^* = \frac{y}{(cw + d)(c\bar{w} + d)}. \tag{19}$$

In particular, if w is contained in the fundamental region $\mathfrak{B}$, the absolute value of $cw + d$ is at least 1, and hence $y^* \leqslant y$. Therefore, for all points w^* equivalent to w, the imaginary part assumes its maximal value at w. In the sequel the absolute value of the determinant of a square matrix L is simply denoted by abs L. Applying (20) and (21) of Section 3 to an arbitrary symplectic substitution (6) or, in particular, to a modular substitution of degree p, we obtain as a generalization of (19) the two formulas

$$(Y^*)^{-1} = (C\overline{W} + D)Y^{-1}(CW + D)' = Y^{-1}[XC' + D'] + Y[C'], \tag{20}$$

$$|Y^*| = |Y| \operatorname{abs}(CW + D)^{-2} \tag{21}$$

where $W^* = X^* + iY^*$. Now, we call W *semireduced* if the relation $|Y^*| \leqslant |Y|$ is satisfied for every modular substitution. According to (21), this condition is equivalent to

$$\text{abs}\,(CW + D) \geqslant 1. \tag{22}$$

Furthermore, we may assume $C \neq 0$ since, otherwise, D is unimodular and equality holds identically in (22). The matrix $(C \quad D)$ will also be called the second row of

$$M = \begin{pmatrix} A & B \\ C & D \end{pmatrix}.$$

Theorem 2. Every class of points in $\mathfrak{H}$ which are equivalent with respect to Γ contains a semireduced point.

Proof: In view of (21) we must show that the value $|Y^*|$ assumes a maximum for a suitable modular substitution. If M is a modular matrix and U any $p \times p$ unimodular matrix, the product

$$\begin{pmatrix} (U^{-1})' & 0 \\ 0 & U \end{pmatrix} M = \begin{pmatrix} (U^{-1})'A & (U^{-1})'B \\ UC & UD \end{pmatrix} \tag{23}$$

is also a modular matrix. Hence if $(C \quad D)$ is the second row of a modular matrix, then so is $(UC \quad UD)$. In this case we call the matrix pairs C, D and UC, UD *associated*. Now, replacing M by the modular matrix appearing on the right-hand side of (23), the matrix $(Y^*)^{-1}$ changes to $(Y^*)^{-1}[U']$ in view of (20); whereas, the value abs $(CW + D)$ remains unchanged. Therefore, in order to prove the theorem, we may assume that the positive symmetric matrix $(Y^*)^{-1}$ is reduced in the sense of Minkowski for all modular substitutions under consideration. Hence it lies in the fundamental region $\mathfrak{R}$ formed for $m = p$.

Now, let $\mathfrak{c}_k, \mathfrak{d}_k$ $(k = 1, \ldots, p)$ be the columns of C' and $XC' + D'$ respectively and let $u_1, \ldots, u_p$ be the diagonal elements of $(Y^*)^{-1}$. Then (20) yields the formula

$$u_k = Y^{-1}[\mathfrak{d}_k] + Y[\mathfrak{c}_k] \geqslant \begin{cases} Y[\mathfrak{c}_k] \\ Y^{-1}[\mathfrak{d}_k] \end{cases} \qquad (k = 1, \ldots, p). \tag{24}$$

If, for any k, both columns $\mathfrak{c}_k$ and $\mathfrak{d}_k$ vanish, the kth row of the matrix $(C \quad D)$ is also zero. But this is impossible in view of $|M| = 1$. Since $Y > 0$, the value $Y[\mathfrak{c}_k]$ assumes a positive minimum when W is fixed and the integral column $\mathfrak{c}_k \neq 0$ is arbitrary. On the other hand, if $\mathfrak{c}_k = 0$, then $\mathfrak{d}_k$ is the kth colume of D' and hence it is integral and $\neq 0$. Since $Y^{-1} > 0$, the values $Y^{-1}[\mathfrak{d}_k]$ also assume a positive minimum. From (24) we derive that

$u_1, \ldots, u_p$ have a positive lower bound independent of M. Relations (14) and (21) imply the inequality

$$u_1 u_2 \cdots u_p \leqslant \gamma_p \, |Y^*|^{-1} = \gamma_p \, |Y|^{-1} \operatorname{abs} (CW + D)^2. \tag{25}$$

If we now assume that the condition

$$\operatorname{abs} (CW + D) \leqslant \rho \tag{26}$$

is satisfied with any fixed arbitrarily large number ρ, then relation (25) yields upper bounds for $u_1, \ldots, u_p$. Then from (24) we obtain upper bounds for the elements of $\mathfrak{c}_k$, $\mathfrak{d}_k$ and hence for the elements of C, D. Therefore, relation (26) is satisfied only for finitely many pairs C, D if W is fixed and ρ is any fixed arbitrarily large number. Thus, we have proved that abs $(CW + D)$ assumes a minimum. For this minimum, in view of (21), the value $|Y^*|$ is maximal. Hence W^* is semireduced, and the proof of Theorem 2 is complete.

The modular substitution

$$W^* = (AW + B)(CW + D)^{-1}$$

is called an *integral modular substitution* if each element $w^*_{kl} = x^*_{kl} + iy^*_{kl}$ of W^* is a polynomial with respect to all variables w_{kl}. The determinant $|Y|$ is an irreducible polynomial in the n variables y_{kl} $(1 \leqslant k \leqslant l \leqslant p)$. On the other hand, in the case of an integral modular substitution, we conclude from relation (21) that abs $(CW + D)^2$, as a polynomial in x_{kl} and y_{kl}, must divide $|Y|$. Therefore, in view of

$$\operatorname{abs} (CW + D)^2 = |CW + D| \, |C\overline{W} + D|,$$

the determinant $|CW + D|$ must be independent of W. If σ again denotes the trace of a matrix, we have

$$d \log |CW + D| = \sigma((CW + D)^{-1} C \, dW). \tag{27}$$

Here the matrix $(CW + D)^{-1}C$ is symmetric since

$$C(CW + D)' = CWC' + CD' = CWC' + DC' = (CW + D)C'.$$

The left-hand side of (27) vanishes identically in dW. Therefore, it follows that $C = 0$ and hence $A = U'$, $D = U^{-1}$, $B = SU^{-1}$ where U is unimodular and S integral symmetric. Every integral modular substitution thus has the form

$$W^* = W[U] + S \tag{28}$$

with U and S as indicated. Conversely, every modular substitution of this form is clearly integral. If, on the other hand, a modular substitution satisfies the condition that abs $(CW + D)$ is constant identically in W, then the determinant $|CW + D|$ is also constant since it is a regular function of the variables w_{kl}. We have thus proved that abs $(CW + D)$ is constant if and only if $C = 0$. In this case, D is unimodular; hence abs $(CW + D) = 1$ identically in W.

For the integral modular substitution (28) we obtain

$$Y^* = Y[U], \qquad |Y^*| = |Y|, \qquad X^* = X[U] + S. \tag{29}$$

Hence W^* is semireduced if W itself is semireduced. Now, for semireduced W, let the unimodular matrix U be selected so that Y^* is contained in $\mathfrak{R}$. Then, let the integral symmetric matrix S be determined by the conditions

$$-\tfrac{1}{2} \leqslant x^*_{kl} \leqslant \tfrac{1}{2} \qquad (1 \leqslant k \leqslant l \leqslant p). \tag{30}$$

If W^* satisfies all these conditions, it will be called *reduced*. The reduced points W of $\mathfrak{H}$ form a closed region $\mathfrak{F}$ whose properties will be investigated in the sequel. More precisely, $\mathfrak{F}$ is defined by the infinite collection of inequalities abs $(CW + D) \geqslant 1$ given by (22), together with the n inequalities (30), and the condition that Y is contained in $\mathfrak{R}$. The last condition is expressed by finitely many inequalities in (11) and (12), where S has to be replaced by Y.

Theorem 3: The region $\mathfrak{F}$ formed by the reduced points of $\mathfrak{H}$ is a fundamental region for the modular group. For the definition of $\mathfrak{F}$ it suffices to impose only finitely many of the inequalities abs $(CW + D) \geqslant 1$, which then imply all the others. A point of $\mathfrak{F}$ is an interior point if and only if the inequalities which serve to define $\mathfrak{F}$ are all strict inequalities at this point. Only finitely many images $\mathfrak{F}_M$ of $\mathfrak{F}$ are adjacent to $\mathfrak{F}$.

Proof: Since every point of $\mathfrak{H}$ is equivalent to a reduced point with respect to Γ, the domain $\mathfrak{H}$ is, indeed, completely covered by the images $\mathfrak{F}_M$. Furthermore, if a point W of $\mathfrak{F}$ satisfies any of the conditions (11), (12), (22), and (30) with the equality sign, then there exist points arbitrarily close to W which do not satisfy that condition. Hence, these points are located outside of $\mathfrak{F}$, and W is seen to be a boundary point of $\mathfrak{F}$.

Denote the diagonal entries of W, X, Y by w_k, x_k, y_k $(k = 1, \ldots, p)$, and the unit matrix of degree $p - 1$ by E. Then with

$$A = \begin{pmatrix} 0 & 0 \\ 0 & E \end{pmatrix}, \quad B = \begin{pmatrix} -1 & 0 \\ 0 & 0 \end{pmatrix}, \quad C = \begin{pmatrix} 1 & 0 \\ 0 & 0 \end{pmatrix}, \quad D = \begin{pmatrix} 0 & 0 \\ 0 & E \end{pmatrix},$$

we obtain a modular matrix which, in view of (22), yields the inequalities

$$\text{abs}\, z_1 \geqslant 1, \qquad x_1^2 + y_1^2 \geqslant 1.$$

Hence, observing (30), we obtain the inequality

$$y_1^2 \geqslant \tfrac{3}{4}.$$

From condition (13) we derive

$$y_k \geqslant \tfrac{1}{2}\sqrt{3} \qquad (k = 1, \ldots, p)$$

for all diagonal elements, and from (14) we then obtain a positive lower bound for $|Y|$ depending only upon p. Replacing the matrix S in (15) by Y, we have $t_1 = y_1$; then (16) implies the boundedness of T^{-1} and Y^{-1}.

Let V_0 denote the matrix of the linear substitution $x_k \to x_{p-k+1}$ $(k = 1, \ldots, p)$ which results from the previously employed matrix V by replacing $2p$ by p. Putting

$$\begin{pmatrix} T^{-1}[V_0] & 0 \\ 0 & T \end{pmatrix} = T_1, \qquad \begin{pmatrix} V_0(Q^{-1})'V_0 & -V_0(Q^{-1})'X \\ 0 & Q \end{pmatrix} = Q_1$$

and observing (5), we obtain

$$S\begin{bmatrix} V_0 & 0 \\ 0 & E \end{bmatrix} = T_1[Q_1]. \tag{31}$$

In view of (16) and (30), this matrix (31) is contained in $\mathfrak{R}(c)$ for suitable c and $m = 2p$. Now, let W^* and W be two points of $\mathfrak{F}$ which are equivalent by means of the mapping (6). According to the previously indicated property of $\mathfrak{R}(c)$, it follows from (7) and (31) that the modular matrix M is contained in a finite set $\mathfrak{M}$ independent of W and W^*. In particular, since both W and W^* are semireduced, the equations

$$\operatorname{abs}(CW + D) = 1, \qquad \operatorname{abs}(-C'W^* + A') = 1$$

must be satisfied in view of (21). If $C \neq 0$, both points W and W^* turn out to be boundary points of $\mathfrak{F}$. On the other hand if $C = 0$, we are given an integral modular substitution (28). By (29), there are three possibilities: (1) $U \neq \pm E$ and Y and Y^* are boundary points of the Minkowski cone $\mathfrak{R}$, (2) $U = \pm E$, $S \neq 0$ and at least one of the conditions (30) is satisfied with the equality sign, and (3) equation (6) represents the identity mapping. Thus, in particular, it is proved that no two different interior points of $\mathfrak{F}$ can be equivalent. Hence $\mathfrak{F}$ is a fundamental region of Γ. An additional result is that for all neighbors $\mathfrak{F}_M$ of $\mathfrak{F}$ the modular matrix M is contained in the finite set $\mathfrak{M}$. Now, we consider the pairs of matrices C, D corresponding to the elements M of $\mathfrak{M}$. From each class of associated pairs we select precisely one pair, thus obtaining a finite set $\mathfrak{P}$ of nonassociated pairs C, D.

Now consider only those of the infinitely many inequalities (22) obtained from pairs C, D which belong to $\mathfrak{P}$. Hence only finitely many inequalities are involved. Let W be any point of $\mathfrak{H}$ which satisfies these inequalities as well as the inequalities (30), and let Y be located in $\mathfrak{R}$. We will show that W lies in $\mathfrak{F}$. If not, then the inequality $\operatorname{abs}(CW + D) < 1$ holds for suitable M.

This situation, however, can occur only for finitely many nonassociated pairs C, D. Putting $W_0 = W + i\lambda E$ with variable scalar $\lambda \geqslant 0$, we obtain

$$\frac{d}{d\lambda}\log(|CW_0 + D||C\overline{W}_0 + D|)$$
$$= i\sigma((CW_0 + D)^{-1}C - C'(\overline{W}_0C' + D')^{-1})$$
$$= i\sigma((CW_0 + D)^{-1}C(\overline{W}_0 - W_0)C'(\overline{W}_0C' + D')^{-1})$$
$$= 2\sigma((CW_0 + D)^{-1}C(Y + \lambda E)C'(\overline{W}_0C' + D')^{-1}) > 0$$

since $C \neq 0$ and $Y + \lambda E > 0$. Therefore, abs $(CW_0 + D)^2$ is a monotonically increasing function of λ, hence a polynomial in λ which really depends on λ. For sufficiently large values of λ it is greater than 1. Then λ may be selected so that, for a suitable pair C, D not belonging to $\mathfrak{P}$ with $C \neq 0$, the equation abs $(CW_0 + D) = 1$ is satisfied and W_0 is semireduced. Moreover, the matrix W_0 is reduced since (30) remains true for W_0, and $Y_0 = Y + \lambda E$ belongs to $\mathfrak{R}$ because Y and E are also in $\mathfrak{R}$. If M is a modular matrix with second row $(C \quad D)$, then the image W_0^* of W_0 defined by (6) is also semireduced. Combining (6) with a suitable integral modular substitution, we may even establish that W_0^* is reduced. But M does not belong to $\mathfrak{M}$, contradicting our result obtained in the preceding paragraph.

Therefore, all inequalities (22) follow from the assumption that these inequalities are satisfied for those pairs C, D which belong to $\mathfrak{P}$ where, in addition, Y lies in $\mathfrak{R}$ and (30) is valid. Furthermore, if abs $(CW + D) > 1$ is satisfied for all pairs C, D in $\mathfrak{P}$, if in addition Y lies in the interior of $\mathfrak{R}$, and if none of the equality signs hold in (30), then all these conditions remain valid for all points in a sufficiently small neighborhood of W. Hence W is an interior point of $\mathfrak{F}$, and Theorem 3 is completely proved.

In view of Theorem 3, the boundary of $\mathfrak{F}$ consists of finitely many hypersurfaces of real dimension $2n - 1$, each of which is defined by an algebraic equation in the $2n$ variables x_{kl} and y_{kl} $(1 \leqslant k \leqslant l \leqslant p)$. In particular, the region $\mathfrak{F}$ has a well-defined symplectic volume $v(\mathfrak{F})$. The line element of symplectic geometry is given by (22) in Section 3. Denoting the elements of Y^{-1} by Y_{kl}, it follows from

$$d(Y^{-1}) = -Y^{-1}(dY)Y^{-1}$$

that the Jacobian determinant of the functions Y_{kl} $(1 \leqslant k \leqslant l \leqslant p)$ with respect to the variables y_{kl} is given by $(-1)^n\,|Y|^{-p-1}$. Hence, in view of Theorem 4 in Section 3, the symplectic volume element equals 2^{n-p} times the Euclidean volume element in the space of the variables x_{kl} and Y_{kl}. But, since both X and Y^{-1} are bounded on $\mathfrak{F}$, it follows that $v(\mathfrak{F})$ is finite. We may even write down this volume $v(\mathfrak{F})$ explicitly for arbitrary p in terms of a simple formula. However we will not carry out the computations.

Finally, it is easily seen that $\mathfrak{F}$ is not compact. First, we verify that iE belongs to $\mathfrak{F}$. Namely, the inequalities (30) are satisfied, E is contained in $\mathfrak{R}$, and the positive number

$$\text{abs}\,(iC + D)^2 = |(iC + D)(-iC' + D')| = |CC' + DD'|$$

is integral and hence $\geqslant 1$. Our previous consideration then proves that $i\lambda E$ also belongs to $\mathfrak{F}$ for scalar $\lambda > 1$. Hence $\mathfrak{F}$ is not compact. Moreover, we will show that Γ possesses no compact fundamental region at all. Otherwise, consider $W = i\lambda E$ for $\lambda \to \infty$. For every λ there exists a modular matrix $M = M(\lambda)$ such that, in view of (20), the matrices

$$(Y^*)^{-1} = \lambda^{-1}DD' + \lambda CC'$$

as well as Y^* stay bounded. For sufficiently large λ it follows that $C = 0$ and D is unimodular. Hence $|Y^*| = \lambda^p$, which is a contradiction.

We notice that the fundamental region $\mathfrak{F}$ coincides with $\mathfrak{B}$ in the case $p = 1$.

6. *Modular forms*

First, we will deal with the theory of elliptic modular forms. According to the definition given in Section 4, a function $\phi(w)$ of the complex variable $w = x + iy$ is called a *modular form* if it is regular in the upper halfplane $\mathfrak{H}$, bounded in the fundamental region $\mathfrak{B}$, and satisfies the functional equation

$$\phi(w^*) = (cw + d)^k \phi(w) \tag{1}$$

for all modular substitutions $w^* = (aw + b)(cw + d)^{-1}$. Here k is an even integer called the *weight* of the modular form. In particular, in view of (1), this modular form has period 1, and as was already shown, the corresponding Laurent expansion

$$\phi(w) = \sum_{n=0}^{\infty} a_n q^n \qquad (q = e^{2\pi i w}) \tag{2}$$

does not contain negative powers of q. Hence $\phi(w)$, as a function of q, is regular in the unit disk $|q| < 1$ and, in particular, at $q = 0$.

Clearly, if the point $w = w_0$ is a zero of $\phi(w)$, then w^* has the same property. Therefore, in order to investigate the zeros of modular forms, we may restrict ourselves to the fundamental region of the modular group. Moreover, in so doing we omit that part of the boundary of $\mathfrak{B}$ which belongs to the right-half plane $x > 0$. We denote the region thus obtained by $\mathfrak{B}_0$. Now $\mathfrak{B}_0$ contains exactly one representative of each equivalence class. The multiplicities of zeros, w_0, of modular forms will play an important role in the sequel. We adopt the convention that the multiplicity of a zero at $w_0 = i$ is $\frac{1}{2}$ times the order of this zero. Similarly, the multiplicity of a zero at

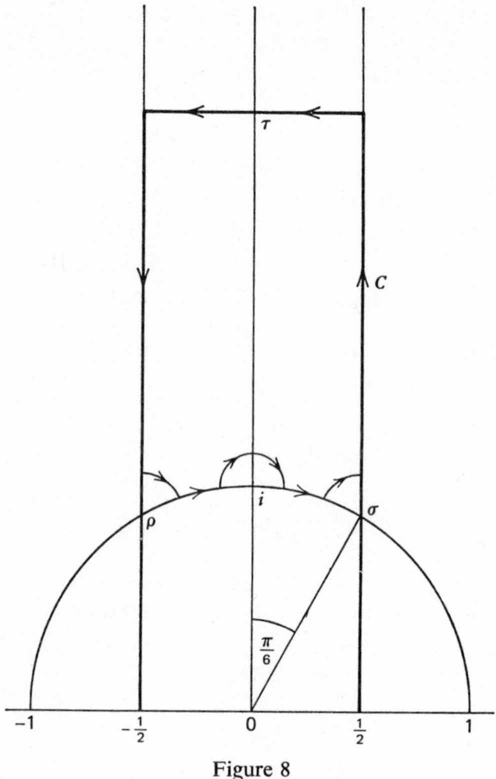

Figure 8

$w_0 = \rho = e^{2\pi i/3}$ will be ⅓ times the order. This is motivated by the fact that the point i is left fixed by precisely two modular substitutions while ρ is left fixed by precisely three modular substitutions. Moreover, $\phi(w)$ has a zero of order g at the point of $\mathfrak{V}_0$ at infinity if $\phi(w)$, as a function of q, vanishes with order g at $q = 0$; that is, if the coefficients in (2) satisfy $a_n = 0$ $(n = 0, 1, \ldots, g - 1)$ and $a_g \neq 0$.

Theorem 1: Let $\phi(w)$ be a nonidentically vanishing elliptic modular form of weight k. Then the number of zeros of $\phi(w)$ in $\mathfrak{V}_0$, counted with their multiplicities, is equal to $k/12$.

Proof: We integrate $d \log \phi(w)$ along the simple curve C indicated in Figure 8, where τ is a point on the imaginary axis and the three little arcs are parts of the circles $|t| = \varepsilon$ with

$$t = \left(\frac{w - \rho}{w - \bar{\rho}}\right)^3, \qquad t = \left(\frac{w - i}{w + i}\right)^2, \qquad t = \left(\frac{w - \sigma}{w - \bar{\sigma}}\right)^3.$$

Then, if C does not contain any zeros of $\phi(w)$, the integral

$$\frac{1}{2\pi i}\int_C d\log\phi(w) = h \tag{3}$$

equals the number of zeros of $\phi(w)$ within the region bounded by C.

At first we assume that, except for the fixed points $w = \rho, i, \sigma$, there are no zeros of $\phi(w)$ on the boundary of $\mathfrak{B}$. Whereas, for $w = \rho$ and $w = i$, we assume $\phi(w)$ to have zeros of the respective multiplicities $a/3$ and $b/2$. The expansion of $\phi(w)$ with respect to powers of $(w - \rho)/(w - \bar{\rho})$ then starts with the ath power, and a corresponding statement is true for i in place of ρ. From this we conclude that the contribution of the three little circles to the left-hand side of (3) tends to $-(a/3) - (b/2)$ if $\varepsilon \to 0$. Furthermore, if $\phi(w)$ vanishes with order g at the point of $\mathfrak{B}_0$ at infinity, then, in view of (2), the contribution of the horizontal part of C tends to $-g$ if $\tau \to \infty$. Because of the periodicity of $\phi(w)$, the contributions of the vertical parts of C cancel each other. Finally, on the unit circle $|w| = 1$ we set $w = e^{i\alpha}$ and obtain the equations

$$\phi\left(-\frac{1}{w}\right) = w^k\phi(w), \quad \log\phi(e^{i(\pi-\alpha)}) = ik\alpha + \log\phi(e^{i\alpha}) \qquad (0 < \alpha < \pi).$$

Therefore, the contributions of the remaining parts of C have the limit

$$\frac{1}{2\pi i}\left(ik\frac{\pi}{6}\right) = \frac{k}{12}$$

if $\varepsilon \to 0$. Thus, the assertion is proved in the case at hand.

If there are further zeros on the boundary of $\mathfrak{B}$, they are correspondingly avoided by half circles of radius ε penetrating the interior of $\mathfrak{B}$. Thus the general case of Theorem 1 is established.

As an application of Theorem 1, we take for $\phi(w)$ the Eisenstein series $E_4(w)$ and $E_6(w)$ obtained from g_2 and g_3 by the normalization $a_0 = 1$. The weights of $E_4(w)$ and $E_6(w)$ are 4 and 6. In view of Theorem 1, the series $E_4(w)$ must vanish with first order at $w = \rho$ and nowhere else on $\mathfrak{B}_0$; according to our convention, the multiplicity of this zero is $1/3$. Similarly, the series $E_6(w)$ must vanish on $\mathfrak{B}_0$ only at $w = i$ and again with first order.

From Theorem 1 we immediately derive that modular forms of negative weight are identically zero, and modular forms of weight 0 are constants. By the way, this is seen more easily by use of the argument in Section 5 of Chapter 3 (Volume II) when, in addition (1) and the formula

$$y^* = y \operatorname{abs}(cw + d)^{-2}$$

are applied.

Now let k be an even natural number and let μ, ν range over all solutions of $4\mu + 6\nu = k$ in nonnegative integers. Putting

$$\sum_{\mu} c_\mu E_4^\mu E_6^\nu = \psi(w) \tag{4}$$

with constant complex coefficients c_μ, we obtain an isobaric polynomial in E_4 and E_6 which obviously is a modular form $\psi(w)$ of weight k.

Theorem 2: Every elliptic modular form has a unique representation as an isobaric polynomial in E_4 and E_6.

Proof: If a modular form $\phi(w)$ vanishes at $w = \rho$, then the quotient $\phi(w)/E_4(w)$ is also a modular form, and the corresponding statement holds for $w = i$. Hence we need only treat the case when $\phi(w)$ differs from zero at $w = \rho$ and $w = i$. In this case, in view of Theorem 1, the weight k is divisible by 12 and $k \geqslant 0$. Furthermore, the number of solutions of $4\mu + 6\nu = k$ in nonnegative integers μ, ν is equal to the integer

$$m = \frac{k}{12} + 1.$$

According to Theorem 1, the number of zeros of $\phi(w)$ in $\mathfrak{B}_0$, including the point at infinity, is equal to $m - 1$. Hence, the m coefficients c_μ in (4) can be determined by use of (12) in Section 4 so that not all of them vanish, and the modular form $\psi(w)$ of weight k vanishes with at least the same order as $\phi(w)$ at all zeros of $\phi(w)$. Then the quotient $\psi(w)/\phi(w)$ is a modular form of weight 0 and hence is constant.

Since $E_4(i) \neq 0$, $E_6(i) = 0$ and $E_6(w)$ does not vanish identically, the function $\psi(w)$ defined by (4) is identically zero if and only if each coefficient c_μ vanishes. Combining this with the previously obtained result, it follows that the modular form under consideration, $\phi(w)$, has a unique representation of the form (4). Hence, Theorem 2 is proved.

As an example of Theorem 2, we consider the Eisenstein series

$$E_8(w) = \sum{}'' (cw + d)^{-8},$$

where c, d ranges over all pairs of relatively prime integers which satisfy either $c > 0$, d arbitrary or $c = 0$, $d = 1$. We obtain

$$E_8(w) = 1 - \frac{16}{B_8} q + \cdots \qquad (q = e^{2\pi i w}),$$

which is a modular form of weight 8. The difference

$$E_4^2 + E_8 = \left(480 + \frac{16}{B_8}\right) q + \cdots$$

is a modular form of the same weight which vanishes at $q = 0$. Hence, in view of Theorem 1, it vanishes identically. Therefore, we have $B_8 = -\frac{1}{30}$ and

$$E_8 = E_4^2$$

Theorems 1 and 2 cover the main points of the theory of elliptic modular forms if we require, as we have until now, condition (1) for all modular substitutions. However, we are led to important new questions if we do not maintain this condition for the entire elliptic modular group Γ, but only for a subgroup of finite index in Γ. Here a particular role is played by the so-called *congruence subgroups*. However, we will not pursue this refined theory of elliptic modular forms.

Now we turn to modular forms of degree p where we assume $p > 1$. Hence, we are dealing with functions $\phi(W)$ which are regular in the generalized upper half plane $\mathfrak{H}$ and which, for all modular substitutions of degree p

$$W^* = (AW + B)(CW + D)^{-1}, \tag{5}$$

satisfy the equation

$$\phi(W^*) = |CW + D|^k \, \phi(W). \tag{6}$$

Here k is an integer called the *weight* of $\phi(W)$, and we will assume k to be even in the sequel. Now, it turns out that the Ansatz (11) in Section 4 can be generalized to the construction of modular forms of degree p. The restrictions on summation which are then necessary require some preparatory work.

According to (28) in the preceding section, the integral modular substitutions defined by $C = 0$ have the form

$$W^* = W[U] + S \tag{7}$$

with unimodular U and integral symmetric S. Hence, the matrices in (5) are given by $A = U'$, $B = SU^{-1}$, $D = U^{-1}$. A subgroup Δ of the modular group Γ of degree p is hereby defined. Let M and M_0 be the matrices of the substitutions (5) and (7). Setting

$$M_1 = M_0 M = \begin{pmatrix} A_1 & B_1 \\ C_1 & D_1 \end{pmatrix},$$

we obtain an element of Γ belonging to the right coset ΔM of Δ with respect to Γ, and we have

$$C = UC_1, \qquad D = UD_1, \qquad C_1 D' = C_1 D_1' U' = D_1 C_1' U' = D_1 C'.$$

Conversely, let the condition

$$C_1 D' = D_1 C' \tag{8}$$

be satisfied for any element M_1 of Γ. Then it follows that

$$M_1M^{-1} = \begin{pmatrix} A_1 & B_1 \\ C_1 & D_1 \end{pmatrix} \begin{pmatrix} D' & -B' \\ -C' & A' \end{pmatrix} = \begin{pmatrix} A_0 & B_0 \\ 0 & D_0 \end{pmatrix} = M_0$$

is an element of Δ. Hence $D_0 = U^{-1}$ with unimodular U and $C = UC_1$, $D = UD_1$. The pairs of matrices C, D and C_1, D_1 are called *associated* if they satisfy condition (8). This condition is satisfied if and only if $C = UC_1$, $D = UD_1$ holds with unimodular U, and also, if and only if M and M_1 belong to the same right coset ΔM. Therefore, if we select a representative from each class of associated pairs C, D, these representatives are in one-to-one correspondence with the right cosets of Δ with respect to Γ.

If $C = UC_1$, $D = UD_1$, it follows that

$$|CW + D| = |U(C_1W + D_1)| = |U|\,|C_1W + D_1| = \pm|C_1W + D_1|.$$

Hence,

$$|CW + D|^k = |C_1W + D_1|^k \tag{9}$$

for every even exponent k. Conversely, if for fixed integral $k \neq 0$ condition (9) is satisfied identically in W by given modular matrices M and M_1, then we set

$$M_1M^{-1} = M_0 = \begin{pmatrix} A_0 & B_0 \\ C_0 & D_0 \end{pmatrix}$$

and obtain

$$|C_0W + D_0|^k = 1$$

identically in W. Hence $C_0 = 0$. Therefore, the right cosets are also in one-to-one correspondence with the different polynomials $|CW + D|^k$ for even $k > 0$.

With the reservation that convergence must be proved, we now define

$$E_k(W) = \sum_M{}' \, |CW + D|^{-k}, \tag{10}$$

where k is even and M ranges over all representatives of the right cosets of Δ with respect to Γ. Hence C, D ranges over a complete system of non-associated pairs of matrices. We call $E_k(W)$ *the Eisenstein series of weight* k *and degree* p.

Theorem 3: If $k > p + 1$, then the Eisenstein series $E_k(W)$ is uniformly and absolutely convergent on every compact subset of $\mathfrak{H}$ as well as on the whole of the fundamental region $\mathfrak{F}$. Furthermore, it represents a modular form of weight k.

Proof: Let $\mathfrak{K}$ be a compact subset of $\mathfrak{H}$. We consider all pairs of $p \times p$ real matrices P, Q which satisfy both conditions

$$PQ' = QP', \qquad PP' + QQ' = E; \tag{11}$$

and we will show that the real function abs $(PW + Q)$ on $\mathfrak{K}$ lies between two positive bounds independent of P, Q. The existence of such an upper bound is trivial since W, P, Q are bounded. Assuming that P, Q only satisfies $PQ' = QP'$ and putting

$$J(PQ)' = F,$$

it follows that

$$S[F] = (P\overline{W} + Q)Y^{-1}(PW + Q)', \tag{12}$$

where S is defined by (5) in Section 5. Now, suppose (11) holds and there is no positive lower bound of abs $(PW + Q)$ independent of P, Q, W. Then, in view of the compactness of $\mathfrak{K}$, there exists a W in $\mathfrak{K}$ and a pair P, Q satisfying (11) such that $|PW + Q| = 0$. Hence, from (12) we conclude that $|S[F]| = 0$. Since $S > 0$ the matrix F must have rank less than p. However, (11) implies

$$F'F = (PQ)(PQ)' = E.$$

Thus the existence of the desired positive lower bound is proved.

More generally, we now only subject the $p \times p$ real matrices P, Q to the conditions $PQ' = QP'$ and $|iP + Q| \neq 0$. If we then set $iP + Q = R$, the Hermitian matrix

$$\bar{R}R' = (-iP + Q)(iP' + Q') = PP' + QQ' \tag{13}$$

is real and positive. Hence an invertible real matrix K can be found satisfying $\bar{R}R' = KK'$. In view of (13), we may apply what we already proved with $K^{-1}P$, $K^{-1}Q$ in place of P, Q. This yields

$$\alpha \leqslant \text{abs}\,(K^{-1}PW + K^{-1}Q) \leqslant \beta.$$

Observing abs R = abs K, we obtain the inequalities

$$\alpha\, \text{abs}\,(iP + Q) \leqslant \text{abs}\,(PW + Q) \leqslant \beta\, \text{abs}\,(iP + Q). \tag{14}$$

where α, β are positive numbers independent of P, Q, W.

In order to prove Theorem 3, we apply a generalization of the argument utilized in Section 5 of Chapter 3 (Volume II) to prove the convergence of Poincaré series. Let $\mathfrak{G}$ be a compact subset of the fundamental region $\mathfrak{F}$. The determinant $|Y|$ has an upper bound, s, on $\mathfrak{G}$ and the estimate $|Y| \leqslant s$ also holds on the image $\mathfrak{G}_M$ of $\mathfrak{G}$ for every modular matrix M. According to (7), for every point W of $\mathfrak{G}_M$ we can determine an element M_0 of Δ such that the imaginary part Y^* of the image point W^* in $\mathfrak{G}_{M_0 M}$ belongs to the

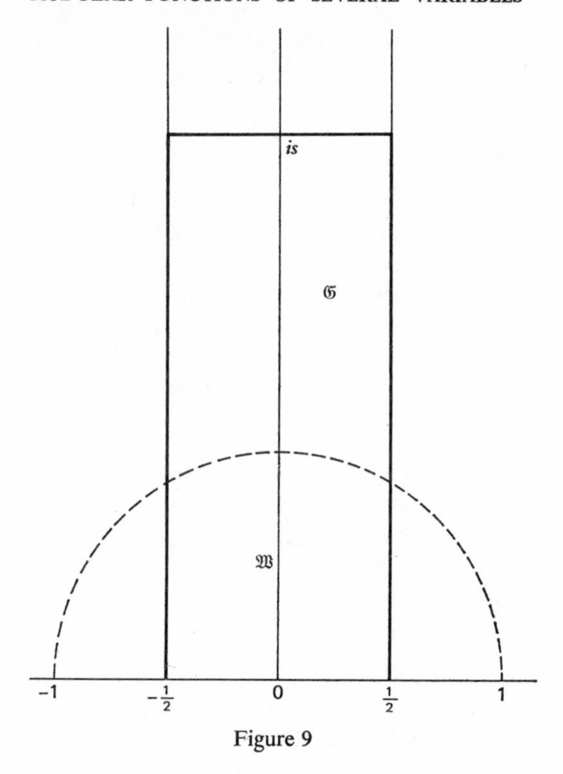

Figure 9

fundamental region $\mathfrak{R}$ of Minkowski, and the real part X^* belongs to the cube $\mathfrak{Q}$ defined by (30) in Section 5. The inequality $|Y| \leqslant s$ defines a subset $\underline{\mathfrak{R}}$ of $\mathfrak{R}$. Let $\mathfrak{W}$ be the Cartesian product of $\underline{\mathfrak{R}}$ and $\mathfrak{Q}$ (Figure 9). Furthermore, let $\{dX\}$ and $\{dY\}$ denote the Euclidean volume elements in $\mathfrak{Q}$ and $\mathfrak{R}$ formed by the coordinates x_{kl} and y_{kl} $(1 \leqslant k \leqslant l \leqslant p)$. Later we will prove that the integral

$$I_r = \int_{\mathfrak{G}} |Y|^r \{dX\}\{dY\} \tag{15}$$

converges for all real $r > \dfrac{-(p+1)}{2}$.

For every right coset ΔM we obtain finitely many subsets of $\mathfrak{W}$ such that the images of these subsets under integral modular substitutions yield a covering of $\mathfrak{G}_M$ without gaps and overlaps. On the other hand, the subsets corresponding to different right cosets do not overlap in $\mathfrak{W}$. Now, we observe that for every symplectic transformation, the Jacobian determinant of X^*, Y^* with respect to X, Y equals abs $(CW + D)^{-2(p+1)}$. In view of

$$|Y^*| = |Y| \operatorname{abs} (CW + D)^{-2},$$

we obtain

(16) $$|Y^*|^r \{dX^*\}\{dY^*\} = \text{abs}\,(CW + D)^{-2(r+p+1)}\,|Y|^r\,\{dX\}\{dY\}.$$

Furthermore, we set

$$\mu_r(W) = \sum_M{}' \text{abs}\,(CW + D)^{-2(r+p+1)}$$

where the matrix M, as in (10), ranges over all representatives of the right cosets of Δ. Integrating the left-hand side of (16) over $\mathfrak{W}$, we obtain

(17) $$I_r \geqslant \int_{\mathfrak{G}} \mu_r(W)\,|Y|^r\,\{dX\}\{dY\}.$$

By use of (14) the convergence assumption on I_r for $r > \dfrac{-(p+1)}{2}$ now implies first, the convergence of $\mu_r(W)$ at $W = iE$ and second, the uniform convergence of $\mu_r(W)$ on the whole of $\mathfrak{K}$. Setting $k = 2(r + p + 1)$, it follows that the Eisenstein series $E_k(W)$ converges uniformly and absolutely in $\mathfrak{K}$ if $k > p + 1$. Hence the function $E_k(W)$ is regular throughout $\mathfrak{H}$. Furthermore, since right multiplication by any fixed modular matrix only permutes the right cosets ΔM, condition (6) is satisfied by $E_k(W)$ for all modular substitutions. Hence $E_k(W)$ is a modular form of weight k.

Now we must prove the uniform convergence on the totality of $\mathfrak{F}$. For this purpose we utilize (12) and (31) in Section 5. Let $T_1^{1/2}$ denote the diagonal matrix whose diagonal elements are the positive square roots of the diagonal elements of T_1. Since $T_1[Q_1]$ is contained in $\mathfrak{R}(c)$ with $m = 2p$ and suitable c independent of W, both matrices $T_1^{1/2}Q_1T_1^{-1/2}$ and

$$T_1[Q_1T_1^{-1/2}] = S_1$$

are bounded. Furthermore, since $S_1 > 0$ and $|S_1| = 1$, the $2p$ eigenvalues of S_1 are bounded from below and above by positive bounds γ and δ which are independent of W. Therefore, it follows that

(18) $$\delta T_1 > S\begin{bmatrix} V_0 & 0 \\ 0 & E \end{bmatrix} > \gamma T_1.$$

Every product of any p diagonal elements of T_1 is at least $\rho\,|Y|^{-1}$, where ρ is a suitable positive number independent of W. If two nonnegative symmetric matrices G_1, G_2 satisfy $G_1 \geqslant G_2$, that is $G_1 - G_2 \geqslant 0$, then the inequality $|G_1| \geqslant |G_2|$ follows, as is seen by transforming G_1 and G_2 simultaneously into diagonal form. From (18) we derive

$$S[F] \geqslant \gamma\left(T_1\begin{bmatrix} 0 & V_0 \\ -E & 0 \end{bmatrix}\right)\begin{bmatrix} P' \\ Q' \end{bmatrix}.$$

Then, forming determinants and using the Laplace expansion theorem, we obtain the estimate

$$\text{abs}\,(PW+Q)^2 > \gamma^p \rho\, \text{abs}\,(iP+Q)^2$$

in view of (12). Here we only need that W is in $\mathfrak{F}$, and $PQ' = QP'$. In particular, setting $P = C$, $Q = D$, the absolute convergence of $E_k(iE)$, proved previously, implies the uniform and absolute convergence of $E_k(W)$ on the whole of $\mathfrak{F}$.

In order to prove the convergence of the integral I_r, first we utilize the Jacobian transformation $Y = T[Q]$. Setting

$$\{dT\} = \prod_{k=1}^{p} dt_k, \qquad \{dQ\} = \prod_{1 \leq k < l \leq p} dq_{kl},$$

it follows that

$$\{dY\} = \{dT\}\{dQ\} \prod_{k=1}^{p} t_k^{p-k}.$$

From the further substitutions

$$\tau_k = \frac{t_k}{t_{k+1}} \quad (k = 1, \ldots, p-1), \qquad t = |T|,$$

we obtain

$$dt \prod_{k=1}^{p-1} d\tau_k = p\{dT\} \prod_{k=1}^{p-1} \tau_k, \qquad \prod_{k=1}^{p-1} \tau_k^k = t t_p^{-p},$$

$$\prod_{k=1}^{p} t_k^{p-k} = t_p^{p(p-1)/2} \prod_{k=1}^{p-1} \tau_k^{k[p-(k+1)/2]} = t^{(p-1)/2} \prod_{k=1}^{p-1} \tau_k^{k(p-k)/2}.$$

Hence

$$|Y|^r \{dY\} = \frac{1}{p} \{dQ\} t^{r+[(p-1)/2]} dt \prod_{k=1}^{p-1} (\tau_k^{[k(p-k)/2]-1} d\tau_k).$$

In view of (16) and the fact that $0 < t \leqslant s$, the convergence of the integral I_r follows for $r + (p-1)/2 > -1$. Hence the remaining assertion of Theorem 3 is proved.

Furthermore, we may state that I_r diverges for $r \leqslant -(p+1)/2$. However, since $\mathfrak{F}$ is not compact, this does not immediately imply that the Eisenstein series is absolutely divergent for $k \leqslant p + 1$. This divergence can be proved by a more detailed investigation which, however, we will omit.

The second row $(C \quad D)$ of any modular matrix M is *primitive*, where primitiveness is defined by the relative primeness of all $p \times p$ subdeterminants of $(C \quad D)$. This is proved by taking the Laplace expansion of $|M|$ with respect to the last p rows. Now, let $|C| = 0$ and $C \neq 0$, that is, let the rank r of C be positive and less than p. We select two unimodular matrices U_1 and V_1 such that

$$U_1 C = \begin{pmatrix} C_1 & 0 \\ 0 & 0 \end{pmatrix} V_1' \tag{19}$$

holds where C_1 is a square matrix of degree r. Hence $|C_1| \neq 0$. An analogous decomposition of the matrix

$$U_1 D V_1 = \begin{pmatrix} D_1 & D_{12} \\ D_{13} & D_{14} \end{pmatrix} \tag{20}$$

together with the equation $CD' = DC'$ yields the equations

$$C_1 D_1' = D_1 C_1', \qquad C_1 D_{13}' = 0, \qquad D_{13} = 0.$$

Since the matrix

$$U_1(C \quad D)\begin{pmatrix} V_1' & 0 \\ 0 & V_1^{-1} \end{pmatrix}^{-1} = \begin{pmatrix} C_1 & 0 & D_1 & D_{12} \\ 0 & 0 & 0 & D_{14} \end{pmatrix}$$

is also primitive, we conclude that $(C_1 \quad D_1)$ is primitive and D_{14}, unimodular. Now we form the unimodular matrix

$$U_3 = \begin{pmatrix} E & -D_{12}D_{14}^{-1} \\ 0 & D_{14}^{-1} \end{pmatrix}$$

and again replace $U_3 U_1$ by U_1. Then relation (19) remains valid, whereas (20) becomes

$$U_1 D V_1 = \begin{pmatrix} D_1 & 0 \\ 0 & E \end{pmatrix}.$$

Now, we assume that

$$U_2 C = \begin{pmatrix} C_2 & 0 \\ 0 & 0 \end{pmatrix} V_2', \qquad U_2 D V_2 = \begin{pmatrix} D_2 & 0 \\ 0 & E \end{pmatrix}$$

is an analogous decomposition, where U_2, V_2 are unimodular and $|C_2| \neq 0$. Then C_2 is also a square matrix of degree r. We put

$$U_2 U_1^{-1} = U_0, \qquad V_1^{-1} V_2 = V_0$$

and obtain

$$U_0 \begin{pmatrix} C_1 & 0 \\ 0 & 0 \end{pmatrix} = \begin{pmatrix} C_2 & 0 \\ 0 & 0 \end{pmatrix} V_0', \qquad U_0 \begin{pmatrix} D_1 & 0 \\ 0 & E \end{pmatrix} = \begin{pmatrix} D_2 & 0 \\ 0 & E \end{pmatrix} V_0^{-1}.$$

Consequently, the matrices U_0, V_0^{-1} must have the form

$$(21)\qquad U_0 = \begin{pmatrix} U_3 & U_{34} \\ 0 & U_4 \end{pmatrix}, \qquad V_0^{-1} = \begin{pmatrix} V_3^{-1} & V_{34} \\ 0 & V_4 \end{pmatrix},$$

where U_3, U_4, V_3, V_4 are unimodular and U_{34}, V_{34} integral. Furthermore, we obtain

$$(22)\quad U_3C_1 = C_2V_3', \qquad U_3D_1 = D_2V_3^{-1}, \qquad U_{34} = D_2V_{34}, \qquad U_4 = V_4.$$

Conversely, let U_1, V_1 be fixed and let the unimodular matrices U_3, V_3, V_4, as well as the integral matrix V_{34}, be selected arbitrarily. Then, from (21) and (22) it follows that C_2, D_2, U_0, V_0 are uniquely determined; hence $U_2 = U_0U_1$, $V_2 = V_1V_0$ are also uniquely determined. Decomposing the matrices

$$V_1' = \begin{pmatrix} G_1 \\ H_1 \end{pmatrix}, \qquad V_2' = \begin{pmatrix} G_2 \\ H_2 \end{pmatrix}$$

into submatrices of r and $p - r$ rows, we obtain

$$G_2 = V_3'G_1$$

We also call G_1 and G_2 *associated.* If we now fix G_1 within its class of associated matrices, then only $V_3 = E$ is admissible. Next, we fix the pair C_1, D_1 within its class of associated pairs. Thus we also obtain $U_3 = E$, whereas V_4 and V_{34} remain arbitrary.

Introducing the modular matrix

$$(23)\qquad \begin{pmatrix} (U_1')^{-1} & 0 \\ 0 & U_1 \end{pmatrix} \begin{pmatrix} A & B \\ C & D \end{pmatrix} \begin{pmatrix} (V_1')^{-1} & 0 \\ 0 & V_1 \end{pmatrix} = \begin{pmatrix} A^* & B^* \\ C^* & D^* \end{pmatrix}$$

we first obtain

$$C^* = \begin{pmatrix} C_1 & 0 \\ 0 & 0 \end{pmatrix}, \qquad D^* = \begin{pmatrix} D_1 & 0 \\ 0 & E \end{pmatrix}.$$

Then the equations

$$A^*(D^*)' - B^*(C^*)' = E, \qquad (A^*)(B^*)' = B^*(A^*)'$$

imply

$$A^* = \begin{pmatrix} A_1 & 0 \\ A_{13} & E \end{pmatrix}, \qquad B^* = \begin{pmatrix} B_1 & B_{12} \\ B_{13} & B_{14} \end{pmatrix}$$

and

$$A_1B_1' = B_1A_1', \qquad A_1D_1' - B_1C_1' = E.$$

Therefore, $(C_1 \quad D_1)$ becomes the second row of the modular matrix

$$M_1 = \begin{pmatrix} A_1 & B_1 \\ C_1 & D_1 \end{pmatrix}$$

of degree r. Conversely, if M_1 is given, then the special choice

$$A^* = \begin{pmatrix} A_1 & 0 \\ 0 & E \end{pmatrix}, \qquad B^* = \begin{pmatrix} B_1 & 0 \\ 0 & 0 \end{pmatrix}$$

yields, by (23), a modular matrix M whose second row $(C \quad D)$ is related to $(C_1 \quad D_1)$ in the same way as it was in (19) and (20). Hence, the classes of associated pairs $(C \quad D)$ are in one to one correspondence with the pairs whose first components are classes of matrices G_1 and whose second components are classes of pairs $(C_1 \quad D_1)$. Here in the class of G_1 a fixed representative must be chosen.

Consider the decomposition of $C = (c_1 \cdots c_p)$ into columns $c_1, \ldots, c_p$, and assume $c_s \neq 0$ and $c_l = 0$ for $l = s + 1, \ldots, p$. The number s is called the *order* of C. For $C = 0$ we define $s = 0$. Clearly, the rank r of C is at most equal to s. From the relation

$$U_1 C = \begin{pmatrix} C_1 & 0 \\ 0 & 0 \end{pmatrix} \begin{pmatrix} G_1 \\ H_1 \end{pmatrix} = \begin{pmatrix} C_1 G_1 \\ 0 \end{pmatrix},$$

we conclude that C and G_1 have the same order.

Let q be any number of the sequence $1, \ldots, p - 1$. Let W_q be the submatrix of W formed by the first q rows and columns, and let $\mathfrak{F}_q$ be the fundamental region of the modular group of degree q. By use of the definition of $\mathfrak{F}$, we easily see that W_q belongs to $\mathfrak{F}_q$ if W belongs to $\mathfrak{F}$.

Theorem 4: Let W range over a sequence in $\mathfrak{F}$ such that the submatrices W_q converge, whereas the diagonal elements z_{q+1} tend to infinity. Then, for $k > p + 1$ the Eisenstein series $E_k(W)$ tends to the corresponding Eisenstein series $E_k(W_q)$ of degree q.

Proof: We have

$$U_1(CW + D)V_1 = \begin{pmatrix} C_1 & 0 \\ 0 & 0 \end{pmatrix} W[V_1] + \begin{pmatrix} D_1 & 0 \\ 0 & E \end{pmatrix},$$

$$\pm|CW + D| = |C_1 W[G_1'] + D_1| = |C_1|\,|W[G_1'] + C_1^{-1} D_1|\,. \tag{24}$$

If we set

$$Y[G_1'] = Y_0, \qquad X[G_1'] + C_1^{-1} D_1 = X_0,$$

then X_0, Y_0 are symmetric, and $Y_0 > 0$. By simultaneously transforming X_0 and Y_0 into diagonal form, it is seen that

$$\text{abs}\,(X_0 + iY_0) \geqslant |Y_0|,$$

and in view of (24) we also have

$$\text{abs}\,(CW + D) \geqslant \text{abs}\,(X_0 + iY_0) \geqslant |Y_0|. \tag{25}$$

For every W in the sequence under consideration we now select G_1 in its class of associated matrices so that Y_0 is reduced in the sense of Minkowski. Furthermore, Y is also reduced. By (14) in Section 5 and the consideration leading to (18), we then obtain the inequality

$$\prod_{l=1}^{r} (y_1 g_{l1}^2 + \cdots + y_p g_{lp}^2) < \rho\,|Y_0|\,. \tag{26}$$

Here $y_1, \ldots, y_p$ are the diagonal elements of Y; furthermore, $G_1 = (g_{kl})$ and the positive number ρ only depends on p.

According to the assumption we have $y_{q+1} \to \infty$. Hence

$$y_k \to \infty \qquad (k = q + 1, \ldots, p).$$

Since G_1 is primitive, no row of G_1 is zero. Let s be the order of C and $s > q$. Since G_1 is integral and $y_1 > \tfrac{1}{2}\sqrt{3}$, it follows from (25) and (26) that $|CW + D| \to \infty$. If, on the other hand, we have $s \leqslant q$ and

$$G_1 = (G_{1q} \quad 0),$$

with a primitive matrix G_{1q} of r rows and q columns, then

$$W[G_1'] = W_q[G_{1q}'], \qquad |CW + D|^2 = |C_1 W_q[G_{1q}'] + D_1|^2\,.$$

The assertion of Theorem 4 now follows from the uniform convergence of $E_k(W)$ in $\mathfrak{F}$.

Theorem 4, as well as its proof, remains true in case $q = 0$ if we define $E_k(W_0) = 1$. In particular, this implies that none of the Eisenstein series $E_k(W)$ vanishes identically in W for even $k > p + 1$.

Although we proved the fundamental Theorem 2 in case $p = 1$, the theory of modular forms and Eisenstein series for arbitrary p has not been developed to this extent. In the case $p = 2$, a more profound investigation by Igusa yields the fact that all modular forms of even weight are isobaric polynomials in the four Eisenstein series E_4, E_6, E_{10}, E_{12}.

7. *The field of modular functions*

We begin with the theory of elliptic modular functions. Thus we are dealing with functions $f(w)$ of one complex variable, w, which are meromorphic in the upper half plane including the point of $\mathfrak{V}_0$ at infinity and which are invariant under all modular substitutions. More precisely, the condition on the behavior at infinity states that there exists a Laurent expansion

$$f(w) = \sum_{k=-m}^{\infty} a_k q^k \tag{1}$$

which converges for sufficiently small values of $|q|$ and contains only finitely many negative powers of q. Here the variable $q = e^{2\pi i w}$. A particular modular function is given by

$$j(w) = g_2^3(g_2^3 - 27g_3^2)^{-1} = E_4^3(E_4^3 - E_6^2)^{-1} = \tfrac{1}{1728}q^{-1} + \cdots. \tag{2}$$

More generally, every rational function of $j(w)$ with constant coefficients is also a modular function. However, the function $e^{j(w)}$ is not a modular function although, as a function of w, it is even regular throughout the upper half plane and is invariant under all modular substitutions. Namely, this function does not satisfy condition (1). Now we will show that the field of elliptic modular functions coincides with the field of rational functions of $j(w)$.

Theorem 1: Every modular function can be expressed rationally in terms of $j(w)$.

Proof: The modular form $E_4^3 - E_6^2$ of weight 12 vanishes to first order at the point of $\mathfrak{V}_0$ at infinity given by $q = 0$. In view of Theorem 1 of Section 6, this modular form has no zero in $\mathfrak{H}$. Hence by (2), the function $j(w)$ is regular on $\mathfrak{H}$ and has a pole of first order at $q = 0$. This result was previously obtained in another way. Furthermore, we know that $E_4(w)$ and $E_6(w)$ vanish to first order at the respective points $w = \rho$ and $w = i$ and that they have no other zeros elsewhere on $\mathfrak{V}_0$. If we count multiplicities of modular functions at the points $w = \rho$ and $w = i$ in the same way as we did for modular forms in the preceding section, it follows by (2) that $j(w)$ assumes the value 0 on $\mathfrak{V}_0$ only at $w = \rho$ with multiplicity 1. The same is true for the value 1 at $w = i$. Finally, if c is any complex number different from 0 and 1, the equation $j(w) = c$ is equivalent to the equation

$$(1 - c)E_4^3 + cE_6^2 = 0$$

which has exactly one solution on $\mathfrak{V}_0$ in view of Theorem 1 in Section 6. This result, in not quite this precise form, was also obtained in a different way in Section 9 of Chapter 3 (Volume II).

Now let $f(w)$ be any modular function which is not identically zero. In order to prove the theorem, we may assume that the function f has neither a zero nor a pole at each of the points $w = \rho$, $w = i$, and $q = 0$. This assumption is justified by means of a suitable fractional linear transformation of f.

Now let $a_1, \ldots, a_r$ be the zeros and $b_1, \ldots, b_s$ be the poles of $f(w)$ in $\mathfrak{B}_0$, where the respective multiplicities are to be taken into consideration. Furthermore, replacing f by f^{-1} if necessary, we may assume that $r \geqslant s$. Since $j(w)$ assumes every value on $\mathfrak{B}_0$ exactly once with multiplicity 1, the derivative of $j(w)$ does not vanish at any zero or pole of $f(w)$. Consequently, the function

$$g(w) = \frac{\prod_{k=1}^{r} (j(w) - j(a_k))}{\prod_{l=1}^{s} (j(w) - j(b_l))}$$

assumes on $\mathfrak{H}$ exactly the same zeros and poles as $f(w)$ with their respective multiplicities. Obviously the quotient $f/g = h(w)$ is a modular function and regular everywhere on $\mathfrak{H}$; moreover, in view of $r \geqslant s$, $h(w)$ is also regular with respect to q at the point of $\mathfrak{B}_0$ at infinity. In case $r > s$ the function h vanishes at $q = 0$. If $\mathfrak{B}_0$ is compactified by adjoining the point $q = 0$ and if the maximum principle is applied to the regular function $h(w)$, it follows that $h(w)$ is constant and $r = s$. Hence, the proof of Theorem 1 is complete.

By use of (2), Theorem 1 implies that every modular function f is the quotient of two modular forms ϕ_1, ϕ_2 of equal weight which are locally relatively prime on $\mathfrak{B}_0$ as well as the point at infinity. On the other hand, for a given function f, these properties determine the modular forms ϕ_1, ϕ_2 uniquely except for a common constant factor. In view of (2), the modular forms ϕ_1, ϕ_2 are isobaric polynomials in E_4 and E_6. This also follows from Theorem 2 in Section 6.

As an application of the theory of elliptic modular functions, we will outline the classical proof of Picard's theorem. This theorem states that every nonconstant meromorphic function $h(z)$ of a complex variable z actually assumes every complex value, including ∞, with at most two exceptions. Namely, if $h(z)$ omits three different values, then by a preliminary fractional linear transformation of h we may assume that these exceptional values are given by $0, 1, \infty$. We then form the equation

$$j(w) = h(z) \tag{3}$$

in order to determine w as a function of z. For every finite value $z = z_0$, this equation (3) has precisely one solution, $w = w_0$ on $\mathfrak{B}_0$, and this solution is different from ρ, i, ∞. We start from a fixed z_0 and the corresponding w_0. Since the derivative of $j(w)$ does not vanish at this point, equation (3) has a uniquely determined regular solution $w = \psi(z)$ within a neighborhood of z_0 satisfying $w_0 = \psi(z_0)$. If we now consider any curve C in the complex z-plane emerging from z_0, it follows from (3) that $\psi(z)$ can be analytically continued along C since $h(z)$ is different from $0, 1, \infty$ on the whole of C.

The monodromy theorem implies that $\psi(z)$ is an entire function. Since the values of this function are located in $\mathfrak{H}$, the function

$$\chi(z) = \frac{\psi(z) - i}{\psi(z) + i}$$

is also an entire function whose absolute value is less than 1 everywhere. Liouville's theorem then implies that $\chi(z)$ is constant. Hence $w = \psi(z)$ and $j(w) = h(z)$ are also constant.

A slightly different formulation of Picard's theorem states that every non-constant entire function assumes every finite value with at most one exception. The example of the exponential function, which is different from zero everywhere, shows that one exceptional value can actually occur without implying that the function is constant.

Further investigations in the theory of elliptic modular functions concern questions arising from the subgroups of the modular group. In particular, these questions are of importance for complex multiplication. We will not enter into this field now, but will treat exclusively the theory of modular functions of degree p for $p > 1$. Thus we are dealing with functions $f(W)$ which are meromorphic in the generalized upper half plane $\mathfrak{H}$ and which remain invariant under all modular substitutions of degree p. In contradistinction to the case $p = 1$, the behavior of $f(W)$ is not subjected to additional conditions at infinity. Clearly, the modular functions of degree p form a field which will be denoted by K.

Just as in the case $p = 1$, the connection between modular forms and modular functions is established by the fact that two modular forms ϕ and $\psi \neq 0$ of equal weight yield a modular function upon formation of the quotient ϕ/ψ. Later we will see that all modular functions of degree p are obtained in this way. We already know this in the case $p = 1$. Let k be an even number larger than $p + 1$. In view of Theorem 3 of the preceding section, the Eisenstein series $E_k(W)$ is a modular form of weight k. We also know that it does not vanish identically in W for any given k. Then for every fixed even $k > p + 1$, we will show that the Ansatz

$$f_l(W) = E_{lk} E_k^{-l} \qquad (l = 1, 2, \ldots) \tag{4}$$

yields certain modular functions in terms of which every modular function can be rationally expressed. Before investigating the functions $f_l(W)$ in more detail, we must first deal with a statement about simultaneous zeros of several Eisenstein series. Clearly, if we exclude the zeros of the Eisenstein series $E_k(W)$, the values $f_l(W)$ depend only on the equivalence class of the matrix W with respect to the modular group Γ.

Theorem 2: Given any natural number r there exists a natural number s depending only on p and r such that for any r points $W_1, \ldots, W_r$ of $\mathfrak{H}$

and any even number $k > p + 1$ at least one of the s functions $E_{\nu k}(W)$ $(\nu = 1, \ldots, s)$ differs from zero at each of the points $W_1, \ldots, W_r$.

Proof: Without loss of generality we may assume that all r points under consideration lie in the fundamental domain $\mathfrak{F}$. In view of Theorem 3 of Section 6, given any even $k > p + 1$ and any natural number ν, the Eisenstein series $E_{\nu k}(W)$ is uniformly and absolutely convergent in $\mathfrak{F}$. Furthermore, since each term of the series satisfies abs $(CW + D) \geqslant 1$ everywhere in $\mathfrak{F}$, the convergence is also uniform with respect to ν and k. Therefore, finitely many pairs $C = C^*$, $D = D^*$ can be selected independently of W, ν, k so that the absolute value of the difference

$$(5) \quad E_l(W) - \sum |C^*W + D^*|^{-l} = R_l(W) \qquad (l = \nu k;\ \nu = 1, 2, \ldots)$$

is less than $\frac{1}{2}$ for all ν and arbitrary W in $\mathfrak{F}$. Here we assume that the pair 0, E actually occurs among the pairs C^*, D^*. Let h be the number of remaining pairs C^*, D^* and let $hr = m$. Let $2\pi\alpha_l$ $(l = 1, \ldots, m)$ denote the arguments of the m complex numbers $|C^*W + D^*|^k$, where C^*, D^* ranges over the h pairs and W ranges over $W_1, \ldots, W_r$. We normalize $0 \leqslant \alpha_l < 1$. According to a special case of Minkowski's theorem about linear forms, which was established by Dirichlet, there exist for every natural number t an integer g in the interval $1 \leqslant g \leqslant t^m$ and m additional integers $g_1, \ldots, g_m$ such that the m inequalities

$$\text{abs}\,(g\alpha_l - g_l) < \frac{1}{t} \qquad (l = 1, \ldots, m)$$

are simultaneously satisfied. An application of this theorem with $t = 4$ yields a natural number $g \leqslant 4^m$ such that each power $|C^*W + D^*|^{-gk}$ for $W = W_1, \ldots, W_r$ and all pairs C^*, D^* occurring in (5) have a positive real part. Now we observe that the term 1 corresponding to the pair 0, E also occurs under the summation sign on the left-hand side of (5). Hence the absolute value of this sum for $\nu = g$ and all $W = W_1, \ldots, W_r$ is at least 1; it follows that

$$\text{abs}\, E_{gk}(W) > \tfrac{1}{2} \qquad (W = W_1, \ldots, W_r).$$

Therefore, the number $s = 4^{hr}$ has the required property, and Theorem 2 is proved.

In the sequel we only need the cases $r = 1, 2$. Hence we may assume that s depends only on p. Now the problem is to construct nonconstant modular functions using expressions (4) and to show that among these there even exist n analytically independent functions. Here $n = p(p + 1)/2$ is the number of the independent variables w_{kl} $(1 \leqslant k \leqslant l \leqslant p)$. For this purpose we need the following theorem which is analogous to Theorem 3 in Section 6 of Chapter 3 (Volume II).

Theorem 3: Let there be given two arbitrary points W and W^* of $\mathfrak{H}$, and select an even number $k > p + 1$ so that $E_k(W)E_k(W^*) \neq 0$. If W is not located on certain algebraic hypersurfaces, only finitely many of which intersect any given compact subset of $\mathfrak{H}$, then the equations

$$f_l(W) = f_l(W^*) \qquad (l = 1, 2, \ldots) \tag{6}$$

imply the equivalence of W and W^* with respect to the modular group of degree p.

Proof: Without loss of generality we may assume that W^* is contained in $\mathfrak{F}$ and W belongs to a compact subset $\mathfrak{K}$ of $\mathfrak{F}$. Now we apply the same argument as in Sections I and II of the proof of Theorem 3 in Section 6 of Chapter 3 (Volume II). According to this argument, equations (6) imply the infinitely many equations

$$E_k(W)\,|CW + D|^k = E_k(W^*)\,|C^*W^* + D^*|^k, \tag{7}$$

where $(C \quad D)$ ranges over a complete system of nonassociated second rows of modular substitutions and where $(C^* \quad D^*)$ ranges over a suitable permutation of this system. This permutation may depend on W and W^*. Now we have abs $(CW + D) \geqslant 1$ and abs $(C^*W^* + D^*) \geqslant 1$. First choosing $C = 0$, $D = E$ and then choosing $C^* = 0$, $D^* = E$, we obtain

$$\text{abs}\,(0^*W^* + E^*) = 1. \tag{8}$$

Let

$$W^{**} = (A_1W^* + B_1)(C_1W^* + D_1)^{-1}$$

be a modular substitution with $C_1 = 0^*$, $D_1 = E^*$. Because of (8), the point W^{**} is semireduced in any case; for suitable choice of A_1, B_1, it is even contained in $\mathfrak{F}$. Since $E_k(W)$ is a modular form of weight k, equation (7) for $C = 0$, $D = E$ implies that

$$E_k(W) = E_k(W^{**}).$$

It suffices to prove the assertion for W^{**} in place of W^*. Again denoting W^{**} by W^*, we obtain from (7) the equations

$$|CW + D| = \varepsilon\,|C^*W^* + D^*|, \tag{9}$$

where ε is a kth root of unity which may depend on C, D, W, W^*. Furthermore, for $C \neq 0$ we also have $C^* \neq 0$.

Since W is contained in $\mathfrak{K}$, the left-hand side of (9) is bounded for fixed C, D. Hence, in view of Theorem 3 of the preceding section, the pair C^*, D^* in (9) can only range over a finite set of nonassociated pairs which is independent of W and W^*. In particular, we choose C of rank 1 and denote the rank of C^* by r. Hence, r is a natural number satisfying $1 \leqslant r \leqslant p$. In view of (24) in the preceding section, equations (9) now read as follows:

$$(10) \qquad cW[q] + d = \varepsilon\,|C_1^* W_1^* + D_1^*|, \qquad W_1^* = W^*[Q_1].$$

Here c, d is any pair of relatively prime integers, and q is any column with p relatively prime elements. Furthermore, $(C_1^* \;\; D_1^*)$ is the second row of a modular matrix of degree r, $|C_1^*| \neq 0$, and Q_1 is a primitive matrix with p rows and r columns. It should be remarked that c, d, q can be chosen arbitrarily, whereas ε, C_1^*, D_1^*, Q_1 depend on c, d, q, W, W^*. In particular, choosing c, d to be at first 1, 1 and then 1, 0, while q is kept fixed, subtraction of the corresponding equations (10) yields an equation

$$(11) \qquad \varepsilon_1\,|C_1^* W_1^* + D_1^*| - \varepsilon_2\,|C_2^* W_2^* + D_2^*| = 1.$$

Here ε_1, ε_2 are kth roots of unity, and $W_2^* = W^*[Q_2]$, where Q_2 is a primitive matrix with p rows and s columns and $1 \leqslant s \leqslant p$. Furthermore, $(C_2^* \;\; D_2^*)$ is the second row of a modular matrix of degree s, and $|C_2^*| \neq 0$. In particular, let q_{kl} be the column with kth and lth elements 1 and all other elements zero. Then it follows that

$$(12) \qquad W[q_{kk}] = w_{kk}, \qquad W[q_{kl}] = w_{kk} + 2w_{kl} + w_{ll} \qquad (k \neq l).$$

Setting $c = 1$, $d = 0$ we obtain from (10) and (12) that each element of W is a polynomial in the elements of W^*. The coefficients of these polynomials belong to a finite set of values which is independent of W and W^*. Now if (11) is not satisfied identically in W^*, then the point W^*, hence also W, lies on one of finitely many algebraic hypersurfaces of complex dimension $n - 1$ whose equations are given by (11). By our assumption we may exclude this case, that is, we may assume that (11) is satisfied identically in W^*.

Let U be a unimodular matrix whose first r columns coincide with Q_1. Again replacing $W^*[U]$ by W^* and denoting the submatrix of W^* which is formed by the first r rows and columns by W_1^*, identity (11) now can be written as

$$(13) \quad |W_1^* + S_1| - \alpha\,|W_2^* + S_2| = \beta, \qquad \alpha = \frac{\varepsilon_2\,|C_2^*|}{\varepsilon_1\,|C_1^*|}, \qquad \beta = \frac{1}{\varepsilon_1\,|C_1^*|}$$

with symmetric matrices $S_1 = (C_1^*)^{-1}D_1^*$, $S_2 = (C_2^*)^{-1}D_2^*$, $W_2^* = W^*[Q_0]$ and a primitive matrix Q_0 with s columns. First we conclude that $r = s$ and then we conclude that $|W_1^*| = \alpha\,|W_2^*|$ identically in W^*. In particular, choosing for the matrix W^* a matrix whose entries are all zero except for r arbitrary diagonal entries, it follows that each $r \times r$ subdeterminant of Q_0 vanishes except for the subdeterminant formed by the first r rows. Therefore, the last $p - r$ rows of Q_0 must all be zero. Thus we see that

$$W_2^* = W_1^*[U_1], \qquad \alpha = 1$$

with unimodular U_1. Furthermore,

$$|W_1^* + S_1| - |W_1^* + S_2[U_1^{-1}]| = \beta$$

identically in W_1^*. Now if $r > 1$, we derive that $S_1 = S_2[U_1^{-1}]$ by comparison of the terms of degree $r - 1$. Hence we arrive at the contradiction $\beta = 0$. Therefore, we conclude that $r = s - 1$ and β is rational. In view of (13) both roots of unity, ε_1 and ε_2, are then also rational, that is ± 1.

Thus the elements w_{kl} of W turn out to be linear functions of the elements ζ_{kl} of W^*. We may write explicitly

$$w_{kl} = t_{kl} + \sum_{\kappa,\lambda=1}^{p} a_{kl,\kappa\lambda}\zeta_{\kappa\lambda} \qquad (k, l = 1, \ldots, p), \tag{14}$$

where all coefficients t_{kl}, $a_{kl,\kappa l}$ belong to a finite set of rational numbers and remain invariant if k and l or κ and λ are interchanged. We apply (10) with $c = 1$, $d = 0$ and denote the elements of q by $q_1, \ldots, q_p$. Then for every system of p bounded, relatively prime numbers $q_1, \ldots, q_p$, we derive a relation

$$\sum_{k,l=1}^{p} t_{kl}q_kq_l + \sum_{k,l,\kappa,\lambda=1}^{p} a_{kl,\kappa\lambda}\zeta_{\kappa\lambda}q_kq_l = c\sum_{\kappa,\lambda=1}^{p} \zeta_{\kappa\lambda}r_\kappa r_\lambda + d$$

with suitable relatively prime $r_1, \ldots, r_p$ and suitable relatively prime $c \neq 0$, d. Again we may assume that this relation holds identically in the variables $\zeta_{\kappa\lambda}$. Therefore we have

$$\sum_{k,l=1}^{p} a_{kl,\kappa\lambda}q_kq_l = cr_\kappa r_\lambda \qquad (\kappa, \lambda = 1, \ldots, p), \tag{15}$$

and hence also

$$\left(\sum_{k,l=1}^{p} a_{kl,\kappa\kappa}q_kq_l\right)\left(\sum_{k,l=1}^{p} a_{kl,\lambda\lambda}q_kq_l\right) = \left(\sum_{k,l=1}^{p} a_{kl,\kappa\lambda}q_kq_l\right)^2 \qquad (\kappa, \lambda = 1, \ldots, p). \tag{16}$$

We may assume that (16) is valid for all $q_k = 0, \pm 1, \pm 2$ $(k = 1, \ldots, p)$. As an equation of degree 4 with respect to each quantity q_k, equation (16) then holds identically in $q_1, \ldots, q_p$.

If for all pairs κ, λ the corresponding quadratic forms on the left-hand side of (16) only differ by constant factors, we derive from (15) a decomposition of the form

$$a_{kl,\kappa\lambda} = h_{kl}c_\kappa c_\lambda \qquad (k, l, \kappa, \lambda = 1, \ldots, p).$$

By (14), this decomposition implies in the case $p > 1$ that the variables w_{kl} are linearly dependent. Again excluding W from certain hyperplanes, we may assume that the coefficients $a_{kl,\kappa\lambda}$ are not decomposed in the above way.

Finally, since the argument concerning linear dependence of the w_{kl} remains valid for any number less than or equal to p but greater than 1, the only case which must be considered is the case where each quadratic form on the left-hand side of (16) is the square of a linear form. Now equations (15) and (16) imply the decomposition

$$(17) \qquad 2a_{kl,\kappa\lambda} = h_{k\kappa}h_{l\lambda} + h_{k\lambda}h_{l\kappa} \qquad (k, l, \kappa, \lambda = 1, \ldots, p).$$

Hereby, relation (14) becomes

$$(18) \qquad W = W^*[H] + T, \qquad H = (h_{kl}), \qquad T = (t_{kl}),$$

and H and T belong to a finite set of matrices. Moreover, from (15) and (17) we derive that

$$(19) \qquad H = \lambda G,$$

where G is an integral matrix, λ is a scalar, and λ^2 is an integer.

Again, excluding a suitable hypersurface for W, we may assume that $|H| \neq 0$ in (18). Hence W^* also belongs to a compact subset of $\mathfrak{F}$, and what we have proved for W holds correspondingly for W^*. By (18) and (19) we then obtain an equation

$$\lambda\lambda_1 GG_1 = E,$$

where G_1 is an integral matrix and λ_1^2 an integral scalar. Therefore G is unimodular, and $\lambda^2 = \pm 1$. Since W and W^* both lie in $\mathfrak{H}$, we conclude from (18) and (19) that $\lambda^2 = 1$. Hence H is unimodular. Finally, substituting into (9) the pair $C = E$, $D = 0$ which corresponds to the modular substitution $W \to -W^{-1}$, we obtain the equation

$$|W^* + T[H^{-1}]| = \varepsilon\, |C^*W^* + D^*|,$$

and by a suitable assumption this holds identically in W^*. Then we derive

$$\varepsilon\, |C^*| = 1, \qquad T[H^{-1}] = (C^*)^{-1}D^*.$$

Hence $\varepsilon = \pm 1$, C^* is unimodular, and T is integral. Thus relation (18) is seen to be a modular substitution, and Theorem 3 is proved.

In view of Theorem 2, for every two points W and W^* in $\mathfrak{H}$, there exists a number k satisfying the required property in Theorem 3. If $k_0 > p + 1$ is an arbitrary even number, we may set $k = \nu k_0$ with at least one element ν of the sequence $1, \ldots, s$.

Theorem 4: For every fixed even number $k > p + 1$ the sequence

$$f_l = E_{lk}E_k^{-l} \qquad (l = 1, 2, \ldots)$$

contains at least n analytically independent functions.

Proof: In view of Theorem 3, we can determine a domain $\mathfrak{G}$ in $\mathfrak{H}$ such that $E_k(W)$ differs from 0 at every point of $\mathfrak{G}$, and for any two distinct points W, W^* of $\mathfrak{G}$, the condition $f_l(W) \neq f_l(W^*)$ is satisfied for at least one index l. We suppose that there are not n analytically independent functions $f_l(W)$. Let $r < n$ be the maximum number of analytically independent functions f_l and let the functions $F_\nu = f_{l_\nu}$ $(\nu = 1, \ldots, r)$ be analytically independent, where $l_1, \ldots, l_r$ are certain distinct natural numbers. Select a point $W = W_0$ in $\mathfrak{G}$ where the Jacobian matrix of $F_1, \ldots, F_r$ with respect to the n variables w_{kl} $(1 \leqslant k \leqslant l \leqslant p)$ is of rank r. There is a neighborhood of W_0 where certain suitable r of these variables can be expanded with respect to powers of the r differences

$$d_\nu = F_\nu(W) - F_\nu(W_0) \qquad (\nu = 1, \ldots, r)$$

and where the $n - r$ remaining variables can be chosen arbitrarily. Inserting these power series expansions into f_l $(l = 1, 2, \ldots)$, we obtain the functions f_l as power series in the r variables d_ν which converge in a sufficiently small neighborhood of the origin and are independent of the $n - r$ remaining variables w_{kl}.

The r conditions $d_\nu = 0$ $(\nu = 1, \ldots, r)$ define, locally, an analytic manifold of $n - r$ complex dimensions in a subdomain of $\mathfrak{G}$. This manifold contains W_0, and every point W of this manifold satisfies all the equations $f_l(W) = f_l(W_0)$. Thus we obtain a contradiction to Theorem 3. Hence we have $r \geqslant n$, and Theorem 4 is proved.

Since n is the number of the independent variables w_{kl}, we obviously also have $r \leqslant n$. Whence r equals n.

8. Algebraic dependence

The most important result in the theory of modular functions is the following theorem for which we will give a simplified proof due to Andreotti and Grauert.

Theorem 1: Let $\phi_1(W), \ldots, \phi_n(W)$ be any n fixed algebraically independent modular functions of degree p. Then every modular function $f(W)$ satisfies an algebraic equation

$$f^m + g_1 f^{m-1} + \cdots + g_m = 0$$

of bounded degree m whose coefficients $g_1, \ldots, g_m$ belong to the field of rational functions of $\phi_1, \ldots, \phi_n$.

Proof: The proof consists mainly of generalizing the proofs of Theorems 2 and 3 in Section 2. Now the essential difficulty is that the fundamental region $\mathfrak{F}$ is not compact in $\mathfrak{H}$. The beautiful idea of Andreotti and Grauert is based on the investigation of the compact subset $\mathfrak{F}_\lambda$ of $\mathfrak{F}$ defined by $|Y| \leqslant \lambda$ for sufficiently large positive λ and utilizes an important property of the subset of the boundary given by $|Y| = \lambda$.

We consider the $n + 1$ given modular functions, $f, \phi_1, \ldots, \phi_n$, and represent them at every point a of $\mathfrak{F}_\lambda$ as quotients of power series

$$(1) \qquad f(W) = \frac{p_a(W)}{q_a(W)}, \qquad \phi_k(W) = \frac{p_{ka}(W)}{q_{ka}(W)} \qquad (k = 1, \ldots, n),$$

which converge in a closed n-disk $\mathfrak{K}_a$ of radius r_a and which are relatively prime throughout this disk. We distinguish three cases pertaining to the position of a. First, if a is located in the interior of $\mathfrak{F}_\lambda$, we select the radius r_a so that the entire n-disk $\mathfrak{K}_a$ also belongs to $\mathfrak{F}_\lambda$. Second, assume that a lies on the boundary of $\mathfrak{F}_\lambda$ and that the condition $|Y| < \lambda$ is satisfied at a. Then the point a belongs to the boundary of $\mathfrak{F}$, and we may select r_a so small that each image, $\mathfrak{F}_{M^{-1}}$, of $\mathfrak{F}$ under modular substitutions which intersects $\mathfrak{K}_a$ also contains a as boundary point. In this case $|Y|$ has the same value $<\lambda$ at both points a and a_M. Therefore, by further diminishing of r_a if necessary, we may assume that the entire intersection of $(\mathfrak{K}_a)_M$ and $\mathfrak{F}$ is contained in $\mathfrak{F}_\lambda$. Third, let the equation $|Y| = \lambda$ be satisfied at a and let $\mathfrak{F}_{M^{-1}}$ be any image of $\mathfrak{F}$ which contains a. Here M may also be the identity. Then the equation $\operatorname{abs}(CW + D) = 1$ again holds at the point a. Recalling an argument from the proof of Theorem 4 in Section 6 it follows, for sufficiently large values of λ, that at least the last column of C vanishes. Hence the determinant $|CW + D|$, as a function of W, does not depend on the p independent elements $w_{kp} = w_{pk}$ $(k = 1, \ldots, p)$ of the last row and column of W. Utilizing this result, we will subsequently show that the following lemma holds.

Lemma: There exists a smaller n-disk $\mathfrak{K}_a^*$ with radius r_a^* concentric to $\mathfrak{K}_a$ such that for any regular function on $\mathfrak{K}_a$ the maximum of its absolute value on $\mathfrak{K}_a^*$ is not larger than the maximum of its absolute value on the intersection of $\mathfrak{K}_a$ with the union of all images $(\mathfrak{F}_\lambda)_{M^{-1}}$.

We set $\rho_a = e^{-1}r_a^*$ in the third case and $\rho_a = e^{-1}r_a$ in the first two cases. Finally, let $\mathfrak{L}_a$ be the n-disk of radius ρ_a concentric to $\mathfrak{K}_a$.

Now we are able to carry over the proof of Theorem 2 in Section 2. Only at the end are there some arguments to be added. For the sake of clarity we repeat the previous considerations too. First, according to the covering property for compact sets, we determine finitely many points $a_1, \ldots, a_m$ of $\mathfrak{F}_\lambda$ so that the corresponding n-disks, $\mathfrak{L}_a$ $(a = a_1, \ldots, a_m)$, form a covering of the entire compact region $\mathfrak{F}_\lambda$. Let a, b be any two of these points and suppose the intersection $\mathfrak{D}(M, a, b)$ of $\mathfrak{K}_a$ and $(\mathfrak{K}_b)_M$ is not void for a certain modular matrix M. Since $f(W)$ is a modular function, we obtain from (1) the equations

$$(2)\qquad \frac{p_a(W)}{q_a(W)} = \frac{p_b(V)}{q_b(V)}, \qquad q_b(V) = j_{Mab}(W)q_a(W), \qquad W - V_M$$

on the intersection $\mathfrak{D}(M, a, b)$, where $j_{Mab}(W)$ is a unit on every connected subset of $\mathfrak{D}(M, a, b)$. Analogously we have

$$(3)\qquad q_{kb}(V) = j_{kMab}(W)q_{ka}(W), \qquad (k = 1, \ldots, n).$$

Let two natural numbers, μ and τ, be selected so that the inequalities

$$(4)\qquad \operatorname{abs} j_{Mab} < e^{\mu}, \qquad \operatorname{abs}\left(\prod_{k=1}^{n} j_{kMab}\right) < e^{\tau}$$

are valid on all intersections $\mathfrak{D}(M, a, b)$. We put

$$(5)\qquad s = m\tau^n + 1,$$

and for every $l = 1, 2, \ldots$ we denote by $t = t_l$ the largest nonnegative integer satisfying

$$(6)\qquad st^n < ml^n.$$

It follows that

$$(7)\qquad ml^n \leqslant s(t + 1)^n < (s + 1)(t + 1)^n.$$

Let l be selected so large that, in view of (5) and (7), the condition

$$s > \left(\frac{\mu s}{t} + \tau\right)^n m$$

is satisfied. By (6) we have the inequality

$$(8)\qquad \mu s + \tau t < l.$$

Furthermore, we set

$$(9)\qquad Q_a = q_a^s \prod_{k=1}^{n} q_{ka}^t, \qquad I_{Mab} = j_{Mab}^s \prod_{k=1}^{n} j_{kMab}^t.$$

Let $g(x, x_1, \ldots, x_n)$ be a polynomial in the $n + 1$ variables $x, x_1, \ldots, x_n$ which has degree s with respect to x and degree t with respect to each of the other variables and which has indeterminate coefficients. Then in view of (1) and (9), the function

$$(10)\qquad P_a = Q_a g(f, \phi_1, \ldots, \phi_n)$$

is regular on $\mathfrak{K}_a$. We require that this function, as well as all its partial derivatives of orders $1, \ldots, l - 1$, vanish at the point a. Thus for every fixed a, this condition yields C homogeneous linear equations for the coefficients of g where

$$(11)\qquad C = \sum_{k=0}^{l-1} \binom{n+k-1}{k} = \binom{n+l-1}{n} = \prod_{k=1}^{n} \frac{l+k-1}{k} \leqslant l^n.$$

If a ranges over $a_1, \ldots, a_m$, we obtain mC such equations, whereas the number of coefficients of g is equal to $(s+1)(t+1)^n$. Therefore by (7) and (11), we can determine a polynomial g which does not vanish identically in $x, x_1, \ldots, x_n$ and which satisfies all the required conditions.

Now we consider the maximum, γ, of the absolute values of $P_a(W)$ for $a = a_1, \ldots, a_m$ where W ranges over $\mathfrak{R}_a$ in the first two cases pertaining to the location of a and where W ranges over $\mathfrak{R}_\lambda^*$ in the third case. Let the maximum be attained at $W = V$ and $a = b$. Hence, we have

$$\operatorname{abs} P_b(V) = \gamma, \qquad \operatorname{abs} P_a(W) \leqslant \gamma$$

in the corresponding n-disks. At first we suppose that the location of the point b is given by one of the first two cases. Then for a suitable modular matrix M, the point $V_M = W_0$ lies in $\mathfrak{F}_\lambda$, and hence also in one of the n-disks $\mathfrak{L}_a$. Thus the estimate

$$(12)\qquad \operatorname{abs} P_a(W_0) \leqslant \gamma e^{-l}$$

follows from Theorem 1 in Section 2. Now let b be situated as in the third case. Hence V is a point of $\mathfrak{R}_b^*$. In view of the lemma mentioned above, there exists a modular matrix M and a point V^* in the intersection of $\mathfrak{R}_b$ and $(\mathfrak{F}_\lambda)_{M^{-1}}$ such that

$$\operatorname{abs} P_b(V^*) \geqslant \gamma.$$

Moreover, since the point $V_M^* = W_0$ again lies in $\mathfrak{F}_\lambda$, equation (12) is also valid in the third case. Again writing V instead of V^*, we conclude by (2), (3), (4), (9), and (10) in each case that

$$P_b(V) = I_{Mab}(W_0)P_a(W_0), \qquad \gamma \leqslant \gamma e^{\mu s + \tau t - l}.$$

By (8) it follows that $\gamma = 0$. Hence,

$$g(f, \phi_1, \ldots, \phi_n) = 0.$$

Thus it is shown that the $n+1$ functions $f, \phi_1, \ldots, \phi_n$ are algebraically dependent. The stronger statement of Theorem 1 is then obtained by repeating precisely the method of the proof of Theorem 3 in Section 2. We will not repeat even the outline here.

It remains for us to prove the essential point, namely the previously utilized lemma. Let $W_0 = X_0 + iY_0$ be any point of $\mathfrak{H}$ and $W - W_0 = W_1$, $Y - Y_0 = Y_1$. By a simultaneous transformation of Y_0 and Y_1 into diagonal form, we see that the power series

$$\log |Y| = \log |Y_0| + \sigma(Y_0^{-1}Y_1) - \tfrac{1}{2}\sigma(Y_0^{-1}Y_1Y_0^{-1}Y_1) \tag{13}$$
$$+ \sum_{l=3}^{\infty} \frac{(-1)^{l-1}}{l} \sigma((Y_0^{-1}Y_1)^l)$$

converges in a real neighborhood of $Y = Y_0$. Introducing the quadratic polynomial

$$\psi = \psi(W) = \frac{1}{2i}\sigma(Y_0^{-1}W_1) + \frac{1}{8}\sigma(Y_0^{-1}W_1Y_0^{-1}W_1) \tag{14}$$

with $W_1 = W - W_0$, we obtain

$$\sigma(Y_0^{-1}Y_1) - \tfrac{1}{4}\sigma(Y_0^{-1}Y_1Y_0^{-1}Y_1) = \psi + \bar{\psi} - \tfrac{1}{4}\sigma(Y_0^{-1}W_1Y_0^{-1}\overline{W}_1).$$

In view of (13), the inequality

$$|Y| < |Y_0| \tag{15}$$

is certainly satisfied if $\psi = 0$ and

$$0 < \sigma(W_1\overline{W}_1) < \varepsilon$$

with a sufficiently small positive ε.

In particular, we assume that all elements of W_1 outside the last row and column are zero. In view of (14), the equation $\psi = 0$ is solvable for the last diagonal element ω_p of W_1 for sufficiently small ε, and we obtain ω_p as a power series in the remaining $p - 1$ elements ω_{kp} $(k = 1, \ldots, p - 1)$ of the last column of W_1. Thus we have determined an analytic manifold with $p - 1$ independent complex coordinates ω_{kp} in a neighborhood of W_0, and inequality (15) is satisfied throughout this manifold for $W \neq W_0$. Furthermore, we put $\omega_{1p} = \omega$, $\omega_{kp} = 0$ $(k = 2, \ldots, p - 1)$, whereby ω_p becomes a power series in the single complex variable ω.

Now let $\mathfrak{P}$ be an n-disk with center W_0. Then the point $W = W_0 + W_1$ lies in $\mathfrak{P}$ if the variable ω is restricted to a disk abs $\omega \leqslant \rho$ of sufficiently small positive radius ρ, and inequality (15) is valid on the entire boundary circle abs $\omega = \rho$. This remains true if we define $W = W_2 + W_1$, where for fixed $\mathfrak{P}$ and ρ we select W_2 in a sufficiently small neighborhood $\mathfrak{U}$ of W_0. In particular, let W_0 be a point of $\mathfrak{F}$ satisfying $|Y_0| = \lambda$. Hence W_0 is a boundary point of $\mathfrak{F}_\lambda$. For the second row $(C \quad D)$ of modular matrices M we have abs $(CW_0 + D) \geqslant 1$. Consider a complete system of nonassociated pairs, $C = C_k$, $D = D_k$ $(k = 1, \ldots, r)$ satisfying the condition abs $(C_kW_0 + D_k) = 1$. As was previously stated, the function $|C_kW + D_k|$ is then independent of the p variables w_{kp} $(k = 1, \ldots, p)$. Furthermore, for every pair C, D not associated with one of the pairs, C_k, D_k, the inequality abs $(CW_0 + D) > 1$

holds. Let the point W_2 also be located in $\mathfrak{F}$. Then we have

$$\operatorname{abs}(C_k W_2 + D_k) \geqslant 1 \quad (k = 1, \ldots, r).$$

Hence $\operatorname{abs}(C_k W + D_k) \geqslant 1$ for $W = W_2 + W_1$ under the previously established conditions. If $\mathfrak{P}$ is chosen sufficiently small, then in view of Theorem 3 of Section 6 we also have $\operatorname{abs}(CW + D) > 1$ for all the remaining pairs C, D. Therefore the point W is semireduced. There exists an integral modular substitution with matrix G such that W_G lies in $\mathfrak{F}$ and hence in $\mathfrak{F}_\lambda$ for $\operatorname{abs}\omega = \rho$.

Now if $h(W)$ is any function regular on $\mathfrak{P}$, we apply the substitution $W = W_2 + W_1$. Then for every fixed W_2 in $\mathfrak{U}$ this function $h(W)$ becomes a regular function of the single variable ω in the disk $\operatorname{abs}\omega \leqslant \rho$. Hence the absolute value of this function assumes its maximum on the periphery $\operatorname{abs}\omega = \rho$. Therefore, there exists a point W in $\mathfrak{P}$ and a modular matrix G such that W_G is in $\mathfrak{F}_\lambda$ and

$$\operatorname{abs} h(W_2) \leqslant \operatorname{abs} h(W). \tag{16}$$

Finally, let a be a boundary point of $\mathfrak{F}_\lambda$ satisfying $|Y| = \lambda$. Select the radius r_a of the n-disk $\mathfrak{K}_a$ so small that every image $\mathfrak{F}_{M^{-1}}$ of $\mathfrak{F}$ under a modular substitution which also intersects $\mathfrak{K}_a$ contains the point a. For every such M we take the corresponding image a_M for the point W_0. Then we choose an n-disk contained in the image $(\mathfrak{K}_a)_M$ as the domain $\mathfrak{P}$. Next, the radius r_a^* of the n-disk $\mathfrak{K}_a^*$ is selected so that for every admissible M the image $(\mathfrak{K}_a^*)_M$ is entirely contained in the neighborhood $\mathfrak{U}$ introduced above. Now if $d(W)$ is any regular function on $\mathfrak{K}_a$, the result (16) may be applied with $h(W) = d(W_{M^{-1}})$. Then, for every point W^* located in the intersection of $\mathfrak{K}_a^*$ and $\mathfrak{F}_{M^{-1}}$, we can find a suitable point W in $\mathfrak{F}_\lambda$ such that

$$\operatorname{abs} d(W^*) \leqslant \operatorname{abs} d(W_{M^{-1}G^{-1}})$$

is satisfied and $W_{M^{-1}G^{-1}}$ lies in $\mathfrak{K}_a$. Thus the lemma used above has been proved. We note explicitly that the assumption $p > 1$ was essential to the proof. Now Theorem 1 is completely proved.

According to Theorem 4 of the preceding section, for even $k > p + 1$ there always exist n analytically independent modular functions of the form

$$F_r = f_l = E_{lk} E_k^{-l} \quad (l = l_r;\ r = 1, \ldots, n). \tag{17}$$

These functions are also algebraically independent. Let k be fixed in the sequel. In view of Theorem 1, there exists another modular function F_0 connected with $F_1, \ldots, F_n$ by an algebraic equation such that the adjunction of F_0 to the field of rational functions of $F_1, \ldots, F_n$ yields the field K of all modular functions of degree p. Here it is not asserted that F_0 is also expressible in the form (17). However, again utilizing Theorem 3 of the preceding section, this time in a more profound fashion, we will show that F_0 can be represented as a quotient of isobaric polynomials of equal weight in suitably selected finitely many Eisenstein series. More precisely, an expression of the form

$$F_0 = \lambda_1 f_1 + \cdots + \lambda_t f_t \tag{18}$$

for suitable t and certain complex constants $\lambda_1, \ldots, \lambda_t$ will serve our purposes.

Theorem 2: There exists a modular function of the particular form (18) such that every modular function can be expressed rationally in terms of $F_0, F_1, \ldots, F_n$.

Proof: Let Λ be the field of all rational functions of $f_1, f_2, \ldots$. Then Theorem 1 implies that Λ is a simple algebraic extension of the field of rational functions of $F_1, \ldots, F_n$. By Abel's theorem we may generate Λ by adjoining a function F_0 of the form (18) to the field of rational functions of $F_1, \ldots, F_n$. Let

$$P_0(F_0, F_1, \ldots, F_n) = 0 \tag{19}$$

be the irreducible algebraic equation for F_0 whose coefficients are polynomials in $F_1, \ldots, F_n$. Denote the coefficient of the highest power of F_0 occurring in P_0 by g_0 and let Δ_0 be the discriminant of P_0 with respect to F_0. Furthermore, let F be a modular function whose adjunction to the field of rational functions of $F_1, \ldots, F_n$ yields the entire field K. In analogy with (19), let

$$P(F, F_1, \ldots, F_n) = 0 \tag{20}$$

be the irreducible algebraic equation for F. We define g and Δ the same way we define g_0 and Δ_0.

We form the Jacobian determinant $\phi = \phi(W)$ of $F_1(W), \ldots, F_n(W)$ with respect to the n independent variables w_{kl} $(1 \leqslant k \leqslant l \leqslant p)$. For every modular substitution

$$W_M = (AW + B)(CW + D)^{-1}$$

we have

$$\phi(W_M) = |CW + D|^{p+1}\, \phi(W). \tag{21}$$

Hence, the product

$$\psi(W) = \phi^k E_k^{-p-1} \tag{22}$$

is a modular function. Denote the irreducible algebraic equation for ψ by

$$P_1(\psi, F_1, \ldots, F_n) = 0$$

and let h be the term which is independent of ψ.

Let $W^{(1)}$ be the submatrix of W formed by the first $p - 1$ rows and columns of W and denote the corresponding Eisenstein series of weight k and degree $p - 1$ by $E_k(W^{(1)})$. Suppose W runs over a sequence in the fundamental region $\mathfrak{F}$ such that the last diagonal element w_p tends to infinity. Then, in view of Theorem 4 in Section 6, the difference $E_k(W) - E_k(W^{(1)})$ tends to zero. Because $\frac{1}{2}p(p - 1) = n - p < n$, the n modular functions

$$F_r^* = F_r^*(W^{(1)}) = E_{lk}(W^{(1)})(E_k(W^{(1)}))^{-l} \qquad (l = l_r;\ r = 1, \ldots, n)$$

of degree $p - 1$ satisfy an algebraic equation

$$\chi(F_1^*, \ldots, F_n^*) = 0, \tag{23}$$

whereas the polynomial $\chi(F_1, \ldots, F_n)$ does not vanish identically in W.

If μ is the precise degree of the polynomial P_0 in (19) with respect to F_0, we have the representations

$$f_l = Q_{l1}F_0^{\mu-1} + Q_{l2}F_0^{\mu-2} + \cdots + Q_{l\mu} \qquad (l = 1, 2, \ldots) \tag{24}$$

with uniquely determined rational functions $Q_{l1}, \ldots, Q_{l\mu}$ of $F_1, \ldots, F_n$. Here we take the values $l = 1, \ldots, s$, where s is defined as in the discussion following Theorem 2 of the preceding section. For each of these values $l = 1, \ldots, s$ we form the common denominator of the μ rational functions $Q_{l1}, \ldots, Q_{l\mu}$ and denote the product of the s common denominators by $\tau = \tau(F_1, \ldots, F_n)$. Finally, let $\gamma = \gamma(F_1, \ldots, F_n)$ be the product of the seven polynomials $g_0, \Delta_0, g, \Delta, h, \chi, \tau$ in the variables $F_1, \ldots, F_n$.

Applying the limit point theorem, we can construct a point W_0 in $\mathfrak{H}$ which satisfies each of the following conditions: the n functions $F_1, \ldots, F_n$ are regular at W_0; the value of γ at W_0 is different from zero; none of the denominators of the countably many functions $Q_{l\nu}$ ($\nu = 1, \ldots, \mu$; $l = 1, 2, \ldots$), which are rational in $F_1, \ldots, F_n$, vanishes at W_0; and W_0 is not located on any of the algebraic hypersurfaces to be excluded according to Theorem 3 of Section 7. We consider the space $\mathfrak{R}$ of the n complex finite coordinates $F_1, \ldots, F_n$ and a closed oriented curve C in this space starting from the point $\mathfrak{p}_0$ with coordinates $F_1(W_0), \ldots, F_n(W_0)$ such that $\gamma(F_1, \ldots, F_n) \neq 0$ is satisfied throughout this curve. In view of (19) and (20), the functions $F_0(W)$ and $F(W)$ are regular at $W = W_0$. By means of equations (19) and (20), both $F_0(W)$ and $F(W)$, as functions of $F_1, \ldots, F_n$ in $\mathfrak{R}$, can be continued regularly along C since $g_0\Delta_0\, g\Delta \neq 0$ remains valid. Traversing C once, we again obtain at the point $\mathfrak{p}_0$ certain solutions F_0, F of the equations (19), (20) which might differ from the starting values $F_0(W_0)$,

$F(W_0)$. Now suppose Λ is a proper subfield of K. Hence, F cannot be represented in the form of the right-hand side of (24). Then we apply an important theorem about algebraic functions of several variables whose proof will not be presented here. We only note that the proof is obtained by generalizing the arguments in Section 1 of Chapter 2 (Volume I). A detailed explanation would be rather extensive. According to this theorem, the curve C can be selected so that F_0, indeed, returns to its initial value $F_0(W_0)$, but F does not. We are now going to derive, by analytic methods, a contradiction to this statement.

Starting at $\mathfrak{p}_0$ we invert the system of n regular functions $F_1(W), \ldots, F_n(W)$, thus transplanting the curve C from $\mathfrak{R}$ to $\mathfrak{H}$. First, we show that this is possible in a certain neighborhood of $\mathfrak{p}_0$. Since by our assumptions $F_1(W), \ldots, F_n(W)$ are regular at $W = W_0$, the Jacobian determinant $\phi(W)$ is also regular at this point. If we suppose $\phi(W_0) = 0$, then by (22) and $h(W_0) \neq 0$ it follows that $E_k(W_0) = 0$. In view of Theorem 2 of the preceding section, at least one of the s values $E_{\nu k}(W_0)$ differs from zero. Hence, at least one of the s functions $f_l(W)$ $(l = 1, \ldots, s)$ has a pole at $W = W_0$. However, since τ does not vanish at $\mathfrak{p}_0$, in view of (24) these functions must be regular at W_0. Therefore the w_{kl} are regular functions of $F_1, \ldots, F_n$ on C in a neighborhood of $\mathfrak{p}_0$. Let C^* be a subarc of C which has the same orientation and which starts at $\mathfrak{p}_0$ and ends at $\mathfrak{p}^*$. We assume that every subarc of C^* starting at $\mathfrak{p}_0$ can be transplanted by the above inversion from $\mathfrak{R}$ to $\mathfrak{H}$ so that $F_1(W), \ldots, F_n(W)$ remain regular and both functions $\phi(W)$ and $E_k(W)$ are different from zero. Then we show that the same statement is true for C^* itself. For this purpose we select a sequence of points $\mathfrak{p}_1, \mathfrak{p}_2, \ldots$ on C^* tending to $\mathfrak{p}^*$. Let $W_1, W_2, \ldots$ be the corresponding sequence in $\mathfrak{H}$. For every point W_l $(l = 1, 2, \ldots)$ we determine a suitable modular substitution with matrix M_l such that the point

$$W_l^* = (W_l)_{M_l} \qquad (l = 1, 2, \ldots)$$

belongs to $\mathfrak{F}$.

Again, in view of Theorem 2 of Section 7, for every $l = 1, 2, \ldots$ at least one of the s quantities $E_{\nu k}(W_l^*)$ differs from 0. More precisely, from the previous proof it even follows that one of these s quantities has absolute value at least $\frac{1}{2}$. Since τ does not vanish on C we then derive, as in the preceding paragraph, that the absolute values of the numbers $E_k(W_l^*)$ $(l = 1, 2, \ldots)$ lie above a positive bound. Because of our assumption about γ, the function χ^{-1} is bounded on C. Therefore, the points W_l^* $(l = 1, 2, \ldots)$ must be located in a compact subset of the fundamental region $\mathfrak{F}$, since otherwise a contradiction would occur in view of (23) and Theorem 4 in Section 6. Hence, the points W_l^* have a limit point W^* in $\mathfrak{F}$, and we obtain $E_k(W^*) \neq 0$. It follows that all the functions $F_1(W), \ldots, F_n(W)$ are regular at $W = W^*$. By the assumption about γ, we have $h(W^*) \neq 0$. Hence $\psi(W^*) \neq 0$ and $\phi(W^*) \neq 0$. Consequently, the system of the n functions $F_1, \ldots, F_n$ is uniquely invertible in a neighborhood of $W = W^*$. Since the modular group is discontinuous in $\mathfrak{H}$, the continuity of C^* at the point $\mathfrak{p}^*$ implies that the sequence of modular matrices M_l contains only finitely many different elements. Hence, the entire arc C^* can be transplanted from $\mathfrak{R}$ to $\mathfrak{H}$. Therefore, the same result is true for C itself.

If W_0^* is the endpoint of the image curve in $\mathfrak{H}$, then the functions $F_1(W), \ldots, F_n(W)$, and hence $F_0(W)$ and $F(W)$, are regular at W_0^*. According to the assumption on C, we then have

$$F_r(W_0^*) = F_r(W_0) \qquad (r = 0, 1, \ldots, n), \qquad F(W_0^*) \neq F(W_0).$$

By (24) all the functions $f_l(W)$ are also regular at W_0^*, and it follows that

$$f_l(W_0^*) = f_l(W_0) \qquad (l = 1, 2, \ldots). \tag{25}$$

Since $E_k(W_0)(E_k(W_0^*) \neq 0$, by use of Theorem 3 in the preceding section, equation (25) yields the equivalence of W_0 and W_0^*. Hence,

$$F(W_0^*) = F(W_0).$$

Therefore we have a contradiction, and the proof of Theorem 2 is complete.

In view of Theorem 2 every modular function of degree p can be expressed rationally in terms of t fixed modular functions

$$f_l(W) = E_{lk}(W)(E_k(W))^{-l} \qquad (l = 1, 2, \ldots, t)$$

with arbitrarily given even $k > p + 1$ and sufficiently large t. Therefore, for all modular functions we obtain a representation as a quotient of two isobaric polynomials of equal weight in the t Eisenstein series $E_{lk}(W)$ $(l = 1, \ldots, t)$. In particular, every modular function is a quotient of two modular forms of equal weight. Here it remains undecided whether numerator and denominator can be selected so that they are locally relatively prime throughout $\mathfrak{H}$. This question was answered in the affirmative in the case $p = 1$.

Let $g(W)$ be any modular form of even weight $\nu = 2\mu$. Then $gE_k^{\mu}E_{k+2}^{-\mu}$ is a modular function; hence, g is representable as a quotient of two modular forms u and v with weight difference ν, where u and v are isobaric polynomials in the $t + 1$ Eisenstein series E_{lk} $(l = 1, \ldots, t)$ and E_{k+2}. Here k is any even integer greater than $p + 1$, and t may depend on k. In the case $p = 1$, the stronger statement that we may take $k = 4$, $t = 1$, and $v = 1$ is true because of Theorem 2 in Section 6. Namely, every elliptic modular form is a polynomial in E_4 and E_6. In the case $p = 2$, $v = 1$ is again admissible in view of Igusa's result mentioned at the end of Section 6. Namely, in this case every modular form of even weight and second degree is a polynomial in E_4, E_6, E_{10}, E_{12}. For $p > 2$ no similar result is known.

In order to know the algebraic structure of the field K, we have to determine the irreducible algebraic equation (19) connecting $F_0, F_1, \ldots, F_n$. Here a constant factor different from 0 remains arbitrary. If we choose the constants $\lambda_1, \ldots, \lambda_t$ in (18) as suitable integers, it turns out that the polynomial P_0 can also be written with rational integral coefficients. This important theorem, as well as the explicit determination of the polynomial P_0, is obtained by use of the invariance of the Eisenstein series under integral modular substitutions. Namely, we must utilize some properties of the corresponding Fourier series whose coefficients are all rational numbers. For $p = 1$ these Fourier series are established explicitly in Section 4. The case of an arbitrary p is also of number theoretical interest; but we cannot enter into this subject.

It is an open question whether K is always a rational function field, that is, whether every modular function of degree p can be expressed rationally in terms of n basic functions. The answer is affirmative in the cases $p = 1$ and $p = 2$. This is seen for the elliptic modular functions from Theorem 1 in Section 7 and for modular functions of second degree by Igusa's result.

Finally, we must explain the relation between the field $K(T)$ of modular functions of level T and the modular functions of level E treated thus far. Here we may assume $p > 1$, since otherwise $K(T)$ and K coincide.

Theorem 3: There exists an algebraic function field of transcendence degree n which is a simple algebraic extension of both $K(T)$ and K.

Proof: The modular group $\Gamma(T)$ of level T consists of all symplectic substitutions

$$W^* = (AW + BT)(T^{-1}CW + T^{-1}DT)^{-1}$$

with integral A, B, C, D. Here we put

$$A^* = A, \qquad B^* = BT, \qquad C^* = T^{-1}C, \qquad D^* = T^{-1}DT.$$

Hence,

$$W^* = (A^*W + B^*)(C^*W + D^*)^{-1}. \tag{26}$$

Then we consider all elements of $\Gamma(T)$ for which (26) is a modular substitution in the sense used before, that is, for which A^*, B^*, C^*, D^* are also integral. These elements form the intersection Δ of $\Gamma(T)$ and Γ. Putting $|T| = t$, we define a subgroup Δ_1 of Δ by the stronger conditions

$$A \equiv E \pmod{t}, \qquad B \equiv 0 \pmod{t}, \qquad C \equiv 0 \pmod{t}, \qquad D \equiv E \pmod{t}$$

and, analogously, another subgroup Δ_2 of Δ by

$$A^* \equiv E \pmod{t}, \qquad B^* \equiv 0 \pmod{t},$$
$$C^* \equiv 0 \pmod{t}, \qquad D^* \equiv E \pmod{t}.$$

Since the integers decompose modulo t into t residue classes, it is easily seen that Δ_1 is of finite index in $\Gamma(T)$ and that Δ_2 is of finite index in Γ. Hence Δ is of finite index in $\Gamma(T)$ as well as in Γ.

Now let $M_1, \ldots, M_h$ be a complete set of representatives of the right cosets $\Delta M_1, \ldots, \Delta M_h$ of Δ in $\Gamma(T)$. Furthermore, let $f(W)$ be any function meromorphic in $\mathfrak{H}$ and invariant with respect to Δ. We put

$$W_{M_k} = W_k, \qquad f(W_k) = f_k(W) \qquad (k = 1, \ldots, h) \tag{27}$$

and form the h elementary symmetric polynomials $\sigma_1, \ldots, \sigma_h$ of $f_1, \ldots, f_h$. Then if M is any element of $\Gamma(T)$, the right cosets $(\Delta M_1)M, \ldots, (\Delta M_h)M$ are a permutation of $\Delta M_1, \ldots, \Delta M_h$. Also, since $f(W)$ is invariant under Δ, the h functions $f_k(W_M)$ are a permutation of the functions $f_k(W)$. Hence, the functions $\sigma_1(W), \ldots, \sigma_h(W)$ are modular functions of level T. The function $f(W)$ itself satisfies an algebraic equation

$$f^h - \sigma_1 f^{h-1} + \cdots + (-1)^h \sigma_h = 0. \tag{28}$$

Therefore, if K* denotes the field of automorphic functions corresponding to Δ, then K* is a simple algebraic extension of K(T). On the other hand, K is certainly contained in K*. The previous discussion can also be carried through with Γ in place of $\Gamma(T)$. Hence, K* is also a simple algebraic extension of K. Thus Theorem 3 is proved.

In view of Theorem 3, the field K(T) is an algebraic function field of transcendence degree n for every fixed level T. If we choose for f in (27) any n analytically independent modular functions of degree p, for instance the functions defined by (17), then in view of (28) the set of the nh corresponding expressions $\sigma_1(W), \ldots, \sigma_h(W)$ must also contain n analytically independent functions. Thus we obtain explicit expressions for n analytically independent functions in K(T). But the corresponding generalization of Theorem 2 is not yet worked out. In particular, for this purpose it is necessary to carry over Theorem 3 in Section 7 to an arbitrary level T.

Now we arrive at the end of the long trip which took us from Fagnano's discovery approximately 250 years ago to research of recent date. Our interest in this trip was stimulated by the attempt to penetrate deeper into those analytic, algebraic and geometric relations which lie hidden behind Euler's addition theorem. At first we were led to the theory of elliptic functions, then to the uniformization of algebraic curves by automorphic functions and finally to the Abelian functions and the related modular functions of several variables. The problems dealt with in the final sections lead to open questions whose solutions might provide a challenge for mathematicians of future generations.

Bibliography

ABEL, N. H.

(1) Über die Integration der Differential-Formel $\rho\, dx/\sqrt{R}$, wenn R und ρ ganze Functionen sind. *J. reine angew. Math.* **1** (1826), pp. 185–221.

(2) Recherches sur les fonctions elliptiques. *J. reine angew. Math.* **2** (1827), pp. 101–181; **3** (1828), pp. 160–190.

(3) Remarques sur quelques propriétés générales d'une certaine sorte de fonctions transcendantes. *J. reine angew. Math.* **3** (1828), pp. 313–323.

(4) Démonstration d'une propriété générale d'une certaine classe de fonctions transcendentes. *J. reine angew. Math.* **4** (1829), pp. 200–201.

(5) Précis d'une théorie des fonctions elliptiques. *J. reine angew. Math.* **4** (1829), pp. 236–277, 309–348.

ALBERT, A. A.

(1) Structure of Algebras. *Coll. Publ. Amer. Math. Soc.* **24,** New York, 1939.

(2) On the construction of Riemann matrices I. *Ann. Math.* **35** (1934), pp. 1–28.

(3) A solution of the principal problem in the theory of Riemann matrices. *Ann. Math.* **35** (1934), pp. 500–515.

(4) On the construction of Riemann matrices II. *Ann. Math.* **36** (1935), pp. 376–394.

(5) Involutorial simple algebras and real Riemann matrices. *Ann. Math.* **36** (1935), pp. 886–964.

(6) On involutorial algebras. *Proc. Nat. Acad. Sci. U.S.A.* **41** (1955), pp. 480–482.

ANDREOTTI, A.

(1) Sopra le varietà di Picard di una superficie algebrica. *Rend. Acc. Naz. dei XL*, Ser. IV 2 (1952), pp. 129–137.

(2) On a theorem of Torelli. *Amer. J. Math.* **80** (1958), pp. 801–828.

(3) Complex pseudoconcave spaces and automorphic functions. *Proc. Int. Congr. Math.*, Stockholm, 1962, pp. 306–308.

(4) Théorèmes de dépendance algébrique sur les espaces complexes pseudo-concaves. *Bull. Soc. Math. France* **91** (1963), pp. 1–38.

ANDREOTTI, A., and GRAUERT, H.

(1) Algebraische Körper von automorphen Funktionen. *Nachr. Akad. Wiss. Göttingen, math.-phys. Klasse*, 1961, pp. 39–48.

(2) Théorèmes de finitude pour la cohomologie des espaces complexes. *Bull. Soc. Math. France* **90** (1962), pp. 193–259.

ANDREOTTI, A., and MAYER, A. L.

(1) On period relations for abelian integrals on algebraic curves. *Ann. Scuola Norm. Sup. Pisa* (3) **21** (1967), pp. 189–238.

ANDREOTTI, A., and STOLL, W.

(1) *Analytic and algebraic dependence of meromorphic functions.* Lecture Notes in Mathematics, Vol. 234, Springer-Verlag, Berlin–Heidelberg–New York, 1971.

ANDREOTTI, A., and VESENTINI, E.

(1) On deformations of discontinuous groups. *Acta Math.* **112** (1964), pp. 249–298.

APPELL, P.

(1) Sur l'inversion des intégrales abéliennes. *J. Math. pures appl.* (*4*) **1** (1885).
(2) Sur les fonctions périodiques de deux variables. *J. Math. pures appl.* (*4*) **7** (1891), pp. 157–219.

APPELL, P., and GOURSAT, E.

(1) *Théorie des fonctions algébriques et de leurs intégrales.* Paris, 1895; 2nd edition 1929/30.

ARAKI, S.

(1) On root systems and an infinitesimal classification of irreducible symmetric spaces. *J. Math. Osaka City Univ.* **13** (1962), pp. 1–34.
(2) On Bott-Samelson K-cycles associated with symmetric spaces. *J. Math. Osaka City Univ.* **13** (1962), pp. 87–133.

BAGNERA, G., and de FRANCHIS, M.

(1) Le superficie algebriche le quali ammettono una rappresentazione parametrica mediante funzioni iperellittiche di due argomenti. *Mem. Mat. Fis. Soc. It. Sci.* (*3*) **15** (1908), pp. 253–343.
(2) Les nombres ρ de M. Picard pour les surfaces hyperelliptiques et pour les surfaces irrégulières de genre zéro. *Rend. Circ. Mat. Palermo* **30** (1910), pp. 185–238.

BAILY, W. L., Jr.

(1) On the quotient of an analytic manifold by a group of analytic homeomorphisms. *Proc. Nat. Acad. Sci. U.S.A.* **40** (1954), pp. 804–808.
(2) The decomposition theorem for V-manifolds. *Amer. J. Math.* **78** (1956), pp. 862–888.
(3) On the embedding of V-manifolds in projective space. *Amer. J. Math.* **79** (1957), pp. 403–430.
(4) Satake's compactification of V_n. *Amer. J. Math.* **80** (1958), pp. 348–364.
(5) On the Hilbert-Siegel modular space. *Amer. J. Math.* **81** (1959), pp. 846–874.
(6) On the moduli of Jacobian varieties. *Ann. Math.* **71** (1960), pp. 303–314.
(7) Some results on the moduli of Riemann surfaces. *Proc. Nat. Acad. Sci. U.S.A.* **47** (1961), pp. 325–327.
(8) On the automorphism group of a generic curve of genus >2. *J. Math. Kyoto Univ.* **1** (1961), pp. 101–108; correction, 325.

(9) On the theory of θ-functions, the moduli of Abelian varieties, and the moduli of curves. *Ann. Math.* **75** (1962), pp. 342–381.

(10) *On the orbit spaces of arithmetic groups.* Arithmetical Algebraic Geometry (Proc. Conf. Purdue Univ., 1963), pp. 4–10.

(11) On the moduli of Abelian varieties with multiplications from an order in a totally real number field. *Proc. Int. Congr. Math.*, Stockholm, 1962, pp. 309–313.

(12) On the theory of automorphic functions and the problem of moduli. *Bull. Amer. Math. Soc.* **69** (1963), pp. 727–732.

(13) On the moduli of Abelian varieties with multiplications. *J. Math. Soc. Japan* **15** (1963), pp. 367–386; correction, **16** (1964), p. 182.

(14) On compactifications of orbit spaces of arithmetic discontinuous groups acting on bounded symmetric domains. *Proc. Sympos. Pure Math.*, Vol. 9, American Mathematical Society, Providence, R.I., 1966, pp. 281–295.

(15) Fourier-Jacobi series. *Ibid.*, pp. 296–300.

(16) Classical theory of θ-functions. *Ibid.*, pp. 306–311.

(17) *On Hensel's lemma and exponential sums.* Global Analysis (Papers in Honor of K. Kodaira), pp. 85–100. University of Tokyo Press, Tokyo, 1969.

(18) An exceptional arithmetic group and its Eisenstein series. *Bull. Amer. Math. Soc.* **75** (1969), pp. 402–406.

(19) *Eisenstein series on tube domains.* Problems in Analysis, A Symposium in Honor of S. Bochner, Princeton, N.J., 1970, pp. 139–156.

BAILY, W. L., Jr., and BOREL, A.

(1) On the compactification of arithmetically defined quotients of bounded symmetric domains. *Bull. Amer. Math. Soc.* **70** (1964), pp. 588–593.

(2) Compactification of arithmetic quotients of bounded symmetric domains. *Ann. Math.* **84** (1966), pp. 442–528.

BAKER, H. F.

(1) *Abel's theorem and the allied theory including the theory of theta functions.* Cambridge University Press, 1897.

(2) *An introduction to the theory of multiply periodic functions.* Cambridge University Press, 1907.

BARSOTTI, I.

(1) A note on Abelian varieties. *Rend. Circ. Mat. Palermo (II)* **2** (1953), pp. 236–257.

(2) Structure theorems for group varieties. *Ann. Mat.* **38** (1955), pp. 77–119.

(3) Abelian varieties over fields of positive characteristic. *Rend. Circ. Mat. Palermo (II)* **5** (1956), pp. 145–169.

(4) Repartitions on abelian varieties. *Illinois J. Math.* **2** (1958), pp. 43–70.

(5) On Witt vectors and periodic group varieties. *Illinois J. Math.* **2** (1958), pp. 99–110, 608–610.

BECKER, H.

(1) Poincarésche Reihen zur Hermiteschen Modulgruppe. *Math. Ann.* **129** (1955), pp. 187–208.

BEHNKE, H., and THULLEN, P.

(1) Theorie der Funktionen mehrerer komplexer Veränderlichen. *Erg. Math.*, Vol. 3, No. 3, Springer-Verlag, Berlin 1934; reprinted by Chelsea Publishing Company, New York; 2nd edition, Springer-Verlag, Berlin 1970.

BERGER, M.

(1) Les espaces symétriques non compactes. *Ann. Sci. École Norm. Sup.* (3) **74** (1957), pp. 85–177.

BINGEN, F.

(1) Les domaines bornés symétriques de l'espace complexe à n dimensions. *Bull. Soc. Math. Belgique* **6** (1953), pp. 53–61, published in 1954.

BLUMENTHAL, O.

(1) Über Modulfunktionen von mehreren Veränderlichen I. *Math. Ann.* **56** (1903), pp. 509–548.
(2) Über Modulfunktionen von mehreren Veränderlichen II. *Math. Ann.* **58** (1904), pp. 497–527.

BOCHNER, S.

(1) Algebraic and linear dependence of automorphic functions in several variables. *J. Indian Math. Soc.* **16** (1952), pp. 1–6.
(2) Linear and algebraic dependence of functions on compact complex spaces with singularities. *Proc. Nat. Acad. Sci. U.S.A.* **45** (1959), pp. 47–49.

BOCHNER, S., and GUNNING, R. C.

(1) Existence of functionally independent automorphic functions. *Proc. Nat. Acad. Sci. U.S.A.* **41** (1955), pp. 746–752.

BOCHNER, S., and MARTIN, W. T.

(1) *Several complex variables.* Princeton University Press, Princeton, N.J., 1948.
(2) Complex spaces with singularities. *Ann. Math.* **57** (1953), pp. 490–516.

BOREL, A.

(1) Le plan projectif des octaves et les sphères comme espaces homogènes. *C. R. Acad. Sci. Paris* **230** (1950), pp. 1378–1380.
(2) Les fonctions automorphes de plusieurs variables complexes. *Bull. Soc. Math. France* **80** (1952), pp. 167–182.
(3) *Les espaces hermitiens symétriques.* Séminaire Bourbaki, 1952.
(4) Sur la cohomologie des espaces fibrés principaux et des espaces homogènes de groupes de Lie compacts. *Ann. Math.* **57** (1953), pp. 115–207.
(5) Les bouts des espaces homogènes de groupes de Lie. *Ann. Math.* **58** (1953), pp. 443–457.

(6) La cohomologie mod 2 de certains espaces homogènes. *Comment. Math. Helv.* **27** (1953), pp. 165–191.

(7) Sur l'homologie et la cohomologie des groupes de Lie compacts connexes. *Amer. J. Math.* **76** (1954), pp. 273–342.

(8) Kaehlerian coset spaces of semi-simple Lie groups. *Proc. Nat. Acad. Sci. U.S.A.* **40** (1954), pp. 1147–1151.

(9) *Lectures on symmetric spaces.* Massachusetts Institute of Technology, Lecture Notes, 1958.

(10) On the curvature tensor of the Hermitian symmetric manifolds. *Ann. Math.* **71** (1960), pp. 508–521.

(11) Compact Clifford–Klein forms of symmetric spaces. *Topology* **2** (1963), pp. 111–122.

(12) Introduction to automorphic forms. *Proc. Sympos. Pure Math.*, Vol. 9, American Mathematical Society, Providence, R.I., 1966, pp. 199–210.

BOREL, A., and HIRZEBRUCH, F.

(1) Characteristic classes and homogeneous spaces. I, II, III. *Amer. J. Math.* **80** (1958), pp. 458–538; **81** (1959), pp. 315–382; **82** (1960), pp. 491–504.

BOREL, A., and REMMERT, R.

(1) Über kompakte homogene Kählersche Mannigfaltigkeiten. *Math. Ann.* **145** (1962), pp. 429–439.

BOREL, A., CHOWLA, S., HERZ, C. S., IWASAWA, K., and SERRE, J.-P.

(1) *Seminar on complex multiplication.* Lecture Notes in Mathematics, Vol. 21, Springer-Verlag, Berlin–Heidelberg–New York, 1966.

BOTT, R., and SAMELSON, H.

(1) Applications of the theory of Morse to symmetric spaces. *Amer. J. Math.* **80** (1958), pp. 964–1028.

BRAUN, H.

(1) Zur Theorie der Modulformen n-ten Grades. *Math. Ann.* **115** (1938), pp. 507–517.

(2) Konvergenz verallgemeinerter Eisensteinscher Reihen. *Math. Z.* **44** (1939), pp. 387–397.

(3) Hermitian modular functions. *Ann. Math.* **50** (1949), pp. 827–855.

(4) Hermitian modular functions II. *Ann. Math.* **51** (1950), pp. 92–104.

(5) Hermitian modular functions III. *Ann. Math.* **53** (1951), pp. 143–160.

(6) Der Basissatz für hermitische Modulformen. *Abh. Math. Sem. Univ. Hamburg* **19** (1955), pp. 134–148.

(7) Darstellung hermitischer Modulformen durch Poincarésche Reihen. *Abh. math. Sem. Univ. Hamburg* **22** (1958), pp. 9–37.

BRAUN, H., and KOECHER, M.

(1) *Jordan-Algebren.* Springer-Verlag, Berlin–Heidelberg–New York, 1966.

BRILL, A.

(1) Über das Verhalten einer Funktion von zwei Veränderlichen in der Umgebung einer Nullstelle. *Sitz.-Ber. Bayer. Akad. Wiss., math.-naturw. Klasse*, 1891, pp. 207–220.

(2) Über den Weierstraßschen Vorbereitungssatz. *Math. Ann.* **69** (1910), pp. 538–549.

BUSAM, R.

(1) Eine Verallgemeinerung gewisser Dimensionsformeln von Shimizu. *Inventiones math.* **11** (1970), pp. 110–149.

BUSAM, R., and FREITAG, E.

(1) Das arithmetische Geschlecht des Körpers der symmetrischen Modulfunktionen zur Hilbertschen Modulgruppe eines reell-quadratischen Zahlkörpers der Klassenzahl eins mit Primzahldiskriminante (to appear).

CALABI, E., and VESENTINI, E.

(1) On compact locally symmetric Kähler manifolds. *Ann. Math.* (*2*) **71** (1960), pp. 472–507.

CARTAN, É.

(1) Sur une classe remarquable d'espaces de Riemann. *Bull. Soc. Math. France* **54** (1926), pp. 214–264; **55** (1927), pp. 114–134.

(2) Sur certaines formes Riemanniennes remarquables des géométries à groupe fondamentale simple. *Ann. Sci. École Norm. Sup.* **44** (1927), pp. 345–467.

(3) Groupes simples clos et ouverts et géométrie Riemannienne. *J. Math. pures appl.* **8** (1929), pp. 1–33.

(4) Sur les domaines bornés homogènes de l'espace de *n* variables complexes. *Abh. Math. Sem. Hamburg. Univ.* **11** (1935), pp. 116–162.

CARTAN, H.

(1) Les fonctions de deux variables complexes et le problème de la représentation analytique. *J. Math. pures appl.* (*9*), **10** (1931), pp. 1–114.

(2) Détermination des points exceptionnels d'un système de *p* fonctions analytiques de *n* variables complexes. *Bull. Sci. Math.* **57** (1933), pp. 334–344.

(3) *Sur les groupes de transformations analytiques*. Hermann et Cie, Paris, 1935.

(4) Sur les matrices holomorphes de *n* variables complexes. *J. Math. pures appl.* **19** (1940), pp. 1–26.

(5) Idéaux de fonctions analytiques de *n* variables complexes. *Ann. Sci. École Norm. Sup.* (*3*) **61** (1944), pp. 149–197.

(6) Idéaux et modules de fonctions analytiques de variables complexes. *Bull. Soc. Math. France* **78** (1950), pp. 28–64.

(7) Problèmes globaux dans la théorie des fonctions analytiques de plusieurs variables complexes. *Proc. Int. Congr. Math.*, 1950, I, pp. 152–164, Cambridge, Mass.

(8) *Quotient d'un espace analytique par un groupe d'automorphismes*. Symposium in Honor of S. Lefschetz, Princeton University Press, Princeton, N.J., 1957, pp. 90–102.

(9) *Fonctions automorphes.* Séminaire, Paris, 1957/58.

(10) Prolongement des espaces analytiques normaux. *Math. Ann.* **136** (1958), pp. 97–110.

(11) Fonctions automorphes et séries de Poincaré. *J. Analyse Math.* **6** (1958), pp. 169–175.

(12) Sur les fonctions de plusieurs variables complexes: les espaces analytiques. *Proc. Int. Congr. Math.*, Edinburgh, 1958, pp. 33–52.

(13) Quotients of complex analytic spaces. *Contributions to Function Theory*, Tata Institute, Bombay, 1960, pp. 1–15.

(14) *Faisceaux analytiques cohérents.* Centro Int. Mat. Estivo, Roma, 1963.

(15) *Sur le théorème de préparation de Weierstraß. Festschrift zur Gedächtnisfeier für K. Weierstraß*, pp. 155–168, Arbeitsgemeinschaft für Forschung des Landes Nordrh.-Westf., Westdeutscher Verlag, Köln, 1966.

CARTIER, P.

(1) Isogenies and duality of abelian varieties. *Ann. Math.* **71** (1960), pp. 315–351.

CASSELS, J. W. S.

(1) Diophantine equations with special reference to elliptic curves. *J. London Math. Soc.* **41** (1966), pp. 193–291.

CASTELNUOVO, G.

(1) Sulle funzioni abeliane: I. Le funzioni intermediarie; II. La geometria sulle varietà abeliane; III. Le varietà di Jacobi; IV. Applicazioni alle serie algebriche di gruppi sopra una curva. Roma, *Accad. Lincei Rend.* (5) $\mathbf{30}_1$ (1921), pp. 50–55, 99–103, 195–200, 355–360.

CHEVALLEY, C., and WEIL, A.

(1) Über das Verhalten der Integrale 1. Gattung bei Automorphismen des Funktionenkörpers. *Abh. Math. Sem. Hamburg. Univ.* **10** (1934), pp. 358–361.

CHOW, W. L.

(1) On compact complex analytic varieties. *Amer. J. Math.* **71** (1949), pp. 893–914.

(2) On Picard varieties. *Amer. J. Math.* **74** (1952), pp. 895–909.

(3) The Jacobian variety of an algebraic curve. *Amer. J. Math.* **76** (1954), pp. 453–476.

(4) *On the projective embedding of homogeneous varieties.* Symposium in Honor of S. Lefschetz, Princeton University Press, Princeton, N.J., 1957, pp. 122–128.

(5) On the connectedness theorem in algebraic geometry. *Amer. J. Math.* **81** (1959), pp. 1033–1074.

CHRISTIAN, U.

(1) Zur Theorie der Modulfunktionen n-ten Grades. *Math. Ann.* **133** (1957), pp. 281–297.

(2) Zur Theorie der Modulfunktionen n-ten Grades II. *Math. Ann.* **134** (1958), pp. 298–307.

(3) Über die Multiplikatorensysteme zur Gruppe der ganzen Modulsubstitutionen n-ten Grades. *Math. Ann.* **138** (1959), pp. 363–397.

(4) On the factors of automorphy for the group of integral modular substitutions of second degree. *Ann. Math.* **73** (1961), pp. 134–153.

(5) On certain factors of automorphy for the modular group of degree n. *Monatsh. Math.* **65** (1961), pp. 82–87.

(6) Über die Multiplikatorensysteme gewisser Kongruenzgruppen ganzer Hilbert-Siegelscher Modulsubstitutionen. *Math. Ann.* **144** (1961), pp. 422–459.

(7) Über Hilbert-Siegelsche Modulformen und Poincarésche Reihen. *Math. Ann.* **148** (1962), pp. 257–307.

(8) Zur Theorie der Hilbert-Siegelschen Modulfunktionen. *Math. Ann.* **152** (1963), pp. 275–341.

(9) Bestimmung des Körpergrades der Siegelschen Modulfunktionen über den Eisensteinreihen. *Abh. Math. Sem. Univ. Hamburg* **27** (1964), pp. 171–172.

(10) Über die Modulgruppe zweiten Grades I, II. *Math. Z.* **85** (1964), pp. 1–28, 29–39.

(11) Über die Uniformisierbarkeit der Fixpunkte der Modulgruppe zweiten Grades. *Nachr. Akad. Wiss. Göttingen, math.-phys. Klasse*, 1964, pp. 211–231.

(12) Über die Uniformisierbarkeit nicht-elliptischer Fixpunkte Siegelscher Modulgruppen. *J. reine angew. Math.* **219** (1965), pp. 97–112.

(13) Über die Uniformisierbarkeit elliptischer Fixpunkte Hilbert-Siegelscher Modulgruppen. *J. reine angew. Math.* **223** (1966), pp. 113–130.

(14) Einführung in die Theorie der paramodularen Gruppen. *Math. Ann.* **168** (1967), pp. 59–104

(15) Über die erste Zeile paramodularer Matrizen. *Nachr. Akad. Wiss. Göttingen, math.-phys. Klasse*, 1967, pp. 239–245.

(16) Some remarks on symplectic groups, modular groups and Poincaré series. *Amer. J. Math.* **89** (1967), pp. 319–362.

(17) A reduction theory for symplectic matrices. *Math. Z.* **101** (1967), pp. 213–244.

(18) Siegelsche Modulformen und Integralgleichungen. *Math. Z.* **101** (1967), pp. 299–305.

(19) Hilbert-Siegelsche Modulformen und Integralgleichungen. *Monatsh. Math.* **72** (1968), pp. 412–418.

(20) Über teilerfremde symmetrische Matrizenpaare. *J. reine angew. Math.* **229** (1968), pp. 43–49.

(21) Über die Anzahl der Spitzen Siegelscher Modulgruppen. *Abh. Math. Sem. Univ. Hamburg* **32** (1968), pp. 55–60.

(22) Über elliptische Fixpunkte symplektischer Matrizen. *Monatsh. Math.* **72** (1968), pp. 289–295.

(23) Untersuchung einer Poincaréschen Reihe. I, II. *J. reine angew. Math.* **233** (1968), pp. 37–88; **237** (1969), pp. 12–25.

(24) Über gewisse Gleichungen zwischen symplektischen Matrizen. *J. reine angew. Math.* **243** (1970), pp. 55–65.

CLAUS, G.

(1) Die Randmannigfaltigkeiten und die "tiefsten" Punkte des Fundamentalbereichs für drei Hilbertsche Modulgruppen. *Math. Ann.* **176** (1968), pp. 225–256.

CONFORTO, F.

(1) *Abelsche Funktionen und algebraische Geometrie.* Springer-Verlag, Berlin–Göttingen–Heidelberg, 1956.
(2) Funzioni abeliane e matrici di Riemann, Parte I. *Corsi Ist. Naz. Alta Mat.*, Università di Roma, 1942.

COUSIN, P.

(1) Sur les fonctions de *n* variables complexes. *Acta Math.* **19** (1895), pp. 1–62.
(2) Sur les fonctions périodiques. *Ann. Sci. École Norm. Sup.* **19** (1902), pp. 9–61.
(3) Sur les fonctions triplement périodiques de deux variables. *Acta Math.* **33** (1910), pp. 105–232.

DEURING, M.

(1) Invarianten und Normalformen elliptischer Funktionenkörper. *Math. Z.* **47** (1941), pp. 47–56.
(2) Die Typen der Multiplikatorenringe elliptischer Funktionenkörper. *Abh. Math. Sem. Univ. Hamburg* **14** (1941), pp. 197–272.
(3) Algebraische Begründung der komplexen Multiplikation. *Abh. Math. Sem. Univ. Hamburg* **16** (1949), pp. 32–47.
(4) Die Struktur der elliptischen Funktionenkörper und Klassenkörper der imaginären quadratischen Zahlkörper. *Math. Ann.* **124** (1952), pp. 393–426.
(5) Die Klassenkörper der komplexen Multiplikation. *Enzykl. Math. Wiss.*, Vol. I 2, No. 10, Part II, Stuttgart, 1958.

EICHLER, M.

(1) Über die Darstellbarkeit von Modulformen durch Thetareihen. *J. reine angew. Math.* **195** (1956), pp. 156–171.
(2) Modular correspondences and their representations. *J. Indian Math. Soc.* **20** (1956), pp. 163–206.
(3) Eine Verallgemeinerung der Abelschen Integrale. *Math. Z.* **67** (1957), pp. 267–298.
(4) *On modular correspondences.* Tata Institute, Bombay, 1957.
(5) Quadratische Formen und Modulfunktionen. *Acta Arith.* **4** (1958), pp. 217–239.
(6) *Einführung in die Theorie der algebraischen Zahlen und Funktionen.* Birkhäuser Verlag, Basel, 1963.
(7) Eine Spurformel für Korrespondenzen von algebraischen Funktionenkörpern mit sich selber. *Inventiones math.* **2** (1967), pp. 274–300.
(8) Einige Anwendungen der Spurformel im Bereich der Modularkorrespondenzen. *Math. Ann.* **168** (1967), pp. 128–137.
(9) *Projective varieties and modular forms.* Lecture Notes in Mathematics, Vol. 210, Springer-Verlag, Berlin–Heidelberg–New York, 1970.

ENRIQUES, F., and SEVERI, F.

(1) Mémoire sur les surfaces hyperelliptiques. *Acta Math.* **32** (1909), pp. 283–392; **33** (1910), pp. 321–403.

di FAGNANO, CONTE G. C.

(1) *Produzioni matematiche.* 2 volumes, Pesaro, 1750; *Opere matematiche*, 1–2, Milano-Roma-Napoli, 1911.

(2) Metodo per misurare la lemniscata. *Giornale dei letterati d'Italia* **30** (1718), p. 87; *Opere matematiche*, 2, pp. 304–313.

FISCHER, I.

(1) The moduli of hyperelliptic curves. *Trans. Amer. Math. Soc.* **82** (1956), pp. 64–84.

FREITAG, E.

(1) Zur Theorie der Modulformen zweiten Grades. *Nachr. Akad. Wiss. Göttingen, math.-phys. Klasse*, 1965, pp. 151–157.

(2) Modulformen zweiten Grades zum rationalen und Gaußschen Zahlkörper. *Sitz.-Ber. Heidelberger Akad. Wiss., math.-naturw. Klasse*, 1967, pp. 3–49.

(3) *Der Zentralisator eines Torus.* Séminaire Heidelberg-Strasbourg, 1965/66, Lecture 13, §4, 1967.

(4) Fortsetzung von automorphen Funktionen. *Math. Ann.* **177** (1968), pp. 95–100.

(5) Über die Struktur der Funktionenkörper zu hyperabelschen Gruppen I. *J. reine angew. Math.* **247** (1971), pp. 97–117.

FREITAG, E., and SCHNEIDER, V.

(1) Bemerkung zu einem Satz von J. Igusa und W. Hammond. *Math. Z.* **102** (1967), pp. 9–16.

FRICKE, R.

(1) *Die elliptischen Funktionen und ihre Anwendungen.* Vol. 1, 2nd ed., Teubner, Leipzig, 1930; Vol. 2, Teubner, Leipzig, 1922.

FRICKE, R., and KLEIN, F.

(1) *Vorlesungen über die Theorie der automorphen Funktionen.* Vol. 1, Teubner, Leipzig, 1897; Vol. 2, Teubner, Leipzig, 1901/12; reprint 1965, Johnson Reprint Corporation, New York.

FROBENIUS, G.

(1) Theorie der bilinearen Formen mit ganzen Koeffizienten. *J. reine angew. Math.* **86** (1878), pp. 147–208.

(2) Über das Additionstheorem der Thetafunctionen mehrerer Variabeln. *J. reine angew. Math.* **89** (1880), pp. 185–220.

(3) Über Gruppen von Thetacharakteristiken. *J. reine angew. Math.* **96** (1884), pp. 81–99.

(4) Über die Grundlagen der Theorie der Jacobischen Funktionen. *J. reine angew. Math.* **97** (1884), pp. 16–48 and 188–223.

(5) Über die Charaktere der symmetrischen Gruppe. *Sitz.-Ber. Berlin. Akad. Wiss.*, 1900, pp. 516–534.

FUETER, R.

(1) *Vorlesungen über die singulären Moduln und die komplexe Multiplikation der elliptischen Funktionen*, I (1924), II (1927), Teubner.

GARLAND, H., and RAGHUNATHAN, M. S.

(1) Fundamental domains for lattices in rank one semi-simple Lie groups. *Ann. Math.* **92** (1970), pp. 279–326.

GEL'FAND, I. M.

(1) Spherical functions on symmetric Riemann spaces. *Doklady Akad. Nauk SSSR* **70** (1950), pp. 5–8.

(2) Automorphic functions and theory of representations. *Proc. Int. Congr. Math.*, Stockholm, 1962, pp. 74–85.

GEL'FAND, I. M., and GRAEV, M. I.

(1) Geometry of homogeneous spaces, representations of groups in homogeneous spaces and related problems of integral geometry. *Trudy Moskov. Mat. Obshch.* **8** (1959), pp. 321–390.

(2) Irreducible unitary representations of the group of matrices of the second order with elements from a locally compact field. *Doklady Akad. Nauk SSSR* **149** (1963), pp. 499–501; translation: *Soviet Math. Doklady* **4** (1963), pp. 397–400.

(3) Representations of a group of matrices of the second order with elements from a locally compact field, and special functions on locally compact fields. *Uspekhi Mat. Nauk* **18** (1963), No. 4, pp. 29–99; translation: *Russian Math. Surveys* **18** (1963), No. 4, pp. 29–109.

GEL'FAND, I. M., GRAEV, M. I., and PYATETSKII-SHAPIRO, I. I.

(1) *Representation theory and automorphic functions.* W. B. Saunders Company, Philadelphia–London–Toronto, 1969.

GEL'FAND, I. M., and PYATETSKII-SHAPIRO, I. I.

(1) Representation theory and theory of automorphic functions. *Uspekhi Mat. Nauk* **14** (1959), No. 2, pp. 171–194; translation: *Amer. Math. Soc.* (2) **26** (1963), pp. 173–200.

(2) Unitary representations in homogeneous spaces with discrete stability groups. *Doklady Akad. Nauk SSSR* **147** (1962), pp. 17–20; translation: *Soviet Math. Doklady* **3** (1962), pp. 1528–1531.

(3) Unitary representations in the space G/Γ, where G is the group of real matrices of order n and Γ the subgroup of integral matrices. *Doklady Akad. Nauk SSSR* **147** (1962), pp. 275–278; translation: *Soviet Math. Doklady* **3** (1962), pp. 1574–1577.

(4) Automorphic functions and representation theory. *Trudy Moskov. Mat. Obshch.* **12** (1963), pp. 389–412; translation: *Trans. Moscow Math. Soc.* **12** (1963), pp. 438–464.

GINDIKIN, S. G.

(1) Integral formulas for Siegel domains of second kind. *Doklady Akad. Nauk SSSR* **141** (1961), pp. 531–534; translation: *Soviet Math. Doklady* **2** (1961), pp. 1480–1483.

(2) Integral formulas for complex homogeneous bounded domains. *Uspekhi Mat. Nauk* **17** (1962), No. 3, pp. 209–211 (Russian).

(3) Analysis in homogeneous domains. *Uspekhi Mat. Nauk* **19** (1964), No. 4, pp. 3–92; translation: *Russian Math. Surveys* **19** (1964), No. 4, pp. 1–89.

GINDIKIN, S. G., PYATETSKII-SHAPIRO, I. I., and VINBERG, E. B.

(1) Classification and canonical realization of complex homogeneous bounded domains. *Trudy Moscov. Mat. Obshch.* **12** (1963), pp. 359–388; translation: *Trans. Moscow Math. Soc.* **12** (1963), pp. 404–437.

GINDIKIN, S. G., and VINBERG, E. B.

(1) On some nonassociative algebras occurring in the theory of homogeneous domains. *Uspekhi Mat. Nauk* **17** (1962), No. 6, pp. 229–230 (Russian).

GODEMENT, R.

(1) *Introduction aux travaux de A. Selberg*. Séminaire Bourbaki, 1957.

(2) *Analyse spectrale des fonctions modulaires*. Séminaire Bourbaki 1964/65, Exposé 278.

(3) *Introduction à la theorie de Langlands*. Séminaire Bourbaki, 1966/67.

(4) The decomposition of $L^2(G/\Gamma)$ for $\Gamma = SL(2, \mathbb{Z})$. *Proc. Sympos. Pure Math.*, Vol. 9, American Mathematical Society, Providence, R.I., 1966, pp. 211–224.

(5) The spectral decomposition of cusp-forms. *Ibid.*, pp. 225–234.

GÖTZKY, F.

(1) Über eine zahlentheoretische Anwendung von Modulfunktionen zweier Veränderlicher. *Math. Ann.* **100** (1928), pp. 411–437.

GOTTSCHLING, E.

(1) Explizite Bestimmung der Randflächen des Fundamentalbereiches der Modulgruppe zweiten Grades. *Math. Ann.* **138** (1959), pp. 103–124.

(2) Über die Fixpunkte der Siegelschen Modulgruppe. *Math. Ann.* **143** (1961), pp. 111–149.

(3) Über die Fixpunktuntergruppen der Siegelschen Modulgruppe. *Math. Ann.* **143** (1961), pp. 399–430.

(4) Über Poincarésche Reihen und einen Fundamentalbereich diskontinuierlicher Gruppen. *Math. Ann.* **148** (1962), pp. 125–146.

(5) Die Uniformisierbarkeit der Fixpunkte eigentlich diskontinuierlicher Gruppen von biholomorphen Abbildungen. *Math. Ann.* **169** (1967), pp. 26–54.

(6) Invarianten endlicher Gruppen und biholomorphe Abbildungen. *Inventiones math.* **6** (1969), pp. 315–326; correction, **8** (1969), p. 356.

(7) Reflections in bounded symmetric domains. *Comm. Pure Appl. Math.* **22** (1969), pp. 693–714.

GRAUERT, H., and REMMERT, R.

(1) *Analytische Stellenalgebren*. Springer-Verlag, Berlin-Heidelberg-New York, 1971.

GUNDLACH, K.-B.

(1) Über die Darstellung der ganzen Spitzenformen zu den Idealstufen der Hilbertschen Modulgruppe und die Abschätzung ihrer Fourierkoeffizienten. *Acta Math.* **92** (1954), pp. 309–345.

(2) Poincarésche und Eisensteinsche Reihen zur Hilbertschen Modulgruppe. *Math. Z.* **64** (1956), pp. 339–352.

(3) Modulfunktionen zur Hilbertschen Modulgruppe und ihre Darstellung als Quotienten ganzer Modulformen. *Arch. Math.* **7** (1956), pp. 333–338.

(4) Ganze Nichtspitzenformen der Dimension −1 zu den Hilbertschen Modulgruppen reell-quadratischer Zahlkörper. *Arch. Math.* **7** (1956), pp. 453–456.

(5) Über den Rang der Schar der ganzen automorphen Formen zu hyperabelschen Transformationsgruppen in zwei Variablen. *Nachr. Akad. Wiss. Göttingen, math.-phys. Klasse*, 1958, pp. 59–66.

(6) Dirichletsche Reihen zur Hilbertschen Modulgruppe. *Math. Ann.* **135** (1958), pp. 294–314.

(7) Quotientenraum und meromorphe Funktionen zur Hilbertschen Modulgruppe. *Nachr. Akad. Wiss. Göttingen, math.-phys. Klasse*, 1960, pp. 77–85.

(8) *Some new results in the theory of Hilbert's modular group*. Contributions to Function Theory, Tata Institute, Bombay, 1960, pp. 165–180.

(9) Die Bestimmung der Funktionen zur Hilbertschen Modulgruppe des Zahlkörpers $\mathbb{Q}(\sqrt{5})$. *Math. Ann.* **152** (1963), pp. 226–256.

(10) Die Fixpunkte einiger Hilbertscher Modulgruppen. *Math. Ann.* **157** (1965), pp. 369–390.

(11) Zusammenhänge zwischen Modulformen in einer und in zwei Variablen. *Nachr. Akad. Wiss. Göttingen, math.-phys. Klasse*, 1965. pp. 47–88.

(12) Die Bestimmung der Funktionen zu einigen Hilbertschen Modulgruppen. *J. reine angew. Math.* **220** (1965), pp. 109–153.

GUNNING, R. C.

(1) General factors of automorphy. *Proc. Nat. Acad. Sci. U.S.A.* **41** (1955), pp. 496–498.

(2) The structure of factors of automorphy. *Amer. J. Math.* **78** (1956), pp. 357–382.

(3) Indices of rank and of singularity on Abelian varieties. *Proc. Nat. Acad. Sci. U.S.A.* **43** (1957), pp. 167–169.

(4) Multipliers on complex homogeneous spaces. *Proc. Amer. Math. Soc.* **8** (1957), pp. 394–396.

(5) Factors of automorphy and other formal cohomology groups for Lie groups. *Ann. Math.* **69** (1959), pp. 314–326.

(6) Homogeneous symplectic multipliers. *Illinois J. Math.* **4** (1960), pp. 575–583.

(7) On Cartan's theorems A and B in several complex variables. *Ann. Mat. pura appl.* **55** (1961), pp. 1–11.

(8) The Eichler cohomology groups and automorphic forms. *Trans. Amer. Math. Soc.* **100** (1961), pp. 44–62.

(9) Generalized symplectic differential forms and differential operators. *J. Math. Mech.* **11** (1962), pp. 703–723.

(10) *Lectures on modular forms*. Princeton University Press, Princeton, N.J., 1962.

(11) Differential operators preserving relations of automorphy. *Trans. Amer. Math. Soc.* **108** (1963), pp. 326–352.

(12) *Lectures on Riemann surfaces.* Princeton Mathematical Notes. Princeton University Press, Princeton, N.J., 1966.

(13) *Lectures on vector bundles over Riemann surfaces.* University of Tokyo Press, Tokyo; Princeton University Press, Princeton, N.J., 1967.

(14) Special coordinate coverings of Riemann surfaces. *Math. Ann.* **170** (1967), pp. 67–86.

GUNNING, R. C., and ROSSI, H.

(1) *Analytic functions of several complex variables.* Prentice-Hall, Inc., Englewood Cliffs, N.J., 1965.

HAMMOND, W. F.

(1) On the graded ring of Siegel modular forms of genus two. *Amer. J. Math.* **87** (1965), pp. 502–506.

(2) The modular groups of Hilbert and Siegel. *Amer. J. Math.* **88** (1966), pp. 497–516.

(3) The modular groups of Hilbert and Siegel. *Proc. Sympos. Pure Math.*, Vol. 9, American Mathematical Society, Providence, R.I., 1966, pp. 358–360.

HANO, J.

(1) On Kählerian homogeneous spaces of unimodular Lie groups. *Amer. J. Math.* **79** (1957), pp. 885–900.

HANO, J., and MATSUSHIMA, Y.

(1) Some studies on Kählerian homogeneous spaces. *Nagoya Math. J.* **11** (1957), pp. 77–92.

HARISH-CHANDRA

(1) Representations of semisimple Lie groups, I–VI. *Trans. Amer. Math. Soc.* **75** (1953), pp. 185–243; **76** (1954), pp. 26–65, 234–253; *Amer. J. Math.* **77** (1955), pp. 743–777; **78** (1956), pp. 1–41, 564–628.

(2) The characters of semisimple Lie groups. *Trans. Amer. Math. Soc.* **83** (1956), pp. 98–163.

(3) Differential operators on a semisimple Lie algebra. *Amer. J. Math.* **79** (1957), pp. 87–120.

(4) Fourier transforms on a semisimple Lie algebra. I, II. *Amer. J. Math.* **79** (1957), pp. 193–257, 653–686.

(5) A formula for semisimple Lie groups. *Amer. J. Math.* **79** (1957), pp. 733–760.

(6) Spherical functions on a semisimple Lie group. I, II. *Amer. J. Math.* **80** (1958), pp. 241–310, 553–613.

(7) Automorphic forms on a semi-simple Lie group. *Proc. Nat. Acad. Sci. U.S.A.* **45** (1959), pp. 570–573.

(8) Invariant eigendistributions on semisimple Lie groups. *Bull. Amer. Math. Soc.* **69** (1963), pp. 117–123.

(9) Invariant distributions on Lie algebras. *Amer. J. Math.* **86** (1964), pp. 271–309.

(10) Some results on an invariant integral on a semisimple Lie algebra. *Ann. Math.* **80** (1964), pp. 551–593.

(11) Invariant eigendistributions on a semisimple Lie group. *Trans. Amer. Math. Soc.* **119** (1965), pp. 457–508.

(12) Discrete series for semisimple Lie groups. I, II. *Acta Math.* **113** (1965), pp. 241–318; **116** (1966), pp. 1–111.

(13) Two theorems on semisimple Lie groups. *Ann. Math.* **83** (1966), pp. 74–128.

(14) *Automorphic forms on semisimple Lie groups.* Lecture Notes in Mathematics, Vol. 62, Springer-Verlag, Berlin–Heidelberg–New York, 1968.

HARTOGS, F.

(1) Zur Theorie der analytischen Funktionen mehrerer unabhängiger Veränderlichen insbesondere über die Darstellung derselben durch Reihen, welche nach Potenzen einer Veränderlichen fortschreiten. *Math. Ann.* **62** (1906), pp. 1–88.

(2) Einige Folgerungen aus der Cauchyschen Integralformel bei Funktionen mehrerer Veränderlichen. *Sitz.-Ber. Bayer. Akad. Wiss., math.-phys. Klasse* **36** (1906), pp. 223–241.

(3) Über die aus den singulären Stellen einer analytischen Funktion mehrerer Veränderlichen bestehenden Gebilde. *Acta Math.* **32** (1909), pp. 57–79.

(4) Über die elementare Herleitung des Weierstraßschen "Vorbereitungssatzes." *Sitz.-Ber. Bayer. Akad. Wiss., math.-natur. Klasse*, 1909, No. 3, 12 pp.

HASSE, H.

(1) Neue Begründung der komplexen Multiplikation, I, II. *J. reine angew. Math.* **157** (1927), pp. 115–139; **165** (1931), pp. 64–88.

(2) Theory of cyclic algebras over an algebraic number field. *Trans. Amer. Math. Soc.* **34** (1932), pp. 170–214.

(3) Theorie der relativ-zyklischen algebraischen Funktionenkörper, insbesondere bei endlichem Konstantenkörper. *J. reine angew. Math.* **172** (1934), pp. 37–54.

(4) Abstrakte Begründung der komplexen Multiplikation und Riemannsche Vermutung in Funktionenkörpern. *Abh. Math. Sem. Hamburg. Univ.* **10** (1934), pp. 325–348.

HECKE, E.

(1) Höhere Modulfunktionen und ihre Anwendung auf die Zahlentheorie. *Math. Ann.* **71** (1912), pp. 1–37; *Mathematische Werke*, Vandenhoeck & Ruprecht, Göttingen, 1959, pp. 21–58.

(2) Über die Konstruktion relativ-Abelscher Zahlkörper durch Modulfunktionen von zwei Variablen. *Math. Ann.* **74** (1913), pp. 465–510; *Mathematische Werke*, pp. 69–114.

(3) Über die Perioden vierfach periodischer Funktionen. *Nachr. K. Gesellsch. Wiss. Göttingen, math.-phys. Klasse*, 1915, pp. 81–112; *Mathematische Werke*, pp. 127–158.

(4) Analytische Funktionen und algebraische Zahlen, I, II. *Abh. Math. Sem. Hamburg. Univ.* **1** (1921), pp. 102–126; **3** (1924), pp. 213–236; *Mathematische Werke*, pp. 336–360, 381–404.

(5) Darstellung von Klassenzahlen als Perioden von Integralen 3. Gattung aus dem Gebiet der elliptischen Modulfunktionen. *Abh. Math. Sem. Hamburg. Univ.* **4** (1925), pp. 211–223; *Mathematische Werke*, pp. 405–417.

(6) Über einen neuen Zusammenhang zwischen elliptischen Modulfunktionen und indefiniten quadratischen Formen. *Nachr. K. Gesellsch. Wiss. Göttingen, math.-phys. Klasse*, 1925, pp. 35–44; *Mathematische Werke*, pp. 418–427.

(7) Zur Theorie der elliptischen Modulfunktionen. *Math. Ann.* **97** (1926), pp. 210–242; *Mathematische Werke*, pp. 428–460.

(8) Theorie der Eisensteinschen Reihen höherer Stufe und ihre Anwendung auf Funktionentheorie und Arithmetik. *Abh. Math. Sem. Hamburg. Univ.* **5** (1927), pp. 199–224; *Mathematische Werke*, pp. 461–486.

(9) Über ein Fundamentalproblem aus der Theorie der elliptischen Modulfunktionen. *Abh. Math. Sem. Hamburg. Univ.* **6** (1928), pp. 235–257; *Mathematische Werke*, pp. 525–547.

(10) Über das Verhalten der Integrale 1. Gattung bei Abbildungen, insbesondere in der Theorie der elliptischen Modulfunktionen. *Abh. Math. Sem. Hamburg. Univ.* **8** (1930), pp. 271–281; *Mathematische Werke*, pp. 548–558.

(11) Die Riemannschen Periodenrelationen für die elliptischen Modulfunktionen. *J. reine angew. Math.* **167** (1932), pp. 337–345; *Mathematische Werke*, pp. 559–567.

(12) Die eindeutige Bestimmung der Modulfunktionen q-ter Stufe durch algebraische Eigenschaften. *Math. Ann.* **111** (1935), pp. 293–301; *Mathematische Werke*, pp. 568–576.

(13) Die Primzahlen in der Theorie der elliptischen Modulfunktionen. *Kgl. Danske Videnskabernes Selskab. Matematiskfysiske Meddelelser XIII* **10** (1935), 16 pp.; *Mathematische Werke*, pp. 577–590.

(14) Über Modulfunktionen und die Dirichletschen Reihen mit Eulerscher Produktentwicklung, I, II. *Math. Ann.* **114** (1937), pp. 1–28, 316–351; *Mathematische Werke*, pp. 644–671, 672–707.

(15) Grundlagen einer Theorie der Integralgruppen und der Integralperioden bei den Normalteilern der Modulgruppe. *Math. Ann.* **116** (1939), pp. 469–510; *Mathematische Werke*, pp. 731–772.

HELGASON, S.

(1) *Differential geometry and symmetric spaces*. Academic Press, New York and London, 1962.

(2) Fundamental solutions of invariant differential operators on symmetric spaces. *Amer. J. Math.* **86** (1964), pp. 565–601.

(3) Totally geodesic spheres in compact symmetric spaces. *Math. Ann.* **165** (1966), pp. 309–317.

HELWIG, K.-H.

(1) Automorphismengruppen des allgemeinen Kreiskegels und des zugehörigen Halbraums. *Math. Ann.* **157** (1964), pp. 1–33.

(2) Zur Koecherschen Reduktionstheorie in Positivitätsbereichen, I, II, III. *Math. Z.* **91** (1966), pp. 152–168, 169–178, 355–362.

(3) Eine Verallgemeinerung der formal-reellen Jordan-Algebren. *Inventiones math.* **1** (1966), pp. 18–35.

(4) Halbeinfache reelle Jordan-Algebren. *Math. Z.* **109** (1969), pp. 1–28.

HERMANN, R.

(1) C-W cell decompositions of symmetric homogeneous spaces. *Bull. Amer. Math. Soc.* **66** (1960), pp. 126–128.

(2) A Poisson kernel for certain homogeneous spaces. *Proc. Amer. Math. Soc.* **12** (1961), pp. 892–899.

(3) Geodesics of bounded symmetric domains. *Comment. Math. Helv.* **35** (1961), pp. 1–8.

(4) Geometric aspects of potential theory in bounded symmetric domains, I, II, III. *Math. Ann.* **148** (1962), pp. 349–366; **151** (1963), pp. 143–149; **153** (1964), 384–394.

(5) Complex domains and homogeneous spaces. *J. Math. Mech.* **13** (1964), pp. 243–274.

(6) Compactification of homogeneous spaces, I, II. *J. Math. Mech.* **14** (1965), pp. 655–678; **15** (1966), pp. 667–681.

HERMITE, C.

(1) Sur la théorie des formes quadratiques. *Oeuvres* **1** (1853), pp. 200–233.

(2) Sur la théorie de la transformation des fonctions Abéliennes. *C. R. Acad. Sci. Paris* **40** (1855).

(3) Sur la résolution de l'équation de cinquième degré. *C. R. Acad. Sci. Paris* **46** (1858), pp. 508–515.

(4) Éléments de la théorie des fonctions elliptiques. *Oeuvres* **4**, Gauthier-Villars, Paris, 1898.

HERRMANN, O.

(1) Über Hilbertsche Modulfunktionen und die Dirichletschen Reihen mit Eulerscher Produktentwicklung. *Math. Ann.* **127** (1954), pp. 357–400.

HERTNECK, CH.

(1) Positivitätsbereiche und Jordan-Strukturen. *Math. Ann.* **146** (1962), pp. 433–455.

HERVÉ, M.

(1) Sur les fonctions fuchsiennes de deux variables complexes. *Ann. Sci. École Norm. Sup.* (*3*) **69** (1952), pp. 277–302.

HIRONAKA, H.

(1) Resolution of singularities of an algebraic variety over a field of characteristic zero, I, II. *Ann. Math.* **79** (1964), pp. 109–203, 205–326.

HIRZEBRUCH, F.

(1) Automorphe Formen und der Satz von Riemann-Roch. *Symp. Int. Top. Alg.*, 1956, pp. 129–144. Universidad de Mexico, 1958.

(2) *Characteristic numbers of homogeneous domains*. Seminars on Analytic Functions, Vol. 2, Princeton, 1957, pp. 92–104.

(3) *Elliptische Differentialoperatoren auf Mannigfaltigkeiten. Festschrift zur Gedächtnisfeier für K. Weierstraß*, pp. 583–608, Arbeitsgemeinschaft für Forschung des Landes Nordrh.-Westf., Westdeutscher Verlag, Köln, 1966.

(4) *The Hilbert modular group, resolution of the singularities at the cusps and related problems*. Séminaire Bourbaki, Vol. 1970/71, Exposé 396; Lecture Notes in Mathematics, Vol. 244, Springer-Verlag, Berlin-Heidelberg-New York, 1971, pp. 275–288.

HIRZEBRUCH, U.

(1) Halbräume und ihre holomorphen Automorphismen. *Math. Ann.* **153** (1964), pp. 395–417.

(2) Über Jordan-Algebren und kompakte Riemannsche symmetrische Räume vom Rang 1. *Math. Z.* **90** (1965), pp. 339–354.

(3) Über Jordan-Algebren und beschränkte symmetrische Gebiete. *Math. Z.* **94** (1966), pp. 387–390.

HOYT, W. L.

(1) Some decomposition theorems on abelian varieties, Thesis. The University of Chicago.

(2) On products and algebraic families of Jacobian varieties. *Ann. Math.* **77** (1963), pp. 415–423.

(3) Embeddings of Picard varieties. *Proc. Amer. Math. Soc.* **15** (1964), pp. 26–31.

(4) On the Chow bunches for different projective embeddings of a complete variety. *Amer. J. Math.* **88** (1966), pp. 273–278.

HSU, I., and LOOK, K. H.

(1) A note on transitive domains. *Acta Math. Sinica* **11** (1961), pp. 11–23 (Chinese); translated as *Chinese Math.* **2** (1962), pp. 11–26.

HUA, L. K.

(1) On the theory of automorphic functions of a matrix variable I. Geometrical basis. *Amer. J. Math.* **66** (1944), pp. 470–488.

(2) On the theory of automorphic functions of a matrix variable II. The classification of hypercircles under the symplectic group. *Amer. J. Math.* **66** (1944), pp. 531–563.

(3) On the theory of Fuchsian functions of several variables. *Ann. Math.* **47** (1946), pp. 167–191.

(4) *Harmonic analysis of functions of several complex variables in the classical domains*. American Mathematical Society, Providence, R.I., 1963.

HUA, L. K., and LOOK, K. H.

(1) Theory of harmonic functions in classical domains I, II, III (Chinese). *Acta Math. Sinica* **8** (1958), pp. 531–547; **9** (1959), pp. 295–305, 306–314.

HUA, L. K., and REINER, I.

(1) On the generators of the symplectic modular group. *Trans. Amer. Math. Soc.* **65** (1949), pp. 415–426.

HUMBERT, G.

(1) Théorie générale des surfaces hyperelliptiques. *J. Math.* (*IV*) **9** (1893), pp. 29–170, 361–475.
(2) Sur les fonctions abéliennes singulières. *Oeuvres*, Vol. 2, 1936, pp. 297–498; *J. Math.* (*5*) **5** (1899), pp. 233–350.

HUMBERT, P.

(1) Théorie de la réduction des formes quadratiques définies positives dans un corps algébrique K fini. *Comment Math. Helv.* **12** (1939/40), pp. 263–306.

HURWITZ, A.

(1) Beweis des Satzes, daß eine einwertige Funktion beliebig vieler Variabeln, welche überall als Quotient zweier Potenzreihen dargestellt werden kann, eine rationale Funktion ihrer Argumente ist. *J. reine angew. Math.* **95** (1883), pp. 201–206.
(2) Über algebraische Korrespondenzen und das verallgemeinerte Korrespondenzprinzip. *Math. Ann.* **28** (1887), pp. 561–585.
(3) Über diejenigen algebraischen Gebilde, welche eindeutige Transformationen in sich zulassen. *Math. Ann.* **32** (1888), pp. 290–308.
(4) Über algebraische Gebilde mit eindeutigen Transformationen in sich. *Math. Ann.* **41** (1893), pp. 403–442.
(5) Die unimodularen Substitutionen in einem algebraischen Zahlkörper. *Nachr. Akad. Wiss. Göttingen, math.-phys. Klasse*, 1895, pp. 332–356.
(6) Sur l'intégrale finie d'une fonction entière. *Acta Math.* **20** (1897), pp. 285–312.

IGUSA, J.

(1) Some remarks on the theory of Picard varieties. *J. Math. Soc. Japan* **3** (1951), pp. 345–348.
(2) On the Picard varieties attached to algebraic varieties. *Amer. J. Math.* **74** (1952), pp. 1–22.
(3) On the structure of a certain class of Kaehler varieties. *Amer. J. Math.* **76** (1954), pp. 669–678.
(4) A fundamental inequality in the theory of Picard varieties. *Proc. Nat. Acad. Sci. U.S.A.* **41** (1955), pp. 317–320.
(5) Fibre systems of Jacobian varieties, I–III. *Amer. J. Math.* **78** (1956), pp. 171–199, 745–760; **81** (1959), pp. 453–476.
(6) On the transformation theory of elliptic functions. *Amer. J. Math.* **81** (1959), pp. 436–452.
(7) Kroneckerian model of fields of elliptic modular functions. *Amer. J. Math.* **81** (1959), pp. 561–577.

(8) Arithmetic variety of moduli for genus two. *Ann. Math.* **72** (1960), pp. 612–649.

(9) Betti and Picard numbers of abstract algebraic surfaces. *Proc. Nat. Acad. Sci. U.S.A.* **46** (1960), pp. 724–726.

(10) Structure theorems of modular varieties. *Proc. Int. Congr. Math.*, Stockholm, 1962, pp. 522–525.

(11) On Siegel modular forms of genus two. *Amer. J. Math.* **84** (1962), pp. 175–200.

(12) On Siegel modular forms of genus two (II). *Amer. J. Math.* **86** (1964), pp. 392–412.

(13) On the graded ring of theta-constants. *Amer. J. Math.* **86** (1964), pp. 219–246.

(14) *On the theory of compactifications.* Summer Institute on Algebraic Geometry, 1964, Lecture Notes.

(15) On the desingularization of Satake compactifications. *Proc. Sympos. Pure Math.*, Vol. 9, American Mathematical Society, Providence, R.I., 1966, pp. 301–305.

(16) On the graded ring of theta-constants. II. *Amer. J. Math.* **88** (1966), pp. 221–236.

(17) Modular forms and projective invariants. *Amer. J. Math.* **89** (1967), pp. 817–855.

(18) A desingularization problem in the theory of Siegel modular functions. *Math. Ann.* **168** (1967), pp. 228–260.

(19) On the algebraic theory of elliptic modular functions. *J. Math. Soc. Japan* **20** (1968), pp. 96–106.

IHARA, S.

(1) Holomorphic imbeddings of symmetric domains into a symmetric domain. *Proc. Japan Acad.* **42** (1966), pp. 193–197.

(2) Holomorphic imbeddings of symmetric domains. *J. Math. Soc. Japan* **19** (1967), pp. 261–302.

ISE, M.

(1) Generalized automorphic forms and certain holomorphic vector bundles. *Amer. J. Math.* **86** (1964), pp. 70–108.

(2) Realization of irreducible bounded symmetric domain of type (V). *Proc. Japan Acad.* **45** (1969), pp. 233–237.

(3) Realization of irreducible bounded symmetric domain of type (VI). *Proc. Japan Acad.* **45** (1969), pp. 846–849.

(4) On canonical realizations of bounded symmetric domains as matrix-spaces. *Nagoya Math. J.* **42** (1971), pp. 115–133.

JACQUET, H., and LANGLANDS, R. P.

(1) *Automorphic forms on GL(2).* Lecture Notes in Mathematics, Vol. 114, Springer-Verlag, Berlin–Heidelberg–New York, 1970.

KANEYUKI, S.

(1) On the automorphism groups of homogeneous bounded domains. *J. Fac. Sci. Univ. Tokyo, Sect. I* **14** (1967), pp. 89–130.

KANEYUKI, S., and NAGANO, T.

(1) On the first Betti numbers of compact quotient spaces of complex semi-simple Lie groups by discrete subgroups. *Sci. Papers College Gen. Ed. Univ. Tokyo* **12** (1962), pp. 1–11.

(2) On certain quadratic forms related to symmetric Riemannian spaces. *Osaka Math. J.* **14** (1962), pp. 241–252.

KANEYUKI, S., and SUDO, M.

(1) On Šilov boundaries of Siegel domains. *J. Fac. Sci. Univ. Tokyo, Sect. I* **15** (1968), pp. 131–146.

KATAYAMA, K.

(1) On the Hilbert-Siegel modular group and abelian varieties. *J. Fac. Sci. Univ. Tokyo, Sect. I* **9** (1962), pp. 261–291.

(2) On the Hilbert-Siegel modular group and abelian varieties. II. *J. Fac. Sci. Univ. Tokyo, Sect. I* **9** (1963), pp. 433–467.

KAUFHOLD, G.

(1) Dirichletsche Reihe mit Funktionalgleichung in der Theorie der Modulfunktion 2. Grades. *Math. Ann.* **137** (1959), pp. 454–476.

KLEIN, F., and FRICKE, R.

(1) *Vorlesungen über die Theorie der elliptischen Modulfunktionen.* Vol. 1, Teubner, Leipzig, 1890; Vol. 2, Teubner, Leipzig, 1892; reprint 1966, Johnson Reprint Corporation, New York.

KLINGEN, H.

(1) Diskontinuierliche Gruppen in symmetrischen Räumen, I, II. *Math. Ann.* **129** (1955), pp. 345–369; **130** (1955), pp. 137–146.

(2) Über die analytischen Abbildungen verallgemeinerter Einheitskreise auf sich. *Math. Ann.* **132** (1956), pp. 134–144.

(3) Über die Erzeugenden gewisser Modulgruppen. *Nachr. Akad. Wiss. Göttingen, math.-phys. Klasse*, 1956, pp. 173–185.

(4) Zur Theorie der hermitischen Modulfunktionen. *Math. Ann.* **134** (1958), pp. 355–384.

(5) Bemerkung über Kongruenzuntergruppen der Modulgruppe *n*-ten Grades. *Arch. Math.* **10** (1959), pp. 113–122.

(6) Eisensteinreihen zur Hilbertschen Modulgruppe *n*-ten Grades. *Nachr. Akad. Wiss. Göttingen, math.-phys. Klasse*, 1960, pp. 87–104.

(7) Analytic automorphisms of bounded symmetric complex domains. *Pacific J. Math.* **10** (1960), pp. 1327–1332.

(8) Zur Transformationstheorie von Thetareihen indefiniter quadratischer Formen. *Math. Ann.* **140** (1960), pp. 76–86.

(9) Quotientendarstellung Hermitescher Modulfunktionen durch Modulformen. *Math. Ann.* **143** (1961), pp. 1–18.

(10) Charakterisierung der Siegelschen Modulgruppe durch ein endliches System definierender Relationen. *Math. Ann.* **144** (1961), pp. 64–72.

(11) Volumbestimmung des Fundamentalbereichs der Hilbertschen Modulgruppe n-ten Grades. *J. reine angew. Math.* **206** (1961), pp. 9–19.

(12) Über die Werte der Dedekindschen Zetafunktion. *Math. Ann.* **145** (1962), pp. 265–272.

(13) Über den arithmetischen Charakter der Fourierkoeffizienten von Modulformen. *Math. Ann.* **147** (1962), pp. 176–188.

(14) Über einen Zusammenhang zwischen Siegelschen und Hermiteschen Modulfunktionen. *Abh. Math. Sem. Univ. Hamburg* **27** (1963), pp. 1–12.

(15) Eine potentialtheoretische Methode zur Behandlung von Poincaréschen Reihen. *Nachr. Akad. Wiss. Göttingen, math.-phys. Klasse*, 1965, pp. 17–29.

(16) Bemerkungen zur Konvergenz von Poincaréschen Reihen. *Nachr. Akad. Wiss., Göttingen, math.-phys. Klasse*, 1966, pp. 1–9.

(17) Über Poincarésche Reihen zur Siegelschen Modulgruppe. *Math. Ann.* **168** (1967), pp. 157–170.

(18) Zum Darstellungssatz für Siegelsche Modulformen. *Math. Z.* **102** (1967), pp. 30–43; correction, **105** (1968), pp. 399–400.

KLOOSTERMANN, H. D.

(1) Asymptotische Formeln für die Fourierkoeffizienten ganzer Modulformen. *Abh. Math. Sem. Hamburg. Univ.* **5** (1927), pp. 337–352.

(2) Theorie der Eisensteinschen Reihen von mehreren Veränderlichen. *Abh. Math. Sem. Hamburg. Univ.* **6** (1928), pp. 163–188.

(3) Thetareihen in total-reellen algebraischen Zahlkörpern. *Math. Ann.* **103** (1930), pp. 279–299.

(4) The behaviour of general theta functions under the modular group and the characters of binary modular congruence groups I, II. *Ann. Math.* **47** (1946), pp. 317–375, 376–447.

(5) On the characters of binary modular congruence groups. *Proc. Int. Congr. Math.*, Cambridge, Mass., 1950, pp. 257–280.

KOBAYASHI, S.

(1) Espaces à connexion de Cartan complète. *Proc. Japan Acad. Sci.* **30** (1954), pp. 709–710.

(2) Fixed points of isometries. *Nagoya Math. J.* **13** (1958), pp. 63–68.

(3) On the automorphism group of a certain class of algebraic manifolds. *Tohoku Math. J.* **11** (1959), pp. 184–190.

(4) Geometry of bounded domains. *Trans. Amer. Math. Soc.* **92** (1959), pp. 267–290.

(5) On complete Bergman metrics. *Proc. Amer. Math. Soc.* **13** (1962), pp. 511–518.

KOECHER, M.

(1) Über Dirichlet-Reihen mit Funktionalgleichung. *J. reine angew. Math.* **192** (1953), pp. 1–23.

(2) Über Thetareihen indefiniter quadratischer Formen. *Math. Nachr.* **9** (1953), pp. 51–85.

(3) Zur Theorie der Modulformen n-ten Grades I, II. *Math. Z.* **59** (1954), pp. 399–416; **61** (1955), pp. 455–466.

(4) Positivitätsbereiche im $\mathbb{R}^n$. *Amer. J. Math.* **79** (1957), pp. 575–596.

(5) Analysis in reellen Jordan-Algebren. *Nachr. Akad. Wiss. Göttingen, math.-phys. Klasse*, 1958, pp. 67–74.

(6) Die Geodätischen von Positivitätsbereichen. *Math. Ann.* **135** (1958), pp. 192–202.

(7) Beiträge zu einer Reduktionstheorie in Positivitätsbereichen I, II. *Math. Ann.* **141** (1960), pp. 384–432; **144** (1961), pp. 175–182.

(8) On real Jordan algebras. *Bull. Amer. Math. Soc.* **68** (1962), pp. 374–377.

(9) Eine Charakterisierung der Jordan-Algebren. *Math. Ann.* **148** (1962), pp. 244–256.

(10) *Jordan-Algebras and their Applications.* Lecture Notes, 1962, University of Minnesota.

(11) *An elementary approach to bounded symmetric domains.* Lecture Notes, 1969, Rice University, Houston, Texas.

KOECHER, M., and ROELCKE, W.

(1) Diskontinuierliche und diskrete Gruppen von Isometrien metrischer Räume. *Math. Z.* **71** (1959), pp. 258–267.

KÖHLER, G.

(1) Erweiterungsfähigkeit paramodularer Gruppen. *Nachr. Akad. Wiss. Göttingen, math.-phys. Klasse*, 1967, pp. 229–238.

(2) Ein Trennungssatz für Eisensteinsche Reihen zweiten und dritten Grades der Stufe T. *J. reine angew. Math.* **231** (1968), pp. 47–74.

KOIZUMI, S.

(1) On specialization of the Albanese and Picard varieties. *Mem. Coll. Sci. Univ. Kyoto* (*A*), **32** (1960), pp. 371–382.

KORÁNYI, A.

(1) On some classes of analytic functions of several variables. *Trans. Amer. Math. Soc.* **101** (1961), pp. 520–554.

(2) The Bergman kernel function for tubes over convex cones. *Pacific J. Math.* **12** (1962), pp. 1355–1360.

(3) On the boundary values of holomorphic functions in wedge domains. *Bull. Amer. Math. Soc.* **69** (1963), pp. 475–480.

(4) The Poisson integral for generalized half-planes and bounded symmetric domains. *Ann. Math.* (*2*), **82** (1965), pp. 332–350.

KORÁNYI, A., and WOLF, J. A.

(1) Realization of Hermitian symmetric spaces as generalized half-planes. *Ann. Math.* **81** (1965), pp. 265–288.

(2) Generalized Cayley transformations of bounded symmetric domains. *Amer. J. Math.* **87** (1965), pp. 899–939.

KOSZUL, J. L.

(1) Homologie et cohomologie des algèbres de Lie. *Bull. Soc. Math. France* **78** (1950), pp. 65–127.

(2) Sur la forme Hermitienne canonique des espaces homogènes complexes. *Canad. J. Math.* **7** (1955), pp. 562–576.

(3) *Exposés sur les espaces homogènes symétriques.* Publ. Soc. Math., São Paulo, 1959, 71 pp.

(4) Domaines bornés homogènes et orbites de groupes de transformations affines. *Bull. Soc. Math. France* **89** (1961), pp. 515–533.

(5) Ouverts convexes des espaces affines. *Math. Z.* **79** (1962), pp. 254–259.

KRAZER, A.

(1) *Lehrbuch der Thetafunktionen.* B. G. Teubner, Leipzig, 1903; reprinted Chelsea Publishing Company, New York, 1970.

KRAZER, A., and PRYM, F.

(1) *Neue Grundlagen einer Theorie der allgemeinen Thetafunktionen.* B. G. Teubner, Leipzig, 1892.

KRAZER, A., and WIRTINGER, W.

(1) Abelsche Funktionen und allgemeine Thetafunktionen. *Enzykl. Math. Wiss.* **II B 7** (1920), pp. 604–873.

KUGA, M.

(1) *Fibre varieties over a symmetric space whose fibres are abelian varieties.* Lecture Notes, University of Chicago, Chicago, Ill., 1963–1964.

(2) Fiber varieties over a symmetric space whose fibers are Abelian varieties. *Proc. Sympos. Pure Math.*, Vol. 9, American Mathematical Society, Providence, R.I., 1966, pp. 338–346.

KUGA, M., and SATAKE, I.

(1) Abelian varieties attached to polarized K_3-surfaces. *Math. Ann.* **169** (1967), pp. 239–242; correction, *Math. Ann.* **173** (1967), p. 322.

KUGA, M., and SHIMURA, G.

(1) On vector differential forms attached to automorphic forms. *J. Math. Soc. Japan* **12** (1960), pp. 258–270.

(2) On the zeta-function of a fibre variety whose fibres are abelian varieties. *Ann. Math.* (*2*) **82** (1965), pp. 478–539.

LAGRANGE, J. L.

(1) Démonstration d'un théorème d'arithmétique. *Oeuvres* **3** (1770), pp. 189–201.

(2) Recherches d'arithmétique. *Oeuvres* **3** (1773/75), pp. 695–795.

LAL, SUNDER

(1) On the Fourier coefficients of Hilbert-Siegel modular forms. *Math. Z.* **88** (1965), pp. 207–243.

LANG, S.

(1) *Introduction to Algebraic Geometry*. Interscience Publishers, New York, 1958.

(2) *Abelian Varieties*. Interscience Publishers, New York, 1959.

LANGLANDS, R. P.

(1) The dimension of spaces of automorphic forms. *Amer. J. Math.* **85** (1963), pp. 99–125.

(2) The volume of the fundamental domain for some arithmetical subgroups of Chevalley groups. *Proc. Sympos. Pure Math.*, Vol. 9, American Mathematical Society, Providence, R.I., 1966, pp. 143–148.

(3) Eisenstein series. *Ibid.* pp. 235–252.

(4) Dimension of spaces of automorphic forms. *Ibid.* pp. 253–257.

(5) On the functional equations satisfied by Eisenstein series. Mimeographed Notes.

LAPIN, A. I.

(1) On the theory of Siegel's modular functions of degree two. *Izv. Akad. Nauk USSR* **20** (1956), pp. 325–336 (Russian).

LASKER, E.

(1) Zur Theorie der Moduln und Ideale. *Math. Ann.* **60** (1905), pp. 20–116.

LEFSCHETZ, S.

(1) On certain numerical invariants of algebraic varieties with application to Abelian varieties. *Trans. Amer. Math. Soc.* **22** (1921), pp. 327–482.

(2) *L'analysis situs et la géométrie algébrique*. Gauthier-Villars, Paris, 1924.

(3) Hyperelliptic surfaces and Abelian varieties. *Bull. Nat. Res. Council* **63** (1928), pp. 349–395.

LEGENDRE, A.-M.

(1) *Zahlentheorie*. Teubner, Leipzig, 1886.

LEHNER, J.

(1) *Discontinuous groups and automorphic functions*. American Mathematical Society, Providence, R.I., 1964.

LICHNEROWICZ, A.

(1) Espaces homogènes Kaehlériens. *Coll. Géom. Diff.*, Strasbourg, 1953, pp. 171–184.
(2) Un théorème sur les espaces homogènes complexes. *Arch. Math.* **5** (1954), pp. 207–215.
(3) Opérateurs différentiels invariants sur un espace symétrique. *C.R. Acad. Sci. Paris* **256** (1963), pp. 3548–3550.
(4) Opérateurs différentiels invariants sur un espace homogène. *Ann. Sci. Ècole Norm. Sup.* (*3*) **81** (1964), pp. 341–385.
(5) Variétés complexes, tenseur de Bergmann et espaces homogènes. *C.R. Acad. Sci. Paris* **261** (1965), pp. 2143–2145.
(6) Variétés complexes et tenseur de Bergmann. *Ann. Inst. Fourier* (*Grenoble*) **15** (1965), fasc. 2, pp. 345–407.

LOOK, K. H.

(1) Schwarz Lemma in the theory of functions of several complex variables. *Sci. Record* (*N.S.*) **1** (1957), no. 2, pp. 5–8.
(2) An analytic invariant and its characteristic properties. *Sci. Record* (*N.S.*) **1** (1957), pp. 307–310.
(3) Schwarz Lemma and analytic invariants. *Scientia Sinica* **7** (1958), pp. 453–504.
(4) On a class of homogeneous complex manifolds. *Acta Math. Sinica* **12** (1962), pp. 229–249 (Chinese); translated as *Chinese Math.* **3** (1963), pp. 247–276.

LOOS, O.

(1) Spiegelungsräume und homogene symmetrische Mannigfaltigkeiten. Dissertation, München, 1966.
(2) Spiegelungsräume und homogene symmetrische Räume. *Math. Z.* **99** (1967), pp. 141–170.
(3) *Symmetric spaces*, I: *General theory*, II: *Compact spaces and classification.* W. A. Benjamin, Inc., New York, 1969.

LOWDENSLAGER, D. B.

(1) Potential theory in bounded symmetric homogeneous complex domains. *Ann. Math.* **67** (1958), pp. 467–484.
(2) Potential theory and a generalized Jensen-Nevanlinna formula for functions of several complex variables. *J. Math. Mech.* **7** (1958), pp. 207–218.

MAASS, H.

(1) Konstruktion ganzer Modulformen halbzahliger Dimension mit ϑ-Multiplikatoren in zwei Variablen. *Math. Z.* **43** (1938), pp. 709–738.
(2) Über Gruppen von hyperabelschen Transformationen. *Sitz.-Ber. Heidelberger Akad. Wiss., math.-naturw. Klasse* 1940, No. 2, pp. 1–26.
(3) Zur Theorie der automorphen Funktionen von *n* Veränderlichen. *Math. Ann.* **117** (1940/41), pp. 538–578.

(4) Modulformen und quadratische Formen über dem quadratischen Zahlkörper $R(\sqrt{5})$. *Math. Ann.* **118** (1941/43), pp. 65–84.

(5) Über eine Metrik im Siegelschen Halbraum. *Math. Ann.* **118** (1941/43), pp. 312–318.

(6) Theorie der Poincaréschen Reihen zu den hyperbolischen Fixpunktsystemen der Hilbertschen Modulgruppe. *Math. Ann.* **118** (1941/43), pp. 518–543.

(7) Über automorphe Funktionen von mehreren Veränderlichen und die Bestimmung von Dirichletschen Reihen durch Funktionalgleichungen. *Ber. Math. Tagung Tübingen*, 1946 (1947), pp. 100–102.

(8) Über die Erweiterungsfähigkeit der Hilbertschen Modulgruppe. *Math. Z.* **51** (1948), pp. 255–261.

(9) Automorphe Funktionen und indefinite quadratische Formen. *Sitz.-Ber. Heidelberger Akad. Wiss., math.-naturw. Klasse*, 1949, No. 1, pp. 3–42.

(10) Über eine neue Art von nichtanalytischen automorphen Funktionen und die Bestimmung Dirichletscher Reihen durch Funktionalgleichungen. *Math. Ann.* **121** (1949), pp. 141–183.

(11) Modulformen zweiten Grades und Dirichletreihen. *Math. Ann.* **122** (1950), pp. 90–108.

(12) Über die Darstellung der Modulformen n-ten Grades durch Poincarésche Reihen. *Math. Ann.* **123** (1951), pp. 125–151.

(13) Die Primzahlen in der Theorie der Siegelschen Modulformen. *Math. Ann.* **124** (1951), pp. 87–122.

(14) Die Differentialgleichungen in der Theorie der elliptischen Modulfunktionen. *Math. Ann.* **125** (1953), pp. 235–263.

(15) Die Differentialgleichungen in der Theorie der Siegelschen Modulfunktionen. *Math. Ann.* **126** (1953), pp. 44–68.

(16) *Lectures on Siegel's modular functions.* Tata Institute of Fundamental Research, Bombay, 1954–55.

(17) Die Bestimmung der Dirichletreihen mit Größencharakteren zu den Modulformen n-ten Grades. *J. Indian Math. Soc.* **19** (1955), pp. 1–23.

(18) Spherical functions and quadratic forms. *J. Indian Math. Soc.* **20** (1956), pp. 117–162.

(19) Zetafunktionen mit Größencharakteren und Kugelfunktionen. *Math. Ann.* **134** (1957), pp. 1–32.

(20) Zur Theorie der Kugelfunktionen einer Matrixvariablen. *Math. Ann.* **135** (1958), pp. 391–416.

(21) Über die Verteilung der zweidimensionalen Untergitter in einem euklidischen Gitter. *Math. Ann.* **137** (1959), pp. 319–327.

(22) Über die räumliche Verteilung der Punkte in Gittern mit indefiniter Metrik. *Math. Ann.* **138** (1959), pp. 287–315.

(23) Die Multiplikatorsysteme zur Siegelschen Modulgruppe. *Nachr. Akad. Wiss. Göttingen, math.-phys. Klasse*, 1964, pp. 125–135.

(24) Über die gleichmäbige Konvergenz der Poincaréschen Reihen n-ten Grades. *Nachr. Akad. Wiss. Göttingen, math.-phys. Klasse*, 1964, pp. 137–144.

(25) *Lectures on modular functions of one complex variable.* Tata Institute of Fundamental Research, Bombay, 1964.

(26) Die Fourierkoeffizienten der Eisensteinreihen zweiten Grades. *Mat.-Fys. Medd. Danske Vid. Selsk.* **34** (1964), No. 7.

(27) Modulformen zu indefiniten quadratischen Formen. *Math. Scand.* **17** (1965), pp. 41–55.

(28) *Siegel's modular forms and Dirichlet series*. Lecture Notes in Mathematics, Vol. 216, Springer-Verlag, Berlin–Heidelberg–New York, 1971.

MATSUSAKA, T.

(1) On the algebraic construction of the Picard variety. *Japan. J. Math.* **21** (1951), pp. 217–235.

(2) On the algebraic construction of the Picard variety. II. *Japan. J. Math.* **22** (1952), pp. 51–62.

(3) On a generating curve of an Abelian variety. *Nat. Sci. Rep. Ochanomizu Univ.* **3** (1952), pp. 1–4.

(4) Some theorems on Abelian varieties. *Nat. Sci. Rep. Ochanomizu Univ.* **4** (1953), pp. 22–35.

(5) On the theorem of Castelnuovo-Enriques. *Nat. Sci. Rep. Ochanomizu Univ.* **4** (1953), pp. 164–171.

(6) A remark on my paper "Some theorems on Abelian varieties." *Nat. Sci. Rep. Ochanomizu Univ.* **4** (1954), pp. 172–174.

(7) A note on my paper "Some theorems on Abelian varieties." *Nat. Sci. Rep. Ochanomizu Univ.* **5** (1954), pp. 21–23.

(8) On a theorem of Torelli. *Amer. J. Math.* **80** (1958), pp. 784–800.

(9) Polarized varieties, fields of moduli, and generalized Kummer varieties of polarized abelian varieties. *Amer. J. Math.* **80** (1958), pp. 45–82.

(10) On a characterization of a Jacobian variety. *Mem. Coll. Sci. Univ. Kyoto (A)* **32** (1959), pp. 1–19; correction, **33** (1960/61), p. 350.

MATSUSAKA, T., and MUMFORD, D.

(1) Two fundamental theorems on deformations of polarized varieties. *Amer. J. Math.* **86** (1964), pp. 668–684; correction, *Amer. J. Math.* **91** (1969), p. 851.

MATSUSHIMA, Y.

(1) Un théorème sur les espaces homogènes complexes. *C.R. Acad. Sci. Paris* **241** (1955), pp. 785–787.

(2) Sur la structure du groupe d'homéomorphismes analytiques d'une certaine variété kaehlérienne. *Nagoya Math. J.* **11** (1957), pp. 145–150.

(3) Sur les espaces homogènes kaehlériens d'un groupe de Lie réductif. *Nagoya Math. J.* **11** (1957), pp. 53–60.

(4) Fibrés holomorphes sur un tore complexe. *Nagoya Math. J.* **14** (1959), pp. 1–24.

(5) Espaces homogènes de Stein des groupes de Lie complexes. I, II. *Nagoya Math. J.* **16** (1960), pp. 205–218; **18** (1961), pp. 153–164.

(6) Sur certaines variétés homogènes complexes. *Nagoya Math. J.* **18** (1961), pp. 1–12.

(7) On the first Betti number of compact quotient spaces of higher dimensional symmetric spaces. *Ann. Math.* (2) **75** (1962), pp. 312–330.

(8) On Betti numbers of compact, locally symmetric Riemannian manifolds. *Osaka Math. J.* **14** (1962), pp. 1–20.

(9) *On the cohomology groups of locally symmetric, compact Riemannian manifolds.* Differential Analysis, Bombay Colloquium, 1964, pp. 237–242. Oxford University Press, London, 1964.

(10) A formula for the Betti numbers of compact locally symmetric Riemannian manifolds. *J. Diff. Geom.* **1** (1967), pp. 99–109.

(11) Affine structures on complex manifolds. *Osaka J. Math.* **5** (1968), pp. 215–222.

MATSUSHIMA, Y., and MORIMOTO, A.

(1) Sur certaines espaces fibrés holomorphes sur une variété de Stein. *Bull. Soc. Math. France* **88** (1960), pp. 137–155.

MATSUSHIMA, Y., and MURAKAMI, S.

(1) On vector bundle valued harmonic forms and automorphic forms on symmetric Riemannian manifolds. *Ann. Math.* **78** (1963), pp. 365–416.

(2) On certain cohomology groups attached to hermitian symmetric spaces. I, II. *Osaka J. Math.* **2** (1965), pp. 1–35; **5** (1968), pp. 223–241.

MATSUSHIMA, Y., and SHIMURA, G.

(1) On the cohomology groups attached to certain vector valued differential forms on the product of the upper half planes. *Ann. Math.* **78** (1963), pp. 417–449.

MAUTNER, F. I.

(1) Geodesic flows on symmetric Riemannian spaces. *Ann. Math.* (2) **65** (1957), pp. 416–431.

(2) Spherical functions over p-adic fields. I, II. *Amer. J. Math.* **80** (1958), pp. 441–457; **86** (1964), pp. 171–200.

MINKOWSKI, H.

(1) Grundlagen für eine Theorie der quadratischen Formen mit ganzzahligen Koeffizienten. *Gesammelte Abhandlungen* **1**, pp. 3–144 (1884), Teubner, Leipzig, 1911.

(2) Untersuchungen über quadratische Formen. *Acta Math.* **7** (1885), pp. 201–258; *Gesammelte Abhandlungen* **1**, pp. 157–202.

(3) Diskontinuitätsbereich für arithmetische Äquivalenz. *J. reine angew. Math.* **129** (1905), pp. 220–274; *Gesammelte Abhandlungen* **2**, pp. 53–100.

(4) *Geometrie der Zahlen.* Teubner, Leipzig, 1910; reprint, Chelsea, New York, 1953.

MOORE, C. C.

(1) Compactifications of symmetric spaces. *Amer. J. Math.* **86** (1964), pp. 201–218.

(2) Compactifications of symmetric spaces II: The Cartan domains. *Amer. J. Math.* **86** (1964), pp. 358–378.

(3) Decomposition of unitary representations defined by discrete subgroups of nilpotent groups. *Ann. Math.* **82** (1965), pp. 146–182.

MORIKAWA, H.

(1) On abelian varieties. *Nagoya Math. J.* **6** (1953), pp. 151–170.

(2) Cycles and endomorphisms of abelian varieties. *Nagoya Math. J.* **7** (1954), pp. 95–102.

(3) Cycles and multiple integrals on abelian varieties. *Proc. Japan Acad.* **31** (1955), pp. 317–320.

(4) Generalized Jacobian varieties and separable abelian extensions of function fields. *Nagoya Math. J.* **12** (1957), pp. 231–254.

(5) Theta functions and abelian varieties over valuation fields of rank one. I. *Nagoya Math. J.* **20** (1962), pp. 1–27.

(6) On theta functions and abelian varieties over valuation fields of rank one. II. Theta functions and abelian functions of characteristic p (> 0). *Nagoya Math. J.* **21** (1962), pp. 231–250.

(7) On the explicit defining relations of abelian schemes of level three. *Nagoya Math. J.* **27** (1966), pp. 143–157.

(8) On the defining equations of abelian varieties. *Nagoya Math. J.* **30** (1967), pp. 143–162.

(9) On the relation for two-dimensional theta constants of level three. *J. Math. Soc. Japan* **20** (1968), pp. 248–262.

MORITA, K.

(1) On the kernel functions for symmetric domains. *Sci. Rep. Tokyo Kyoiku Daigaku* (*A*) **5** (1956), pp. 190–212.

MOSTOW, G. D.

(1) On the conjugacy of subgroups of semisimple groups. *Proc. Sympos. Pure Math.*, Vol. 9, American Mathematical Society, Providence, R.I., 1966, pp. 413–419.

(2) On the rigidity of hyperbolic space forms under quasi-conformal mappings. *Proc. Nat. Acad. Sci. U.S.A.* **57** (1967), pp. 211–215.

(3) Quasi-conformal mappings in n-space and the rigidity of hyperbolic space forms. *Inst. Hautes Études Sci. Publ. Math.*, No. 34 (1968), pp. 53–104.

MOSTOW, G. D., and TAMAGAWA, T.

(1) On the compactness of arithmetically defined homogeneous spaces. *Ann. Math.* (*2*) **76** (1962), pp. 446–463.

MOUNTJOY, R. H.

(1) Abelian varieties attached to representations of discontinuous groups. Ph.D. Thesis, University of Chicago, Ill., 1965; *Amer. J. Math.* **89** (1967), pp. 149–224.

MUMFORD, D.

(1) Families of abelian varieties. *Proc. Sympos. Pure Math.*, Vol. 9, American Mathematical Society, Providence, R.I., 1966, pp. 347–351.

(2) On the equations defining abelian varieties. I, II, III. *Inventiones math.* **1** (1966), pp. 287–354; **3** (1967), pp. 75–135; **3** (1967), pp. 215–244.

(3) Abelian quotients of the Teichmüller modular group. *J. Analyse Math.* **18** (1967), pp. 227–244.

(4) A note of Shimura's paper "Discontinuous groups and abelian varieties." *Math. Ann.* **181** (1969), pp. 345–351.

(5) *Abelian varieties.* Oxford University Press, London, 1970.

MUMFORD, D., and NEWSTEAD, P.

(1) Periods of a moduli space of bundles on curves. *Amer. J. Math.* **90** (1968), pp. 1200–1208.

MURAKAMI, S.

(1) Cohomologies of vector-valued forms on compact, locally symmetric Riemann manifolds. *Proc. Sympos. Pure Math.*, Vol. 9, American Mathematical Society, Providence, R.I., 1966, pp. 387–399.

(2) *Cohomology groups of vector-valued forms on symmetric spaces.* Lecture Notes, University of Chicago, Chicago, Ill., 1966.

(3) *Plongements holomorphes de domaines symétriques.* Geometry of Homogeneous Bounded Domains (C.I.M.E., 3° Ciclo, Urbino, 1967), pp. 265–277. Edizioni Cremonese, Rome, 1968.

(4) *Facteurs d'automorphie associés à un espace hermitien symétrique.* Ibid, pp. 281–287.

MYRBERG, P. J.

(1) Untersuchungen über die automorphen Funktionen beliebig vieler Variablen. *Acta Math.* **46** (1925), pp. 215–336.

NAGANO, T.

(1) Transformation groups on compact symmetric spaces. *Trans. Amer. Math. Soc.* **118** (1965), pp. 428–453.

NAGATA, M.

(1) On the normality of the Chow variety of positive o-cycles of degree m in an algebraic variety. *Mem. Coll. Sci. Univ. Kyoto (A)* **29** (1955), pp. 165–176.

NAKAI, Y.

(1) On the divisors of differential forms on algebraic varieties. *J. Math. Soc. Japan* **5** (1953), pp. 184–199.

NISHI, M.

(1) Some results on abelian varieties. *Nat. Sci. Rep. Ochanomizu Univ.* **9** (1958), pp. 1–12.

OKA, K.

(1) *Sur les fonctions analytiques de plusieurs variables*. Iwanami Shoten, Tokyo, 1961. This is a collection of reprints of nine articles under the same general title, which have appeared in the following journals:

I. Domaines convexes par rapport aux fonctions rationnelles. *J. Sci. Hiroshima Univ.* (*A*) **6** (1936), pp. 245–255.

II. Domaines d'holomorphie. *J. Sci. Hiroshima Univ.* (*A*) **7** (1937), pp. 115–130.

III. Deuxième problème de Cousin. *J. Sci. Hiroshima Univ.* (*A*) **9** (1939), pp. 7–19.

IV. Domaines d'holomorphie et domaines rationnellement convexes. *Jap. J. Math.* **17** (1941), pp. 517–521.

V. L'intégrale de Cauchy. *Jap. J. Math.* **17** (1941), pp. 523–531.

VI. Domaines pseudoconvexes. *Tohoku Math. J.* **49** (1942), pp. 15–52.

VII. Sur quelques notions arithmétiques. *Bull. Soc. Math. France* **78** (1950), pp. 1–27.

VIII. Lemme fondamental. *J. Math. Soc. Japan* **3** (1951), pp. 204–214, 259–278.

IX. Domaines finis sans points critique intérieur. *Jap. J. Math.* **23** (1953), pp. 97–155.

Since then, the following paper in the series has also appeared:

X. Une mode nouvelle engendrant les domaines pseudoconvexes. *Jap. J. Math.* **32** (1962), pp. 1–12.

ONO, T.

(1) The Gauss-Bonnet theorem and the Tamagawa number. *Bull. Amer. Math. Soc.* **71** (1965), pp. 345–348.

(2) On algebraic groups and discontinuous groups. *Nagoya Math. J.* **27** (1966), pp. 279–322.

OSGOOD, W. F.

(1) *Lehrbuch der Funktionentheorie*. Vol. II, Teubner, Leipzig, 1929/32; reprinted, Chelsea Publishing Company, New York, 1965.

OTHMER, F-E.

(1) Elementarer Beweis des Hauptsatzes über meromorphe Funktionenkörper. *Math. Ann.* **141** (1960), pp. 99–106.

PETERSSON, H.

(1) Theorie der automorphen Formen beliebiger reeller Dimension und ihre Darstellung durch eine neue Art Poincaréscher Reihen. *Math. Ann.* **103** (1930), pp. 369–436.

(2) Über die Entwicklungskoeffizienten der automorphen Formen. *Acta Math.* **58** (1932), pp. 169–215.

(3) Über die eindeutige Bestimmung und die Erweiterungsfähigkeit von gewissen Grenzkreisgruppen. *Abh. Math. Sem. Hansischen Univ.* **12** (1938), pp. 180–199.

(4) Die linearen Relationen zwischen den ganzen Poincaréschen Reihen von reeller Dimension zur Modulgruppe. *Abh. Math. Sem. Hansischen Univ.* **12** (1938), pp. 415–472.

(5) Zur analytischen Theorie der Grenzkreisgruppen, I–V. *Math. Ann.* **115** (1938), pp. 23–67, 175–204, 518–572, 670–709; *Math. Z.* **44** (1939), pp. 127–155.

(6) Über die Metrisierung der ganzen Modulformen. *Jber. Deutsche Math.-Verein.* **49** (1939), pp. 49–75.

(7) Konstruktion der sämtlichen Lösungen einer Riemannschen Funktionalgleichung durch Dirichletreihen mit Eulerscher Produktentwicklung. I, II, III. *Math. Ann.* **116** (1939), pp. 401–412; **117** (1940/41), pp. 39–64; **117** (1940/41), pp. 277–300.

(8) Über eine Metrisierung der automorphen Formen und die Theorie der Poincaréschen Reihen. *Math. Ann.* **117** (1940/41), pp. 453–537.

(9) Einheitliche Begründung der Vollständigkeitssätze für die Poincaréschen Reihen von reeller Dimension bei beliebigen Grenzkreisgruppen von erster Art. *Abh. Math. Sem. Hansischen Univ.* **14** (1941), pp. 22–60.

(10) Über den Bereich absoluter Konvergenz der Poincaréschen Reihen. *Acta Math.* **80** (1948), pp. 23–63.

(11) Über die lineare Zerlegung der den ganzen Modulformen von höherer Stufe entsprechenden Dirichletreihen in vollständige Eulersche Produkte. *Acta Math.* **80** (1948), pp. 191–221.

(12) Über den Körper der Fourierkoeffizienten der von Hecke untersuchten Eisensteinreihen. *Abh. Math. Sem. Univ. Hamburg* **16** (1949), pp. 101–113.

(13) Ein Konvergenzbeweis für Poincarésche Reihen. *Abh. Math. Sem. Univ. Hamburg* **16** (1949), pp. 127–130.

(14) Über die Berechnung der Skalarprodukte ganzer Modulformen. *Comment. Math. Helv.* **22** (1949), pp. 168–199.

(15) Über die Transformationsfaktoren der relativen Invarianten linearer Substitutionsgruppen. *Monatsh. Math.* **53** (1949), pp. 17–41.

(16) Konstruktion der Modulformen und der zu gewissen Grenzkreisgruppen gehörigen automorphen Formen von positiver reeller Dimension und die vollständige Bestimmung ihrer Fourierkoeffizienten. *Sitz.-Ber. Heidelberger Akad. Wiss., math.-naturw. Klasse*, 1950, pp. 417–494.

(17) Über automorphe Orthogonalfunktionen und die Konstruktion der automorphen Formen von positiver reeller Dimension. *Math. Ann.* **127** (1954), pp. 33–81.

(18) Über Modulfunktionen und Partitionenprobleme. *Abh. Deutsch. Akad. Wiss. Berlin., math.-phys.-tech. Klasse*, 1954, No. 2, 59 pp.

(19) Über automorphe Formen mit Singularitäten im Diskontinuitätsgebiet. *Math. Ann.* **129** (1955), pp. 370–390.

(20) Über Eisensteinsche Reihen und automorphe Formen von der Dimension -1. *Comment. Math. Helv.* **31** (1956), pp. 111–144.

(21) Explizite Konstruktion der automorphen Orthogonalfunktionen in den multiplikativen Differentialklassen. *Math. Nachr.* **16** (1957), pp. 343–368.

(22) Die Systematik der abelschen Differentiale in der Grenzkreisuniformisierung. *Ann. Acad. Sci. Fenn.* (*AI*), No. 276 (1960), 74 pp.

PICARD, É.

(1) Sur une classe de groupes discontinus de substitutions linéaires et sur les fonctions de deux variables indépendantes restant invariables par ces substitutions. *Acta Math.* **1** (1882), pp. 297–320.

(2) Sur les fonctions hyperabéliennes. *J. Liouville* (*IV*) **1** (1885), pp. 87–128.

(3) Sur les intégrales de différentielles totales algébriques de première espèce. *J. Math.* (*IV*) **1** (1885), pp. 281–346.

(4) Mémoire sur la théorie des fonctions algébriques de deux variables. *J. Math.* (*IV*) **5** (1889).

PICARD, É., and POINCARÉ, H.

(1) Sur un théorème relatif aux fonctions de n variables indépendantes admettant $2n$ systèmes de périodes. *C.R. Acad. Sci. Paris* **97** (1883), pp. 1284–1287.

POINCARÉ, H.

(1) Sur les propriétés des fonctions définies par les équations aux différences partielles. *Oeuvres* **1,** IL-CXXXII, 1879; Gauthier-Villars, Paris, 1951.

(2) Mémoire sur les fonctions Fuchsiennes. *Acta Math.* **1** (1882), pp. 193–294; *Oeuvres* **2,** pp. 169–257.

(3) Sur les fonctions de deux variables. *Acta Math.* **2** (1883), pp. 97–113; *Oeuvres* **4,** pp. 147–161.

(4) Sur la réduction des intégrales abéliennes. *Bull. Soc. Math. France* **12** (1884), pp. 124–143; *Oeuvres* **3,** pp. 333–351.

(5) Sur les résidus des intégrales doubles. *Acta Math.* **9** (1887), pp. 321–380; *Oeuvres* **3,** pp. 440–489.

(6) Les fonctions Fuchsiennes et l'arithmétique. *J. Math.* (*IV*) **3** (1887), pp. 405–464; *Oeuvres* **2,** pp. 463–511.

(7) Sur les fonctions abéliennes. *Acta Math.* **26** (1902), pp. 43–98; *Oeuvres* **4,** pp. 473–526.

(8) Les fonctions analytiques de deux variables et la représentation conforme. *Rend. Circ. Mat. Palermo* **23** (1907), pp. 185–220; *Oeuvres* **4,** pp. 244–289.

(9) Fonctions modulaires et fonctions fuchsiennes. *Ann. Fac. Sci. Toulouse* (*III*) **3** (1912), pp. 125–149; *Oeuvres* **2,** pp. 592–618.

PRESTEL, A.

(1) Die elliptischen Fixpunkte der Hilbertschen Modulgruppen. *Math. Ann.* **177** (1968), pp. 181–209.

PRILL, D.

(1) Local classification of quotients of complex manifolds by discontinuous groups. *Duke Math. J.* **34** (1967), pp. 375–386.

(2) The divisor class groups of some rings of holomorphic functions. *Math. Z.* **121** (1971), pp. 58–80.

PYATETSKII-SHAPIRO, I. I.

(1) Abelian modular functions. *Doklady Akad. Nauk SSSR* **95** (1954), pp. 221–224 (Russian).

(2) Analogue of a theorem of Lefschetz. *Doklady Akad. Nauk SSSR* **96** (1954), pp. 917–920 (Russian).

(3) On the theory of abelian modular functions. *Doklady Akad. Nauk SSSR* **106** (1956), pp. 973–976 (Russian).

(4) Singular modular functions. *Izv. Akad. Nauk SSSR, Ser. mat.* **20** (1956), pp. 53–98; translation: *Amer. Math. Soc. Transl.* (2) **10** (1956), pp. 13–58.

(5) Classification of modular groups. *Doklady Akad. Nauk SSSR* **110** (1956), pp. 19–22 (Russian).

(6) On the estimation of the dimension of the space of automorphic forms for certain types of discrete groups. *Doklady Akad. Nauk SSSR* **113** (1957), pp. 980–983 (Russian).

(7) Some questions of harmonic analysis in homogeneous cones. *Doklady Akad. Nauk SSSR* **116** (1957), pp. 181–184 (Russian).

(8) On a problem proposed by É. Cartan. *Doklady Akad. Nauk SSSR* **124** (1959), pp. 272–273 (Russian).

(9) Discrete subgroups of the group of analytic automorphisms of the polycylinder and automorphic forms. *Doklady Akad. Nauk SSSR* **124** (1959), pp. 760–763 (Russian).

(10) The geometry of homogeneous domains and the theory of automorphic functions. The solution of a problem of É. Cartan. *Uspekhi Mat. Nauk* **14** (1959), No. 3, pp. 190–192 (Russian).

(11) The theory of modular functions and related questions of the theory of discrete groups. *Uspekhi Mat. Nauk* **15** (1960), No. 1, pp. 99–136; translation: *Russian Math. Surveys* **15** (1960), No. 1, pp. 97–128.

(12) Classification of bounded homogeneous domains in n-dimensional complex space. *Doklady Akad. Nauk SSSR* **141** (1961), pp. 316–319; translation: *Soviet Math. Doklady* **2** (1961), pp. 1460–1463.

(13) *Geometry of the classical domains and the theory of automorphic functions* (Russian). Fizmatgiz, Moscow, 1961. Translated as *Automorphic functions and the geometry of classical domains*. Gordon and Breach, New York, 1969.

(14) On bounded homogeneous domains in n-dimensional complex space. *Izv. Akad. Nauk SSSR, Ser. Mat.* **26** (1962), pp. 107–124; translation: *Amer. Math. Soc. Transl.* (2) **43** (1964), pp. 299–320.

(15) The structure of j-algebras. *Izv. Akad. Nauk SSSR, Ser. Mat.* **26** (1962), pp. 453–484; translation: *Amer. Math. Soc. Transl.* (2) **55** (1966), pp. 207–241.

(16) Generalized upper halfplanes in the theory of several complex variables. *Proc. Int. Congr. Math.*, Stockholm, 1962, pp. 389–396.

(17) Arithmetic groups in complex domains. *Uspekhi Mat. Nauk* **19** (1964), No. 6 (120), pp. 93–121; translation: *Russian Math. Surveys* **19** (1964), No. 6, pp. 83–109.

(18) The geometry and the classification of bounded homogeneous regions. *Uspekhi Mat. Nauk* **20** (1965), No. 2 (122), pp. 3–51; translation: *Russian Math. Surveys* **20** (1965), No. 2, pp. 1–48.

RAGHAVAN, S.

(1) Modular forms of degree n and representation by quadratic forms. *Ann. Math.* **70** (1959), pp. 446–477; *Contributions to function theory*, Bombay, 1960, pp. 181–183.

(2) On representations by hermitian forms. *Acta Arith.* **8** (1962), pp. 33–96.

RAGHUNATHAN, M. S.

(1) On the first cohomology of discrete subgroups of semisimple Lie groups. *Amer. J. Math.* **87** (1965), pp. 103–139.

(2) Cohomology of arithmetic subgroups of algebraic groups. I, II. *Ann. Math.* **86** (1967), pp. 409–424; **87** (1968), pp. 279–304.

(3) A note on quotients of real algebraic groups by arithmetic subgroups. *Inventiones math.* **4** (1968), pp. 318–335.

RAMANATHAN, K. G.

(1) Units of quadratic forms. *Ann. Math.* **56** (1952), pp. 1–10.

(2) A note on symplectic complements. *J. Indian Math. Soc.* **18** (1954), pp. 115–125.

(3) Units of fixed points in involutorial algebras. *Proc. Int. Symp. Algebraic Number Theory*, Tokyo 1956, pp. 103–106.

(4) Quadratic forms over involutorial division algebras. *J. Indian Math. Soc.* **20** (1956), pp. 227–257.

(5) On orthogonal groups. *Nachr. Akad. Wiss. Göttingen, math.-phys. Klasse*, 1957, pp. 113–121.

(6) Quadratic forms over involutorial division algebras II. *Math. Ann.* **143** (1961), pp. 293–332.

(7) Discontinuous groups. *Nachr. Akad. Wiss. Göttingen, math.-phys. Klasse*, 1963, pp. 293–323.

(8) Discontinuous groups II. *Nachr. Akad. Wiss. Göttingen, math.-phys. Klasse*, 1964, pp. 145–164.

REINER, I.

(1) Automorphisms of the symplectic modular group. *Trans. Amer. Math. Soc.* **80** (1955), pp. 35–50.

(2) Real linear characters of the symplectic modular group. *Proc. Amer. Math. Soc.* **6** (1955), pp. 987–990.

REMMERT, R.

(1) Meromorphe Funktionen in kompakten komplexen Räumen. *Math. Ann.* **132** (1956), pp. 277–288.

(2) Holomorphe und meromorphe Abbildungen komplexer Räume. *Math. Ann.* **133** (1957), pp. 328–370.

(3) Analytic and algebraic dependence of meromorphic functions. *Amer. J. Math.* **82** (1960), pp. 891–899.

REMMERT, R., and STEIN, K.

(1) Über die wesentlichen Singularitäten analytischer Mengen. *Math. Ann.* **126** (1953), pp. 263–306.

RIEMANN, B.

(1) Theorie der Abel'schen Functionen. *Gesammelte Math. Werke*, pp. 88–144, Dover Publications, New York, 1953; *J. reine angew. Math.* **54** (1857), pp. 115–155.

(2) Über das Verschwinden der Theta-Functionen. *Gesammelte Math. Werke*, pp. 212–224, Dover Publications, New York, 1953; *J. reine angew. Math.* **65** (1866), pp. 161–172.

(3) Beweis des Satzes, daß eine einwerthige mehr als $2n$-fach periodische Function von n Veränderlichen unmöglich ist. *Gesammelte Math. Werke*, pp. 294–297, Dover Publications, New York, 1953; *J. reine angew. Math.* **71** (1870), pp. 197–200.

(4) Vorlesungen über die allgemeine Theorie der Integrale algebraischer Differentialien. *Gesammelte Math. Werke, Nachträge*, pp. 1–66, Dover Publications, New York, 1953.

ROELCKE, W.

(1) Über die Wellengleichung bei Grenzkreisgruppen erster Art. *Sitz.-Ber. Heidelberger Akad. Wiss., math.-naturw. Klasse*, 1956, No. 4, pp. 159–267.

(2) Über Fundamentalbereiche diskontinuierlicher Gruppen. *Math. Nachr.* **20** (1959), pp. 329–355.

(3) Über den Laplace-Operator auf Riemannschen Mannigfaltigkeiten mit diskontinuierlichen Gruppen. *Math. Nachr.* **21** (1960), pp. 131–149.

(4) Über diskontinuierliche Gruppen in lokal-kompakten Räumen. *Math. Z.* **75** (1961), pp. 36–52.

(5) Das Eigenwertproblem der automorphen Formen in der hyperbolischen Ebene I, II. *Math. Ann.* **167** (1966), pp. 292–337; **168** (1967), pp. 261–324.

ROSATI, C.

(1) Sulle matrici di Riemann. *Rend. Circ. Mat. Palermo* **53** (1929), pp. 79–134.

ROSATI, M.

(1) Sopra le funzioni abeliane pari e le varietà abeliane di rango due. *Rend. Mat. sue Applic.* (*V*) **11** (1952), pp. 28–61.

(2) Osservazioni su alcuni gruppi finiti di omografie appartenenti ad una varietà di Picard e ad una varietà abeliana di rango due. *Rend. Mat. sue Applic.* (*V*) **11** (1952), pp. 453–469.

(3) Sull' equivalenza birazionale delle due varietà di Picard associate ad una varietà superficialmente irregolare. *Rend. Accad. Naz. Lincei* (*VIII*) **16** (1954), pp. 708–715.

(4) Sulle varietà di equazione $\omega M \omega_{-1} = 0$ con M matrice modulare. *Rend. Mat. sue Applic.* (*V*) **14** (1955), pp. 712–728.

(5) Funzioni abeliane ed abeliane modulari nelle lezioni e nei manoscritti inediti di Fabio Conforto. *Rend. Mat. sue Applic.* (*V*) **14** (1955), pp. 696–711.

(6) Le funzioni e le varietà quasi abeliane dalla teoria del Severi ad oggi. *Pontificiae Academiae Scientiarum Scripta Varia*, No. 23, Roma, 1962.

ROSENHAIN, G.

(1) *Abhandlung über die Functionen zweier Variabler mit vier Perioden*, 1851. Ostwald's Klassiker der Exacten Wissenschaften, No. 65, 1895.

ROTHAUS, O. S.

(1) Domains of positivity. *Bull. Amer. Math. Soc.* **64,** 2 (1958), pp. 85–86.

(2) Domains of positivity. *Abh. Math. Sem. Univ. Hamburg* **24** (1960), pp. 189–235.

RÜCKERT, W.

(1) Zum Eliminationsproblem der Potenzreihenideale. *Math. Ann.* **107** (1933), pp. 259–281.

SATAKE, I.

(1) On a theorem of É. Cartan. *J. Math. Soc. Japan* **2** (1951), pp. 284–305.

(2) On the Fuchsian theta function. *Sûgaku* **5** (1953), pp. 73–81 (Japanese).

(3) A remark on bounded symmetric domains. *Sci. Papers Coll. Gen. Ed. Univ. Tokyo* **3** (1953), pp. 131–144.

(4) On Siegel's modular functions. *Proc. Int. Sympos. Algebraic Number Theory*, Tokyo-Nikko, 1955, pp. 107–129.

(5) On a generalization of the notion of manifold. *Proc. Nat. Acad. Sci. U.S.A.* **42** (1956), pp. 359–363.

(6) On the compactification of the Siegel space. *J. Indian Math. Soc.* **20** (1956), pp. 259–281.

(7) The Gauß-Bonnet theorem for V-manifolds. *J. Math. Soc. Japan* **9** (1957), pp. 464–492.

(8) On modular functions of several variables (the compactification and its application). *Sûgaku* **11** (1959/60), pp. 170–175 (Japanese).

(9) On representations and compactifications of symmetric Riemannian spaces. *Ann. Math.* **71** (1960), pp. 77–110.

(10) On compactifications of the quotient spaces for arithmetically defined discontinuous groups. *Ann. Math.* **72** (1960), pp. 555–580.

(11) On the theory of reductive algebraic groups over a perfect field. *J. Math. Soc. Japan* **15** (1963), pp. 210–235.

(12) *Theory of spherical functions on reductive algebraic groups over* $\mathfrak{p}$*-adic fields.* Publ. Math., No. 18, Inst. Hautes Études Sci., Paris, 1963, pp. 5–69.

(13) *Holomorphic imbeddings of symmetric domains into a Siegel space.* Proc. Conf. Complex Analysis, Minneapolis 1964, pp. 40–48, Springer-Verlag, Berlin–Heidelberg–New York, 1965.

(14) Holomorphic imbeddings of symmetric domains into a Siegel space. *Amer. J. Math.* **87** (1965), pp. 425–461.

(15) Spherical functions and Ramanujan conjecture. *Proc. Sympos. Pure Math.*, Vol. 9, American Mathematical Society, Providence, R.I., 1966, pp. 258–264.

(16) Symplectic representations of algebraic groups. Ibid. pp. 352–357.

(17) Clifford algebras and families of abelian varieties. *Nagoya Math. J.* **27** (1966), pp. 435–446; corrections: *Nagoya Math. J.* **31** (1968), pp. 295–296.

(18) Symplectic representations of algebraic groups satisfying a certain analyticity condition. *Acta Math.* **117** (1967), pp. 215–279.

(19) On a certain invariant of the groups of type E_6 and E_7. *J. Math. Soc. Japan* **20** (1968), pp. 322–335.

(20) A note on holomorphic imbeddings and compactification of symmetric domains. *Amer. J. Math.* **90** (1968), pp. 231–247.

(21) *On modular imbeddings of a symmetric domain of type (IV).* Global Analysis (Papers in Honor of K. Kodaira), pp. 341–354. University of Tokyo Press, Tokyo, 1969.

SCORZA, G.

(1) Intorno alla teoria generale delle matrici di Riemann e ad alcune sue applicazioni. *Rend. Circ. Mat. Palermo* **41** (1916), pp. 263–380.

(2) Le algebre di ordine qualunque e le matrici di Riemann. *Rend. Circ. Mat. Palermo* **45** (1921), pp. 1–204.

SELBERG, A.

(1) Harmonic analysis and discontinuous groups in weakly symmetric Riemannian spaces with applications to Dirichlet series. *J. Indian Math. Soc.* **20** (1956), pp. 47–87.

(2) *Automorphic functions and integral operators.* Seminars on analytic functions, Vol. 2, pp. 152–161. Institute for Advanced Study, Princeton, 1957.

(3) *On discontinuous groups in higher-dimensional symmetric spaces.* Contributions to function theory, Internat. Colloquium on Function Theory, Bombay 1960, pp. 147–164. Tata Institute, Bombay, 1960.

(4) Discontinuous groups and harmonic analysis. *Proc. Int. Congr. Math.*, Stockholm, 1962, pp. 177–189.

SERRE, J.-P.

(1) Quelques propriétés des variétés abéliennes en charactéristique *p*. *Amer. J. Math.* **80** (1958), pp. 715–739.

(2) *Groupes algébriques et corps de classes.* Hermann, Paris, 1959.

(3) *Groupes de Lie l-adiques attachés aux courbes elliptiques.* Colloq. de Clermont-Ferrand, IHES, 1964.

SEVERI, F.

(1) Il teorema d'Abel sulle superficie algebriche. *Ann. Mat.* (*III*) **12** (1905), pp. 55–79.

(2) Intorno al teorema d'Abel sulle superficie algebriche ed alla riduzione a forma normale degl'integrali di Picard. *Rend. Circ. Mat. Palermo* **21** (1906).

(3) Funzioni quasi abeliane. Seconda edizione ampliata. *Pont. Acad. Scient. Scripta varia*, No. 20, Roma, 1961.

SHIMIZU, H.

(1) On traces of Hecke operators. *J. Fac. Sci. Univ. Tokyo* **10** (1963), pp. 1–19.

(2) On discontinuous groups operating on the product of the upper half planes. *Ann. Math.* **77** (1963), pp. 33–71.

(3) On zeta-functions of quaternion algebras. *Ann. Math.* **81** (1965), pp. 166–193.

SHIMURA, G.

(1) On a certain ideal of the center of a Frobeniusean algebra. *Sci. Papers Coll. Gen. Ed. Univ. Tokyo* **2** (1952), pp. 117–124.

(2) A note on the normalization-theorem of an integral domain. *Sci. Papers Coll. Gen. Ed. Univ. Tokyo* **4** (1954), pp. 1–8.

(3) Reduction of algebraic varieties with respect to a discrete valuation of the basic field. *Amer. J. Math.* **77** (1955), pp. 134–176.

(4) On complex multiplications. *Proc. Int. Sympos. Algebraic Number Theory*, Tokyo-Nikko, 1955, pp. 23–30.

(5) La fonction ζ du corps des fonctions modulaires elliptiques. *C.R. Acad. Sci. Paris* **244** (1957), pp. 2127–2130.

(6) Correspondances modulaires et les fonctions ζ de courbes algébriques. *J. Math. Soc. Japan* **10** (1958), pp. 1–28.

(7) Fonctions automorphes et correspondances modulaires. *Proc. Int. Congr. Math.*, Cambridge, 1958, pp. 330–338.

(8) Sur les intégrales attachées aux formes automorphes. *J. Math. Soc. Japan* **11** (1959), pp. 291–311.

(9) Automorphic functions and number theory. I, II. (Japanese). *Sûgaku* **11** (1959/60), pp. 193–205; **13** (1961/62), pp. 65–80.

(10) On the theory of automorphic functions. *Ann. Math.* **70** (1959), pp. 101–144.

(11) On the zeta-functions of the algebraic curves uniformized by certain automorphic functions. *J. Math. Soc. Japan* **13** (1961), pp. 275–331.

(12) On the class-fields obtained by complex multiplication of abelian varieties. *Osaka Math. J.* **14** (1962), pp. 33–44.

(13) On Dirichlet series and abelian varieties attached to automorphic forms. *Ann. Math.* **76** (1962), pp. 237–294.

(14) Arithmetic of alternating forms and quaternion hermitian forms. *J. Math. Soc. Japan* **15** (1963), pp. 33–65.

(15) On modular correspondences for $Sp(n, \mathbb{Z})$ and their congruence relations. *Proc. Nat. Acad. Sci. U.S.A.* **49** (1963), pp. 824–828.

(16) On analytic families of polarized abelian varieties and automorphic functions. *Ann. Math.* **78** (1963), pp. 149–192.

(17) On purely transcendental fields of automorphic functions of several variables. *Osaka J. Math.* **1** (1964), pp. 1–14.

(18) Arithmetic of unitary groups. *Ann. Math.* **79** (1964), pp. 369–409.

(19) On the field of definition for a field of automorphic functions. *Ann. Math.* **80** (1964), pp. 160–189.

(20) Class-fields and automorphic functions. *Ann. Math.* **80** (1964), pp. 444–463.

(21) On the field of definition for a field of automorphic functions: II, III. *Ann. Math.* **81** (1965), pp. 124–165; **83** (1966), pp. 377–385.

(22) Moduli and fibre systems of abelian varieties. *Ann. Math.* **83** (1966), pp. 294–338.

(23) Moduli of abelian varieties and number theory. *Proc. Sympos. Pure Math.*, Vol. 9, American Mathematical Society, Providence, R.I., 1966, pp. 312–332.

(24) A reciprocity law in non-solvable extensions. *J. reine angew. Math.* **221** (1966), pp. 209–220.

(25) Discontinuous groups and abelian varieties. *Math. Ann.* **168** (1967), pp. 171–199.

(26) Construction of class fields and zeta functions of algebraic curves. *Ann. Math.* **85** (1967), pp. 58–159.

(27) Algebraic number fields and symplectic discontinuous groups. *Ann. Math.* **86** (1967), pp. 503–592.

(28) Algebraic varieties without deformation and the Chow variety. *J. Math. Soc. Japan* **20** (1968), pp. 336–341.

(29) *Automorphic Functions and Number Theory.* Lecture Notes in Mathematics, Vol. 54, Springer-Verlag, Berlin–Heidelberg–New York, 1968.

SHIMURA, G., and TANIYAMA, Y.

(1) Complex multiplication of abelian varieties and its applications to number theory. *Publ. Math. Soc. Japan*, No. 6, Tokyo, 1961.

SIEGEL, C. L.

(1) Über die Perioden elliptischer Funktionen. *J. reine angew. Math.* **167** (1932), pp. 62–69; *Gesammelte Abhandlungen*, Vol. I, Springer-Verlag, Berlin–Heidelberg–New York, 1966, pp. 267–274.

(2) Über die analytische Theorie der quadratischen Formen. *Ann. Math.* **36** (1935), pp. 527–606; *Gesammelte Abhandlungen*, Vol. I, pp. 326–405.

(3) Über die analytische Theorie der quadratischen Formen II, III. *Ann. Math.* **37** (1936), pp. 230–263; **38** (1937), pp. 212–291; *Gesammelte Abhandlungen*, Vol. I, pp. 410–443, 469–548.

(4) The volume of the fundamental domain for some infinite groups. *Trans. Amer. Math. Soc.* **39** (1936), pp. 209–218; *Gesammelte Abhandlungen*, Vol. I, pp. 459–468.

(5) Analytische Theorie der quadratischen Formen. *C. R. Congr. Int. Math.*, Oslo, 1937, pp. 104–110; *Gesammelte Abhandlungen*, Vol. II, pp. 1–7.

(6) Formes quadratiques et modules des courbes algébriques. *Bull. Sci. Math.* (2) **61** (1937), pp. 331–352; *Gesammelte Abhandlungen*, Vol. II, pp. 20–40.

(7) Über die Zetafunktionen indefiniter quadratischer Formen. *Math. Z.* **43** (1938), pp. 682–708; *Gesammelte Abhandlungen*, Vol. II, pp. 41–67.

(8) Über die Zetafunktionen indefiniter quadratischer Formen II. *Math. Z.* **44** (1939), pp. 398–426; *Gesammelte Abhandlungen*, Vol. II, pp. 68–96.

(9) Einführung in die Theorie der Modulfunktionen *n*-ten Grades. *Math. Ann.* **116** (1939), pp. 617–657; *Gesammelte Abhandlungen*, Vol. II, pp. 97–137.

(10) Einheiten quadratischer Formen. *Abh. Math. Sem. Hansischen Univ.* **13** (1940), pp. 209–239; *Gesammelte Abhandlungen*, Vol. II, pp. 138–168.

(11) Note on automorphic functions of several variables. *Ann. Math.* **43** (1942), pp. 613–616; *Gesammelte Abhandlungen*, Vol. II, pp. 270–273.

(12) Symplectic geometry. *Amer. J. Math.* **65** (1943), pp. 1–86, and Academic Press, New York and London, 1964; *Gesammelte Abhandlungen*, Vol. II, pp. 274–359.

(13) Discontinuous groups. *Ann. Math.* **44** (1943), pp. 674–689; *Gesammelte Abhandlungen*, Vol. II, pp. 390–405.

(14) On the theory of indefinite quadratic forms. *Ann. Math.* **45** (1944), pp. 577–622; *Gesammelte Abhandlungen*, Vol. II, pp. 421–466.

(15) The average measure of quadratic forms with given determinant and signature. *Ann. Math.* **45** (1944), pp. 667–685; *Gesammelte Abhandlungen*, Vol. II, pp. 473–491.

(16) Some remarks on discontinuous groups. *Ann. Math.* **46** (1945), pp. 708–718; *Gesammelte Abhandlungen*, Vol. III, pp. 67–77.

(17) *Indefinite quadratische Formen und Modulfunktionen.* Courant Anniversary Volume, 1948, pp. 395–406; *Gesammelte Abhandlungen*, Vol. III, pp. 85–91.

(18) Indefinite quadratische Formen und Funktionentheorie I, II. *Math. Ann.* **124** (1951), pp. 17–54; **124** (1952), pp. 364–387; *Gesammelte Abhandlungen*, Vol. III, pp. 105–142, 154–177.

(19) Die Modulgruppe in einer einfachen involutorischen Algebra. *Festschrift Akad. Wiss. Göttingen*, 1951, pp. 157–167; *Gesammelte Abhandlungen*, Vol. III, pp. 143–153.

(20) *Analytic functions of several complex variables.* Lecture Notes, Institute for Advanced Study, Princeton, N.J., 1948, reprinted 1952.

(21) Meromorphe Funktionen auf kompakten analytischen Mannigfaltigkeiten. *Nachr. Akad. Wiss. Göttingen, math.-phys. Klasse*, 1955, No. 4, pp. 71–77; *Gesammelte Abhandlungen*, Vol. III, pp. 216–222.

(22) Zur Theorie der Modulfunktionen n-ten Grades. *Comm. Pure Appl. Math.* **8** (1955), pp. 677–681; *Gesammelte Abhandlungen*, Vol. III, pp. 223–227.

(23) *Zur Vorgeschichte des Eulerschen Additionstheorems.* Sammelband Leonhard Euler, Akademie-Verlag, Berlin 1959, pp. 315–317; *Gesammelte Abhandlungen*, Vol. III, pp. 249–251.

(24) Zur Reduktionstheorie quadratischer Formen. *Publ. Math. Soc. Japan*, No. 5, Tokyo, 1959; *Gesammelte Abhandlungen*, Vol. III, pp. 275–327.

(25) Zur Bestimmung des Volumens des Fundamentalbereichs der unimodularen Gruppe. *Math. Ann.* **137** (1959), pp. 427–432; *Gesammelte Abhandlungen*, Vol. III, pp. 328–333.

(26) Über die algebraische Abhängigkeit von Modulfunktionen n-ten Grades. *Nachr. Akad. Wiss. Göttingen, math.-phys. Klasse*, 1960, No. 12, pp. 257–272; *Gesammelte Abhandlungen*, Vol. III, pp. 350–365.

(27) *Lectures on advanced analytic number theory.* Tata Institute of Fundamental Research, Bombay, 1961.

(28) *Lectures on Riemann matrices.* Tata Institute of Fundamental Research, Bombay, 1963.

(29) *Lectures on the analytic theory of quadratic forms.* Institute for Advanced Study, Princeton, 1935, 2nd edition 1949; 3rd revised edition, Buchhandlung Robert Peppmüller, Göttingen, 1963.

(30) Bestimmung der elliptischen Modulfunktion durch eine Transformationsgleichung. *Abh. Math. Sem. Univ. Hamburg* **27** (1964), pp. 32–38; *Gesammelte Abhandlungen*, Vol. III, pp. 366–372.

(31) Moduln Abelscher Funktionen. *Nachr. Akad. Wiss. Göttingen, math.-phys. Klasse*, 1963, No. 25, pp. 365–427; *Gesammelte Abhandlungen*, Vol. III, pp. 373–435.

(32) Über die Fourierschen Koeffizienten der Eisensteinschen Reihen. *Mat.-Fys. Medd. Danske Vid. Selsk.* **34** (1964), No. 6; *Gesammelte Abhandlungen*, Vol. III, pp. 443–458.

(33) Über die Fourierschen Koeffizienten von Eisensteinschen Reihen der Stufe T. *Math. Z.* **105** (1968), pp. 257–266.

(34) *Zu den Beweisen des Vorbereitungssatzes von Weierstraß. Abhandlungen aus Zahlentheorie und Analysis.* Zur Erinnerung an Edmund Landau (1877–1938), pp. 299–306, VEB Dtsch. Verl. Wissenschaften, Berlin, 1968.

(35) *Topics in complex function theory*, Vol. I: *Elliptic functions and uniformization theory*, Vol. II: *Automorphic functions and abelian integrals*. Wiley-Interscience, New York, 1969/71; translation of: *Vorlesungen über ausgewählte Kapitel der Funktionentheorie* I, II, Göttingen, 1964/65.

(36) Über die Fourierschen Koeffizienten von Modulformen. *Nachr. Akad. Wiss. Göttingen, math.-phys. Klasse*, 1970, No. 3, pp. 15–56.

(37) Über Moduln Abelscher Funktionen. *Nachr. Akad. Wiss. Göttingen, math.-phys. Klasse*, 1971, No. 4, pp. 79–96.

SPÄTH, H.

(1) Der Weierstraßsche Vorbereitungssatz. *J. reine angew. Math.* **161** (1929), pp. 95–100.

SPILKER, J.

(1) Algebraische Körper von automorphen Funktionen. *Math. Ann.* **149** (1963), pp. 341–360.

(2) Kompaktifizierung des Humbertschen Fundamentalbereichs. *Math. Ann.* **161** (1965), pp. 296–314.

(3) Über den Rand der Siegelschen Halbebene. *Math. Z.* **90** (1965), pp. 273–285.

(4) Werte von Modulformen. *Nachr. Akad. Wiss. Göttingen, math.-phys. Klasse*, 1965, No. 9, pp. 125–132.

(5) Darstellung automorpher Formen durch Poincaré-Reihen. *Math. Z.* **99** (1967), pp. 216–234.

(6) Eine Anzahlfunktion bei diskontinuierlichen Gruppen. *Archiv Math.* **18** (1967), pp. 597–602.

STEIN, K.

(1) Maximale holomorphe und meromorphe Abbildungen I, II. *Amer. J. Math.* **85** (1963), pp. 298–313; **86** (1964), pp. 823–868.

STICKELBERGER, L.

(1) Über einen Satz des Herrn Noether. *Math. Ann.* **30** (1887), pp. 401–409.

(2) Über eine Verallgemeinerung der Kreisteilung. *Math. Ann.* **37** (1890), pp. 321–367.

STOLL, W.

(1) Über meromorphe Abbildungen komplexer Räume I, II. *Math. Ann.* **136** (1958), pp. 201–239, 393–429.

SUGAWARA, M.

(1) Über eine allgemeine Theorie der Fuchsschen Gruppen und Theta-Reihen. *Ann. Math.* **41** (1940), pp. 488–494.

TAKEUCHI, M.

(1) On Pontrjagin classes of compact symmetric spaces. *J. Fac. Sci. Univ. Tokyo (I)* **9** (1962), pp. 313–328.
(2) On the fundamental group and the group of isometries of a symmetric space. *J. Fac. Sci. Univ. Tokyo (I)* **10** (1964), pp. 88–123.
(3) Cell decompositions and Morse equalities on certain symmetric spaces. *J. Fac. Sci. Univ. Tokyo* **12** (1965), pp. 81–192.
(4) On orbits in a compact hermitian symmetric space. *Amer. J. Math.* **90** (1968), pp. 657–680.

TAMAGAWA, T.

(1) On Hilbert's modular group. *J. Math. Soc. Japan* **11** (1959), pp. 241–246.
(2) On Selberg's trace formula. *J. Fac. Sci. Univ. Tokyo (I)* **8** (1960), pp. 363–386.

TANIYAMA, Y.

(1) Jacobian varieties and number fields. *Proc. Int. Sympos. Algebraic Number Theory*, Tokyo-Nikko, 1955, pp. 31–45.
(2) *L*-functions of number fields and zeta functions of abelian varieties. *J. Math. Soc. Japan* **9** (1957), pp. 330–366.

THIMM, W.

(1) Über algebraische Relationen zwischen meromorphen Funktionen in abgeschlossenen Räumen. Thesis, Königsberg, 1939, 44 pp.
(2) Über meromorphe Abbildungen von komplexen Mannigfaltigkeiten. *Math. Ann.* **128** (1954), pp. 1–48.
(3) Meromorphe Abbildungen von Riemannschen Bereichen. *Math. Z.* **60** (1954), pp. 435–457.
(4) *Vorlesung über Funktionentheorie mehrerer Veränderlichen.* Aschendorffsche Verlagsbuchhandlung, Münster, 1961.
(5) *Der Weierstraß'sche Satz der algebraischen Abhängigkeit von Abelschen Funktionen und seine Verallgemeinerungen. Festschrift zur Gedächtnisfeier für K. Weierstraß*, Westdeutscher Verlag, Köln, 1966, pp. 123–154.

TORELLI, R.

(1) Sulle varietà di Jacobi. *Rend. Acc. Lincei* (5) **22** (1914), pp. 98–103.

TRAYNARD, E.

(1) Sur les fonctions thêta de deux variables et les surfaces hyperelliptiques. *Ann. l'École Norm.* (3) **24** (1907), pp. 77–177.

VINBERG, E.

(1) Homogeneous cones. *Doklady Akad. Nauk SSSR* **133** (1960), pp. 9–12; translation: *Soviet Math. Doklady* **1** (1960), pp. 787–790.

(2) Morozov-Borel theorem for real Lie groups. *Doklady Akad. Nauk SSSR* **141** (1961), pp. 270–273; translation: *Soviet Math. Doklady* **2** (1961), pp. 1416–1419.

(3) Convex homogeneous domains. *Doklady Akad. Nauk SSSR* **141** (1961), pp. 521–524; translation: *Soviet Math. Doklady* **2** (1961), pp. 1470–1473.

(4) Automorphisms of homogeneous convex cones. *Doklady Akad. Nauk SSSR* **143** (1962), pp. 265–268; translation: *Soviet Math. Doklady* **3** (1962), pp. 371–374.

(5) The theory of homogeneous cones. *Trudy Moscov. Mat. Obshch.* **12** (1963), pp. 303–358; translation: *Trans. Moscow Math. Soc.* **12** (1963), pp. 340–403.

WANG, H. C.

(1) Closed manifolds with homogeneous complex structures. *Amer. J. Math.* **76** (1954), pp. 1–32.

WEDDERBURN, J. H. M.

(1) On division algebras. *Trans. Amer. Math. Soc.* **22** (1921), pp. 129–135.

WEIERSTRASS, K.

(1) *Abhandlungen aus der Functionenlehre.* Verlag von Julius Springer, Berlin, 1886.

(2) Einige auf die Theorie der analytischen Functionen mehrerer Veränderlichen sich beziehende Sätze. *Mathematische Werke*, II, pp. 135–188, Mayer & Müller, Berlin, 1895; reprinted by Johnson Reprint Corporation, New York.

(3) Vorlesungen über die Theorie der Abelschen Transcendenten. *Mathematische Werke*, IV, Mayer & Müller, Berlin, 1902.

(4) Allgemeine Untersuchungen über $2n$-fach periodische Functionen von n Veränderlichen. *Mathematische Werke*, III, pp. 53–114; Mayer & Müller, Berlin, 1903.

WEIL, A.

(1) L'arithmétique sur les courbes algébriques. *Acta math.* **52** (1929), pp. 281–315.

(2) *Arithmétique et géométrie sur les variétés algébriques.* Hermann, Paris, 1935, 16 pp.

(3) L'intégrale de Cauchy et les fonctions de plusieurs variables. *Math. Ann.* **111** (1935), pp. 178–182.

(4) Über Matrizenringe auf Riemannschen Flächen und der Riemann-Rochsche Satz. *Abh. Math. Sem. Hamburg. Univ.* **11** (1935), pp. 110–115.

(5) Généralisation des fonctions Abéliennes. *J. Math. pure appl.* (*IX*) **17** (1938), pp. 47–87.

(6) Zur algebraischen Theorie der algebraischen Funktionen. *J. reine angew. Math.* **179** (1938), pp. 129–133.

(7) Sur les fonctions algébriques à corps de constantes finis. *C.R. Acad. Sci. Paris* **210** (1940), pp. 592–594.

(8) *L'intégration dans les groupes topologiques et ses applications.* Hermann, Paris, 1940.

(9) On the Riemann hypothesis in function fields. *Proc. Nat. Acad. Sci. U.S.A.* **27** (1941), pp. 345–347.

(10) *Foundations of algebraic geometry.* American Mathematical Society, Providence, R.I., 1946, 2nd edition 1962.

(11) Sur la théorie des formes différentielles attachées à une variété analytique complexe. *Comment. Math. Helv.* **20** (1947), pp. 110–116.

(12) *Sur les courbes algébriques et les variétés qui s'en déduisent.* Hermann, Paris, 1948.

(13) *Variétés abéliennes et courbes algébriques.* Hermann, Paris, 1948.

(14) On some exponential sums. *Proc. Nat. Acad. Sci. U.S.A.* **34** (1948), pp. 204–207.

(15) Number of solutions of equations in finite fields. *Bull. Amer. Math. Soc.* **55** (1949), pp. 497–508.

(16) Number-theory and algebraic geometry. *Proc. Int. Congr. Math.*, Cambridge, Mass., 1950, Vol. 2, pp. 90–100.

(17) Arithmetic on algebraic varieties. *Ann. Math.* **53** (1951), pp. 412–444.

(18) Sur la théorie du corps de classes. *J. Math. Soc. Japan* **3** (1951), pp. 1–35.

(19) Criteria for linear equivalence. *Proc. Nat. Acad. Sci. U.S.A.* **38** (1952), pp. 258–260.

(20) On Picard varieties. *Amer. J. Math.* **74** (1952), pp. 865–894.

(21) Jacobi sums as "Grössencharaktere". *Trans. Amer. Math. Soc.* **73** (1952), pp. 487–495.

(22) Sur les théorémes de deRham. *Comment Math. Helv.* **26** (1952), pp. 119–145.

(23) Sur les critères d'équivalence en géométrie algébrique. *Math. Ann.* **128** (1954), pp. 95–127.

(24) On algebraic groups of transformations. *Amer. J. Math.* **77** (1955), pp. 355–391.

(25) On algebraic groups and homogeneous spaces. *Amer. J. Math.* **77** (1955), pp. 493–512.

(26) On a certain type of characters of the idèle-class group of an algebraic number-field. *Proc. Int. Sympos. Algebraic Number Theory*, Tokyo-Nikko, 1955, pp. 1–7.

(27) On the theory of complex multiplication. Ibid., pp. 9–22.

(28) The field of definition of a variety. *Amer. J. Math.* **78** (1956), pp. 509–524.

(29) *On the projective embedding of abelian varieties.* Algebraic Geometry and Topology, A symposium in honor of S. Lefschetz, Princeton, 1957, pp. 177–181.

(30) Zum Beweis des Torellischen Satzes. *Nachr. Akad. Wiss. Göttingen, math.-phys. Klasse*, 1957, No. 2, pp. 33–53.

(31) *Modules des surfaces de Riemann.* Séminaire Bourbaki, 1958, Exposé 168, 7 pp.

(32) *Introduction à l'étude des variétés kählériennes.* Hermann, Paris, 1958.

(33) Algebras with involutions and the classical groups. *J. Indian Math. Soc.* **24** (1960), pp. 589–623.

(34) *Discontinuous subgroups of classical groups.* Chicago, 1960.

(35) On discrete subgroups of Lie groups. *Ann. Math.* (2) **72** (1960), pp. 369–384.

(36) On discrete subgroups of Lie groups, II. *Ann. Math.* (2) **75** (1962), pp. 578–602.

(37) *Adèles and Algebraic Groups.* Notes by M. Demazure and T. Ono, Institute for Advanced Study, Princeton, 1961.

(38) Sur certains groupes d'opérateurs unitaires. *Acta Math.* **111** (1964), pp. 143–211.

(39) Remarks on the cohomology of groups. *Ann. Math.* (2) **80** (1964), pp. 149–157.

(40) Sur la formule de Siegel dans la théorie des groupes classiques. *Acta Math.* **113** (1965), pp. 1–87.

(41) Über die Bestimmung Dirichletscher Reihen durch Funktionalgleichungen. *Math. Ann.* **168** (1967), pp. 149–156.

(42) *Basic Number Theory.* Springer-Verlag, Berlin–Heidelberg–New York, 1967.

(43) *Dirichlet series and automorphic forms*. Lecture Notes in Mathematics, Vol. 189, Springer-Verlag, Berlin–Heidelberg–New York, 1971.

WELKE, H.

(1) Über die analytischen Abbildungen von Kreiskörpern und Hartogsschen Bereichen. *Math. Ann.* **103** (1930), pp. 437–449.

WEYL, H.

(1) Die Idee der Riemannschen Fläche. Teubner, Berlin, 1913; 3rd edition, Teubner, Stuttgart, 1955; English translation: Addison-Wesley Publishing Company, London, 1964.

(2) Theorie der Darstellung kontinuierlicher halb-einfacher Gruppen durch lineare Transformationen. I, II, III. *Math. Z.* **23** (1925), pp. 271–309; **24** (1926), pp. 328–376, 377–395; Supplement: *Math. Z.* **24** (1926), pp. 789–791.

(3) On generalized Riemann matrices. *Ann. Math.* **35** (1934), pp. 714–729.

(4) Generalized Riemann matrices and factor sets. *Ann. Math.* **37** (1936), pp. 709–745.

(5) *The classical groups. Their invariants and representations*. Princeton University Press, Princeton, N.J., 1939.

(6) Theory of reduction for arithmetical equivalence I, II. *Trans. Amer. Math. Soc.* **48** (1940), pp. 126–164; **51** (1942), pp. 203–231.

(7) *Fundamental domains for lattice groups in division algebras. I. Festschrift zum 60. Geburtstag von Prof. Dr. Andreas Speiser*, Füssli, Zürich, 1945, pp. 218–232.

(8) Fundamental domains for lattice groups in division algebras. II. *Comment. Math. Helv.* **17** (1945), pp. 283–306.

WIRTINGER, W.

(1) Algebraische Funktionen und ihre Integrale. *Enzykl. math. Wiss.* II B 2 (1901), pp. 115–175.

(2) *Untersuchungen über Thetafunktionen*. Teubner, Leipzig, 1895.

(3) Zur Theorie der $2n$-fach periodischen Funktionen. Part 1: *Monatsh. Math. Phys.* **6** (1895), pp. 69–98. Part 2: *Monatsh. Math. Phys.* **7** (1896), pp. 1–25.

(4) Über einige Probleme in der Theorie der Abelschen Funktionen. *Acta Math.* **26** (1902), pp. 133–156.

(5) Über den Weierstraßschen Vorbereitungssatz. *J. reine angew. Math.* **158** (1927), pp. 260–267.

(6) Zur formalen Theorie der Funktionen von mehreren komplexen Veränderlichen. *Math. Ann.* **97** (1927), pp. 357–375.

WITT, E.

(1) Riemann-Rochscher Satz und Zetafunktion im Hyperkomplexen. *Math. Ann.* **110** (1934), pp. 12–28.

(2) Eine Identität zwischen Modulformen zweiten Grades. *Abh. Math. Sem. Hansischen Univ.* **14** (1941), pp. 323–337.

(3) Über die Konstruktion von Fundamentalbereichen. *Ann. Mat. pura appl.* **36** (1954), pp. 215–221.

WOLF, J. A.

(1) The manifolds covered by a Riemannian homogeneous manifold. *Amer. J. Math.* **82** (1960), pp. 661–688.

(2) Homogeneous manifolds of constant curvature. *Comment. Math. Helv.* **36** (1961), pp. 112–147.

(3) Discrete groups, symmetric spaces, and global holonomy. *Amer. J. Math.* **84** (1962), pp. 527–542.

(4) The Clifford-Klein space forms of indefinite metric. *Ann. Math.* (*2*) **75** (1962), pp. 77–80.

(5) Homogeneous manifolds of zero curvature. *Trans. Amer. Math. Soc.* **104** (1962), pp. 462–469; correction: **106** (1963), p. 540.

(6) Locally symmetric homogeneous spaces. *Comment. Math. Helv.* **37** (1962/63), pp. 65–101.

(7) On the lattice space forms. *Comment. Math. Helv.* **37** (1962/63), pp. 102–110.

(8) On locally symmetric spaces of non-negative curvature and certain other locally homogeneous spaces. *Comment. Math. Helv.* **37** (1962/63), pp. 266–295.

(9) Geodesic spheres in Grassmann manifolds. *Illinois J. Math.* **7** (1963), pp. 425–446.

(10) Elliptic spaces in Grassmann manifolds. *Illinois J. Math.* **7** (1963), pp. 447–462.

(11) Space forms of Grassmann manifolds. *Canad. J. Math.* **15** (1963), pp. 193–205.

(12) On the classification of hermitian symmetric spaces. *J. Math. Mech.* **13** (1964), pp. 489–495.

(13) Self adjoint function spaces on Riemannian symmetric manifolds. *Trans. Amer. Math. Soc.* **113** (1964), pp. 299–315.

(14) Isotropic manifolds of indefinite metric. *Comment. Math. Helv.* **39** (1964), pp. 21–64.

(15) Complex homogeneous contact manifolds and quaternionic symmetric spaces. *J. Math. Mech.* **14** (1965), pp. 1033–1047.

(16) *Spaces of constant curvature.* McGraw-Hill, New York, 1967.

(17) The geometry and structure of isotropy irreducible homogeneous spaces. *Acta Math.* **120** (1968), pp. 59–148.

(18) Symmetric spaces which are real cohomology spheres. *J. Diff. Geom.* **3** (1969), pp. 59–68.

(19) The action of a real semisimple group on a complex flag manifold. I. Orbit structure and holomorphic arc components. *Bull. Amer. Math. Soc.* **75** (1969), pp. 1121–1237.

Cumulative Index

Roman numerals preceding page numbers indicate the volume.